AF598084

Methods in Molecular Biology™

Series Editor
John M. Walker
School of Life Sciences
University of Hertfordshire
Hatfield, Hertfordshire, AL10 9AB, UK

For further volumes:
http://www.springer.com/series/7651

Organ Regeneration

Methods and Protocols

Edited by

Joydeep Basu

Tengion, Inc., Winston-Salem, NC, USA

John W. Ludlow

Zen-Bio, Inc., Research Triangle Park, NC, USA

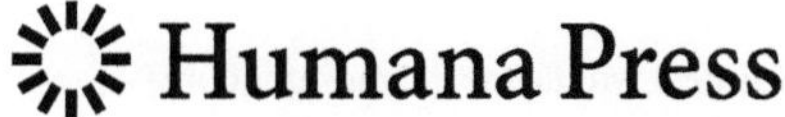

Editors
Joydeep Basu
Tengion, Inc.
Winston-Salem, NC, USA

John W. Ludlow
Zen-Bio, Inc.
Research Triangle Park, NC, USA

ISSN 1064-3745 ISSN 1940-6029 (electronic)
ISBN 978-1-62703-362-6 ISBN 978-1-62703-363-3 (eBook)
DOI 10.1007/978-1-62703-363-3
Springer New York Heidelberg Dordrecht London

Library of Congress Control Number: 2013932712

Printed on acid-free paper

Humana Press is a brand of Springer
Springer is part of Springer Science+Business Media (www.springer.com)

Preface

Tissue engineering and regenerative medicine represents a wide array of cell and biomaterial-based approaches focusing on the repair, augmentation, and regeneration of diseased tissues and organs. To this end, recent successes in clinical outcomes following implantation of tissue-engineered trachea, bladder, and urinary conduit have highlighted the emergence of common methodological frameworks for the development of technical approaches that may be broadly applicable towards the regeneration of multiple, disparate hollow organs. Similarly, recent progress towards regeneration of heart, kidney, liver, pancreas, spleen, and central nervous system is identifying shared methodologies that may underlie the development of foundational platform technologies broadly applicable towards the regeneration of multiple solid organ systems. In all cases, emerging central themes include the use of a biodegradable scaffold to provide structural support for the developing neo-organ and the role of committed or progenitor cell populations in establishing the regenerative microenvironment of key secreted growth factors and extracellular matrix critical for catalyzing de novo organogenesis.

Organ Regeneration: Methods and Protocols has been assembled in response to the growing interest in organ regeneration as a means to treat disease. The goal of this compilation is to provide a detailed guide to aid newcomers and seasoned veterans in their developmental and experimental work in tissue engineering and regenerative medicine.

What you have before you contains contributions by many of the current and emerging leaders in the field. These chapters contain step-by-step information on how to isolate and characterize cells from selected soft tissues and solid organs, preparation and evaluation of natural and synthetic biomaterial scaffolding, implantation of regenerative constructs within experimental animals, and evaluation of regenerative outcomes by molecular and histological methodologies.

Though it is obviously not possible to include contributions by each and every researcher in this field, efforts were made to be inclusive, and avoid being exclusive, regarding methods used to generate tissue-engineered and regenerative medicine products. We hope that you find this body of work both informative and thought provoking.

Winston-Salem, NC, USA *Joydeep Basu*
Research Triangle Park, NC, USA *John W. Ludlow*

Contents

Contributors

STELIOS T. ANDREADIS • *Department of Chemical and Biological Engineering, University at Buffalo, State University of New York, Amherst, NY, USA*

PEDRO M. BAPTISTA • *Wake Forest Institute for Regenerative Medicine, Winston-Salem, NC, USA*

ERIK R. BARTHEL • *Division of Pediatric Surgery, Saban Research Institute, Children's Hospital Los Angeles, Keck School of Medicine of the University of Southern California, Los Angeles, CA, USA*

JOYDEEP BASU • *Tengion Inc., Winston-Salem, NC, USA*

SHANTARAM BHARADWAJ • *Wake Forest Institute for Regenerative Medicine, Winston-Salem, NC, USA*

ANDREW T. BRUCE • *Tengion Inc., Winston-Salem, NC, USA*

BENJAMIN M. BUEHRER • *ZenBio Inc., Research Triangle Park, NC, USA*

TERESA BORA BURNETTE • *Tengion Inc., Winston-Salem, NC, USA*

PETER E. BUTLER • *Department of Plastic and Reconstructive Surgery, Royal Free Hampstead NHS Trust Hospital, London, UK*

BENTLEY CHEATHAM • *ZenBio Inc., Research Triangle Park, NC, USA*

ALAN C. FARNEY • *Department of Surgery, Wake Forest University School of Medicine, Winston-Salem, NC, USA;Wake Forest Institute for Regenerative Medicine, Wake Forest University School of Medicine, Winston-Salem, NC, USA*

JEFF FISH • *Tengion Inc., Winston-Salem, NC, USA*

CHRISTOPHER W. GENHEIMER • *Tengion Inc., Winston-Salem, NC, USA*

PAUL GORES • *Carolinas Medical Center, Charlotte, NC, USA*

TRACY C. GRIKSCHEIT • *Division of Pediatric Surgery, Saban Research Institute, Children's Hospital Los Angeles, Keck School of Medicine of the University of Southern California, Los Angeles, CA, USA*

KELLY I. GUTHRIE • *Tengion Inc., Winston-Salem, NC, USA*

CRAIG HALBERSTADT • *VP Technical Operations and Product Development, Organovo, Inc. Nancy Ridge Drive, San Diego, CA USA*

XIAOGANG HOU • *Division of Pediatric Surgery, Saban Research Institute, Children's Hospital Los Angeles, Keck School of Medicine of the University of Southern California, Los Angeles, CA, USA*

MANUEL J. JAYO • *Tengion Inc., Winston-Salem, NC, USA*

SANDRA L. JOHNSON • *Biomedical Engineering, University of Minnesota, Minneapolis, MN, USA*

RUSTY KELLEY • *Tengion Inc., Winston-Salem, NC, USA*

TOYIN A. KNIGHT • *Tengion Inc., Winston-Salem, NC, USA*

KATHLEEN A. LAMKIN-KENNARD • *Department of Mechanical Engineering, Rochester Institute of Technology, Rochester, NY, USA*

PEDRO LEI • *Department of Chemical and Biological Engineering, University at Buffalo, State University of New York, Amherst, NY, USA*

DANIEL E. LEVIN • *Division of Pediatric Surgery, Saban Research Institute, Children's Hospital Los Angeles, Keck School of Medicine of the University of Southern California, Los Angeles, CA, USA*
GUIHUA LIU • *Wake Forest Institute for Regenerative Medicine, Winston-Salem, NC, USA*
JOHN W. LUDLOW • *Zen-Bio Inc, Research Triangle Park, NC, USA*
JEFFREY M. MACDONALD • *Department of Biomedical Engineering, University of North Carolina, Chapel Hill, NC, USA*
RANDALL MCCLELLAND • *Department of Biomedical Engineering, University of North Carolina, Chapel Hill, NC, USA; SciKon Corporation, Research Triangle Park, NC, USA*
DARELL W. MCCOY • *Tengion Inc., Winston-Salem, NC, USA*
JOHN P. MCQUILLING • *Virginia-tech-Wake Forest school of Biomedical Engineering Sciences, Winston-Salem, NC, USA*
SONYA O. MEHEUX • *Cytonet, Durham, NC, USA*
KIM L. MIHALKO • *Carolinas Medical Center, Charlotte, NC, USA*
EMMA MORAN • *Wake Forest Institute for Regenerative Medicine, Winston-Salem, NC, USA*
EMMANUEL C. OPARA • *Wake Forest Institute for Regenerative Medicine, Wake Forest University School of Medicine, Winston-Salem, NC, USA*
ADELOLA O. OSENI • *Division of Surgery and Interventional Science, UCL Centre for Nanotechnology and Regenerative Medicine, University College London, London, UK*
RICHARD G. PAYNE • *Tengion Inc., Winston-Salem, NC, USA*
SCOTT RAPOPORT • *Organovo Inc., San Diego, CA, USA*
ELIAS A. RIVERA • *Tengion Inc., Winston-Salem, NC, USA*
NEIL ROBBINS • *Tengion Inc., Winston-Salem, NC, USA*
FREDERIC G. SALA • *Division of Pediatric Surgery, Saban Research Institute, Children's Hospital Los Angeles, Keck School of Medicine of the University of Southern California, Los Angeles, CA, USA*
TORI SAMPSELL-BARRON • *Life Technologies, Madison, WI, USA*
NAMRATA SANGHA • *Tengion Inc., Winston-Salem, NC, USA*
ALEXANDER M. SEIFALIAN • *UCL Division of Surgery and Interventional Science, University College London, Royal Free Hampstead NHS Trust Hospital, London, UK*
MONICA A. SERBAN • *Allergan Medical, Medford, MA, USA*
GARY R. SKUSE • *Thomas H. Gosnell School of Life Sciences, Rochester Institute of Technology, Rochester, NY, USA*
SHAY SOKER • *Wake Forest Institute for Regenerative Medicine, Winston-Salem, NC, USA*
ALLISON L. SPEER • *Division of Pediatric Surgery, Saban Research Institute, Children's Hospital Los Angeles, Keck School of Medicine of the University of Southern California, Los Angeles, CA, USA*
KATHERINE TECH • *Department of Biomedical Engineering, University of North Carolina, Chapel Hill, NC, USA*
YASUHIRO TORASHIMA • *Division of Pediatric Surgery, Saban Research Institute, Children's Hospital Los Angeles, Keck School of Medicine of the University of Southern California, Los Angeles, CA, USA*

DIPEN VYAS • *Wake Forest Institute for Regenerative Medicine, Winston-Salem, NC, USA*
ZHAN WANG • *Wake Forest Institute for Regenerative Medicine, Winston-Salem, NC, USA*
DEANA WILLIAMS • *Carolinas Medical Center, Charlotte, NC, USA*
RONGPEI WU • *Wake Forest Institute for Regenerative Medicine, Wake Forest University, Winston-Salem, NC, USA; Department of Urology, First Affiliated Hospital, Sun Yat-Sen University, Guangzhou, China*
HUI YOU • *Department of Chemical and Biological Engineering, University at Buffalo, State University of New York, Amherst, NY, USA*
YUANYUAN ZHANG • *Wake Forest Institute for Regenerative Medicine, Winston-Salem, NC, USA*

Chapter 1

Isolation and Characterization of Human Adipose-Derived Stem Cells for Use in Tissue Engineering

Benjamin M. Buehrer and Bentley Cheatham

Abstract

Human adipose-derived adult stem cells (ASCs) represent a unique population of multipotent stem cells. Their utility in a variety of tissue engineering applications, and as a model system for the study of molecular mechanisms of differentiation, is well established. In addition, their relative abundance, ease of isolation from human subcutaneous lipoaspirates, and functional stability make them an excellent physiologically relevant platform. Here, we describe detailed procedures for handling and purification of ASCs from lipoaspirate, as well as their expansion, cryopreservation, quality control, and functional assays.

Key words Adipose-derived stem cell, Lipoaspirate, Adipocyte, Adipogenesis, Chondrogenesis, Osteogenesis, Myogenesis

1 Introduction

Tissue engineering and regenerative medicine are rapidly growing disciplines that are focused on establishing ex vivo modalities to produce tissues and organs for replacement. Most, if not all, of these approaches involve the use of pluripotent or multipotent stem cells. There are numerous sources of stem cell populations with varying degrees of regenerative capacity and, to an extent, tissue or lineage specificity. These include embryonic stem cells, adult stem cells, and induced pluripotent stem cells (iPS). Each of these sources has distinct advantages and disadvantages primarily related to regenerative capacity, ability to be expanded in culture, utility in vivo, faithful recapitulation of desired phenotype, tumorigenicity, potential utility in both autologous and allogeneic applications, relative abundance, and ease of isolation.

Human adipose-derived adult stem cells (ASCs) are a multipotent population, and have a proven utility in defining molecular mechanisms of adipogenesis and general adipocyte biology, as well as studies in type 2 diabetes. ASCs also possess an extensive portfolio

Joydeep Basu and John W. Ludlow (eds.), *Organ Regeneration: Methods and Protocols*, Methods in Molecular Biology, vol. 1001, DOI 10.1007/978-1-62703-363-3_1,

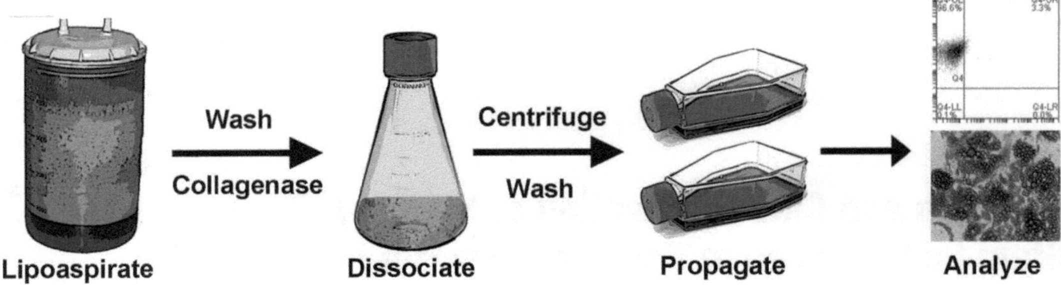

Fig. 1 Schematic overview of ASC isolation and characterization

for in vitro differentiation, including chondrogenesis, osteogenesis, and myogenesis (1–4). In addition, ASCs have been shown to support expansion of hematopoietic stem cells in vitro (5, 6). Furthermore, ASCs have been adapted to high-throughput screening platforms for a variety of robust endpoint assays in several therapeutic areas (7). In addition to the established utility of ASCs they have the advantage of relative abundance, ease of isolation, expansion in vitro, and maintained function post-cryopreservation.

This chapter focuses on the isolation and limited functional characterization of ASCs obtained from lipoaspirates. The steps for handling and processing lipoaspirate, initial plating, expansion, functional characterization, and cryopreservation will be presented. In brief, lipoaspirate material is washed and subjected to digestion with collagenase to release the ASC population. The cells are washed extensively and seeded onto cell culture plates and allowed to expand. Flow cytometry is used to establish a profile of known stem cell markers to determine purity. Finally, the cells are subjected to various protocols to analyze their differentiation potential. An overview of a typical workflow is shown in Fig. 1.

2 Materials

2.1 Equipment

1. Centrifuge.
2. Flow cytometer, such as Accuri Flow Cytometer C6.
3. Microscope.
4. Water baths.
5. Stir plates.
6. Heated stir plate.
7. Sterile stir bars.
8. Sterile 1 l Erlenmeyer flask.
9. 50 ml Centrifuge tubes.

10. 250 ml disposable centrifuge bottles.
11. Cryopreservation tubes.
12. Hemocytometer.
13. Microplate reader (capable of reading absorbance at 540 nm).
14. Control rate freezing apparatus.

2.2 Adipose Tissue Digestion

1. Krebs Ringer's bicarbonate buffer (ZenBio, Research Triangle Park, NC, USA): Sterile solution may be stored at room temperature.
2. 10% (10×) Albumin, bovine (BSA, Fraction V; Sigma-Aldrich, St. Louis, MO, USA) solution in KRB. Suspend 100 g BSA in 900 ml KRB, stirring until completely solubilized. This may require a 37°C water bath to accelerate the process. Bring volume up to 1 l with KRB using graduated cylinder. Sterilize the solution through a 0.2 μm filter into presterilized 500 ml bottles. 50 ml aliquots of the solution may be stored −80°C until required.
3. Phosphate buffered saline. PBS without magnesium or calcium is stored at room temperature until used.
4. Antibiotic solution, Penicillin/Streptomycin/Amphotericin B 100× (ZenBio). Antibiotic solution can be frozen (−20°C) or stored at 4°C until expiration date. Solution contains 10,000 IU Penicillin, 10,000 IU Streptomycin, and 25 μg/ml Amphotericin B.
5. 70% Ethanol. Store in flammables safety cabinet at room temperature.
6. Trypsin/EDTA solution (ZenBio). 0.25% trypsin containing 2.21 mM EDTA in HBSS can be stored at 4°C.
7. 1% (10× stock) Collagenase (Worthington Biochemical, Lakewood, NJ, USA) by w/v in KRB, sterile filtered. 10 g of collagenase is dissolved in 900 ml KRB. Complete solubilization may require heating to 37°C. Bring volume up to 1 l with KRB using graduated cylinder. Sterilize the solution through a 0.2 μm filter into presterilized 500 ml bottles. 50 ml aliquots of the solution may be stored at −80°C until required.

2.3 Growth, Maintenance, and Cryopreservation of ASCs

1. Two-deck Nunc Cell Factory (Thermo Scientific, Waltham, MA, USA).
2. Preadipocyte Medium (ZenBio): Dulbecco's Modified Eagle's Medium (DMEM)/Ham's F12 (1:1 v/v) containing HEPES (pH 7.4), fetal bovine serum, and antibiotics.
3. Cryopreservation medium (ZenBio). Cryopreservation medium can be stored at 4°C until used.

2.4 Analysis of Stem Cell Surface Markers

1. Anti-human Endoglin/CD105-APC conjugate (FAB10971A, R&D systems, Minneapolis, MN, USA).
2. Anti-human CD44-PE conjugate (550989, BD Pharmingen, San Jose, CA, USA).
3. Anti-human CD45-APC conjugate (555485, BD Pharmingen).
4. Anti-human CD31-PE conjugate (555446, BD Pharmingen).
5. PE mouse IgG1 k control (555749, BD Pharmingen).
6. APC mouse IgG1 k control (555751, BD Pharmingen).
7. Block buffer: 0.5% BSA solution in DPBS. Dissolve 0.5 g of BSA in 100 ml of DPBS by stirring. Filter sterilize through a 0.2 μm filter and store at 4°C until ready to use.
8. Mouse Serum (M5905, Sigma-Aldrich).

2.5 Determining Stem Cell Potential

1. Adipocyte Maintenance Medium (ZenBio): Preadipocyte medium containing reduced serum, biotin, pantothenate, insulin, and dexamethasone.
2. Adipocyte Differentiation Medium (ZenBio): Adipocyte medium with IBMX and PPAR gamma agonist.
3. Chondrocyte Differentiation Medium (ZenBio): DMEM (high glucose) containing FBS, TGF-β1, ascorbate-2-phosphate, dexamethasone, ITS, and antibiotics.
4. Osteoblast Differentiation Medium (ZenBio): DMEM/F12 containing FBS, HEPES, β-glycerophosphate, 1,25$(OH)_2$ vitamin D_3, ascorbate-2-phosphate, dexamethasone, and antibiotics.
5. DMEM (high glucose).
6. Initiation medium: DMEM (high glucose) containing FBS.
7. 0.5% Oil Red O: Dissolve 0.5 g Oil Red O (Sigma-Aldrich) in 99.5 ml isopropanol. Store at room temperature away from light.
8. 1.2% alginate solution: Dissolve 1.2 g alginic acid salt (Sigma-Aldrich) in 100 ml of 150 mM NaCl and stir on a hot plate until particles are all in solution. Filter sterilize through a 0.22 μm filter and store at 4°C.
9. 150 mM sodium chloride solution, sterile.
10. 102 mM calcium chloride solution, sterile.
11. 10% formalin solution, neutral buffered (Sigma-Aldrich).
12. 1% Alcian blue solution: Dissolve 1 g of alcian blue (Sigma-Aldrich) in 100 ml 0.1 N HCl.
13. 0.1 N HCl.
14. 70% ice-cold ethanol.

15. 2% alizarin red solution: 0.2 g alizarin red (Sigma-Aldrich) is dissolved in 10 ml of distilled water; pH is adjusted to 4.1–4.3 with dilute NaOH. Filter solution through 0.22 μm filter and store at room temperature away from light.
16. 96-well tissue culture plates.

3 Methods

All of the following procedures, unless otherwise specified, are performed in biosafety hoods using standard laboratory precautions to minimize exposure to human pathogens. Human adipose tissue samples are procured from consenting donors undergoing elective surgeries under IRB approved protocols. Subcutaneous adipose tissue is from surgical waste material derived from subcutaneous lipoaspirate tissue. This protocol assumes 500 ml of tissue and should be scaled accordingly for more or less tissue.

3.1 Preparation of Lipoaspirate

1. Warm all reagents and media in a 37°C water bath and prepare a 37°C water bath on a stir plate.
2. Carefully open liposuction container and aspirate blood if possible. Pour adipose tissue into 1 l beaker and allow blood to separate from tissue prior to aspiration (see Note 1).
3. Wash the tissue with an equal volume of DPBS by gentle stirring at room temperature. Allow the adipose tissue to float to the top and aspirate PBS from the bottom of the container. Repeat this wash step 4–5 times or until PBS wash is a light pink color (see Note 2).

3.2 Collagenase Digestion

1. Prepare collagenase solution by adding 50 ml of 10× BSA and 50 ml of 10× collagenase to 400 ml of KRB. Mix this solution by inversion. The solution can be stored at 4°C for 24 h.
2. Transfer the washed adipose tissue to a sterile 1 l Erlenmeyer flask containing a stir bar and add the prepared collagenase/BSA solution. Cover the flask and wipe the surface with ethanol.
3. Place the flask in the water bath for approximately 15 min, stirring constantly and maintaining temperature at 37°C. Gently swirl the flask and divide contents evenly between four 250 ml disposable centrifuge bottles. Centrifuge at 314 × *g* for 5 min at 20°C.

3.3 Isolation of the Stromal Vascular Fraction

1. Carefully pour off the lipid, primary adipocyte floaters, and collagenase into an empty DPBS bottle leaving 20–50 ml of the brown collagenase solution on top of the stromal vascular fraction at the bottom. Suspend the cells in the remaining solution and transfer to a 50 ml conical centrifuge tube. Bring the

volume up to 40 ml with DPBS and centrifuge the tubes at 314×g for 5 min at 20°C.

2. Aspirate the remaining solution from each tube and suspend the cell pellet in a small volume of DPBS. Pool two cell suspensions and transfer each pool to a new centrifuge tube, bringing the volume to 40 ml with DPBS. Centrifuge tubes as before.
3. Aspirate the supernatant from both tubes and pool the pellets into one clean tube bringing the final volume to 40 ml with DPBS. Centrifuge the tube as before. Aspirate the supernatant and suspend the cell pellet in 50 ml of preadipocyte medium.

3.4 Plating and Expansion of ASCs

1. Pour the cell suspension into a sterile beaker and add an additional 250 ml preadipocyte medium. Gently swirl to mix and transfer the entire contents to a two-deck cell factory and place in incubator (see Note 3).
2. 24–48 h after plating the cell suspension, pour off (or aspirate) medium from the tissue culture flask. Gently wash the cells twice with PBS and add 300 ml preadipocyte medium. Continue to feed the cells every 2 days with preadipocyte medium until 80% confluent.

3.5 Harvest and Cryopreservation of ASCs

1. Pour off (or aspirate) the medium from the cells and wash five times with 50 ml DPBS per deck. Remove the final DPBS wash and add 20 ml trypsin/EDTA per deck, close the caps and incubate at 37°C for 5 min.
2. Dislodge the cells with a sharp rap by hand or using a rubber mallet. Add 30 ml of preadipocyte medium per deck to inactivate the trypsin and gently rock the flask to dissociate the cells. Dispense the cell suspension to an appropriate centrifuge bottle(s). Rinse the cultureware with 20 ml preadipocyte medium per deck and add to the previous cell suspension.
3. Centrifuge to pellet the cells at 314×g for 5 min at 20°C. Aspirate the medium and suspend the pellet in a small volume of preadipocyte medium to count the cells using a hemocytometer (or other method). Calculate the total number of viable cells and determine the number of aliquots to cryopreserved.
4. Pellet the cells by centrifugation at 314×g for 5 min at 20°C and aspirate the medium. Add an appropriate volume of cryopreservation medium to generate the predetermined cell density and number of aliquots. Mix the cell suspension with a pipet until there are no visible cell clumps and transfer the appropriate volume to cryopreservation vials. The cell suspensions are placed immediately into a controlled rate freezing apparatus and place in the −80°C freezer overnight. The next morning, transfer tubes to liquid nitrogen storage tanks for long-term cryopreservation.

3.6 Analysis of Stem Cell Surface Markers

1. Either cryopreserved or freshly cultured cells may be used for cell surface markers. Dilute cells to 0.5×10^6/ml in PBS + 0.5% BSA and add mouse serum to 0.2% final concentration. Incubate on ice or at 4°C for 10 min to block cells.
2. Aliquot 400 μl of blocked cell suspension to three tubes. Add 5 μl of antibodies (see Table 1) as listed below and incubate on ice or 4°C for 30 min.
3. Wash cells twice with 500 μl PBS + 0.5% BSA to remove excess antibody. Suspend cell pellet in 200 μl PBS + 0.5% BSA for data acquisition using C6 flow cytometer or other flow cytometer.
4. Acquire data according to manufacturer's protocol: Label each sample to be collected; isotype controls, CD31/105, and CD44/45. Collect at least 75,000 total events (75,000–100,000). Add histogram plots to follow FL2-A (PE) and FL4-A (APC) signals. Gating can be applied after the first sample, or after all data is collected.
5. Expected outcomes: An example of a typical flow cytometry analysis is shown in Fig. 2. The cells are expected to be positive for both CD44 and CD105, and negative for both CD31 and

Table 1
Antibodies for analysis by flow cytometry

Tube	Ab1	Ab2
1	PE-IgG1k	APC-IgG1k
2	CD31-PE	CD105-APC
3	CD44-PE	CD45-APC

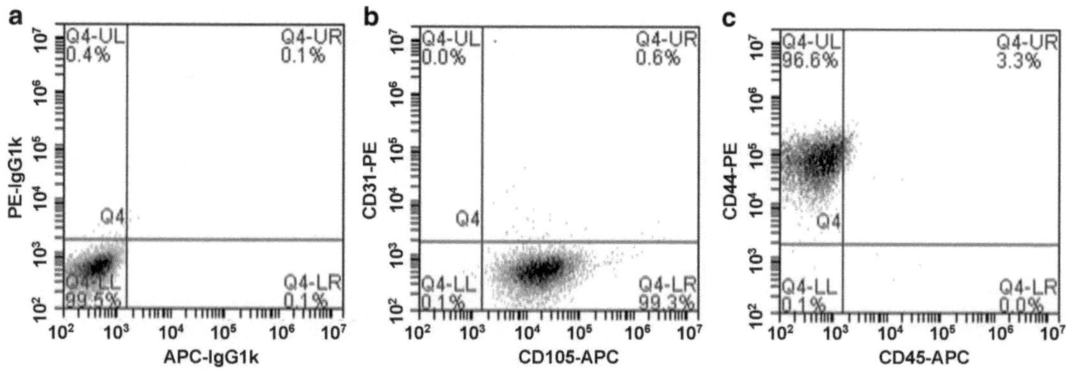

Fig. 2 Typical expected flow cytometry profile of ASCs. (**a**) Isotype controls. (**b**) ASCs showing CD31 negative and CD105 positive staining. (**c**) ASCs showing CD45 negative and CD44 positive staining

CD45 (the latter two markers are specific for endothelial cells and monocytes/macrophages, respectively).

3.7 Adipogenesis

1. Cells are seeded at a density of 40,625 cells/cm^2 in preadipocyte medium on appropriate tissue culture plates (13,000 cells per well of a 96-well plate in 150 μl), using enough test wells to include four different treatments in replicate. The cells are allowed to adhere to the culture ware stratum overnight in an incubator (37°C, 5% CO_2).
2. All of the plating medium is aspirated and replaced with either 150 μl Adipocyte Differentiation Medium (positive control); 150 μl Adipocyte Differentiation Medium containing 10 ng/ml TNF-α (negative control); 150 μl initiation medium + 0.1% DMSO (vehicle control); or 150 μl initiation medium alone (solvent control). The cells are placed back into the incubator (37°C, 5% CO_2) for 7 more days.
3. Remove 90 μl from each well and add 120 μl Adipocyte Maintenance Medium. Place the cells into the incubator (37°C, 5% CO_2) for 7 days to complete the differentiation process (see Note 4).
4. At the end of the 14 day differentiation period, lipid accumulation can be assessed by staining with Oil Red. Remove medium from all wells and wash cells with 150 μl DPBS. Aspirate the DPBS and fix cells with 100 μl of 10% formalin solution at 4°C for at least 1 h (see Note 5).
5. Prepare a working solution of Oil Red O by adding 6 ml of the Oil Red O stock solution to 4 ml of distilled water. Let this solution stand at room temperature for 20 min before use and then filter through PFTE filter to remove any particles. This solution must be prepared fresh the day of the assay and cannot be stored for later use.
6. Remove fixative solution from cells and wash twice with 200 μl of distilled water. Remove all of the water from each well and add 50 μl of the working Oil Red O solution. Incubate for 15 min at room temperature.
7. Aspirate the Oil Red O stain from each well and wash three times with 80 μl of water. The lipid droplets will stain an intense red (Fig. 3b). Photos may be taken for records. In addition, the optical density can be measured using 540 nm wavelength plate reader.

3.8 Osteogenesis

1. Cells are seeded at a density of 40,625 cells/cm^2 in preadipocyte medium on appropriate tissue culture plates (13,000 cells per well of a 96-well plate in 150 μl), using enough test wells to include two different treatments in replicate. The cells are allowed to adhere to the culture ware stratum overnight in an incubator (37°C, 5% CO_2).

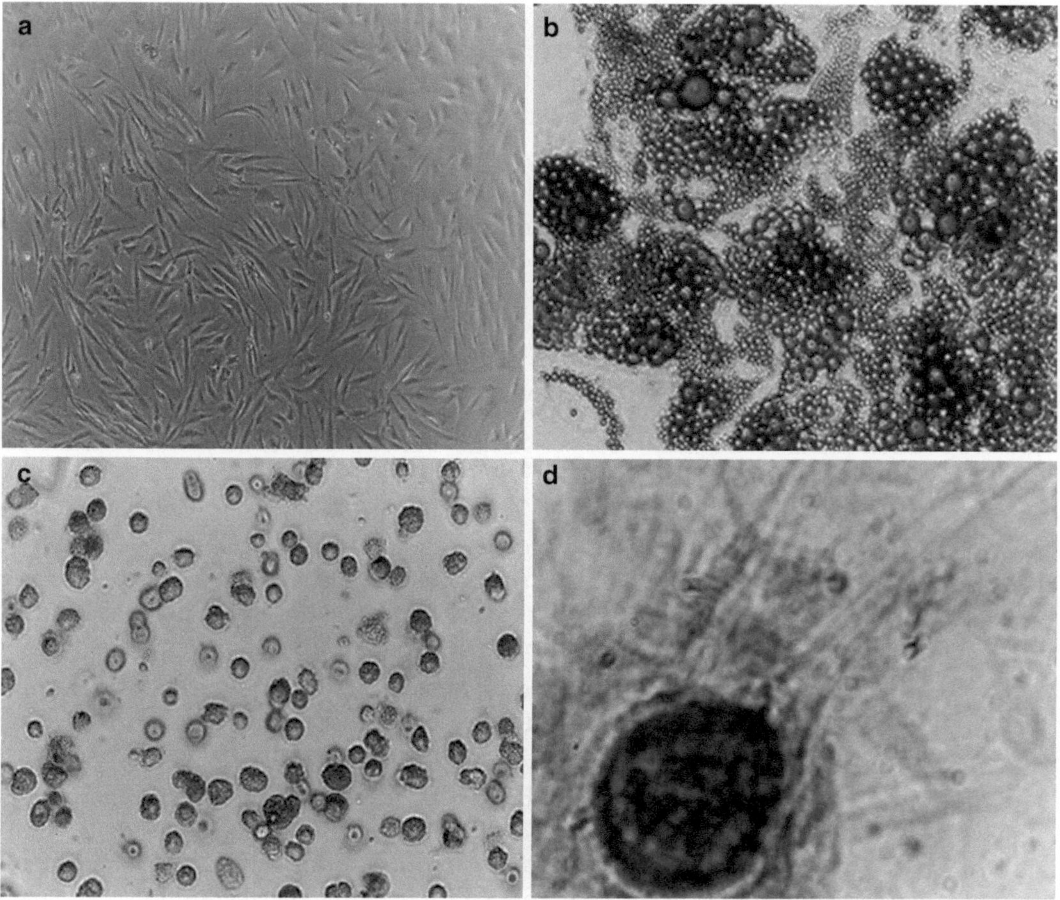

Fig. 3 Differentiation of ASCs. (**a**) Undifferentiated ASCs. (**b**) Cells stained with Oil Red O following adipogenesis. (**c**) Cells stained with alizarin red following osteogenesis. (**d**) Cells stained with alcian blue following chondrogenesis

2. Remove the medium from each well and replace with either 150 μl of Osteoblast Differentiation Medium (positive control) or 150 μl of initiation medium. The cells are placed back into the incubator (37°C, 5% CO_2) and fed the same medium every 3 days for 14 days.
3. During the course of the 14-day differentiation period the morphology of the Osteoblast Differentiation Medium-treated cells will change and appear similar to that shown in Fig. 3c. At this point the cells may be stained with alizarin red. Remove the medium from all of the wells and wash the cells three times with 150 mM NaCl solution.
4. Fix the cells using ice-cold 70% ethanol for 1 h at 4°C. Remove the ethanol from each well and wash cells three times with distilled water.

5. Remove the water from each well and cover the cells with 30 μl of the 2% alizarin red solution and incubate at room temperature for 10 min. Observe the extent of staining using a microscope. Differentiated cells will display dark red staining.
6. Remove the stain from each well and wash five times with distilled water. Wash a final time with water including a 15-min room temperature incubation step. Photograph immediately to document the amount of staining in the positive and negative control wells.

3.9 Chondrogenesis

1. Suspend cells in 1.2% alginate solution at 4×10^6 cells/ml, carefully mixing with a pipette without creating any bubbles.
2. Draw the cell-seeded alginate suspension into a sterile 1 cc syringe using a 26-gauge needle.
3. Add 3 ml of 102 mM $CaCl_2$ per well of a 6-well culture plate. Carefully and slowly dispense equal-sized droplets of cell-seeded alginate solution into the $CaCl_2$ solution. Dispense 10–30 beads per well avoiding clumping. Cure the beads in the $CaCl_2$ solution for 10 min at room temperature.
4. Using a glass pipette, aspirate the $CaCl_2$ solution from around the beads. Wash the cell beads three times with 150 mM NaCl solution and one time with DMEM (high glucose) medium.
5. Remove the DMEM-HG medium and add 3 ml of Chondrocyte Differentiation Medium to positive control wells and 3 ml of initiation medium to negative control wells. Incubate the cell beads at 37°C, 5% CO_2 changing the medium every 2 days. Keep the cell beads in culture up to 4 weeks prior to assessing differentiation.
6. Transfer 10–12 cell beads from each condition to wells on a new plate and add 2 ml of 10% buffered formalin solution. Incubate for 15 min at room temperature to fix the cells. Wash the cell beads five times with 3 ml DPBS.
7. After the final wash, remove all of the DPBS and add 3 ml of 1% alcian blue solution per well and incubate for 30 min at 37°C.
8. Remove the staining solution and add 3 ml of 0.1 N HCl to each well and incubate for 5 min at room temperature. Remove the HCl solution and wash the cell beads twice with DPBS, leaving the DPBS on the cells. Incubate at room temperature for 20 min at which time the cells will be released from the beads and can be photographed to assess staining. Cells that have undergone chondrogenesis will stain blue (Fig. 3d).

4 Notes

1. Adipose tissue floats to the top of vessel. The blood phase is aspirated from the bottom of the beaker or liposuction container.
2. Save an empty DPBS bottle for use later in the procedure.
3. If less tissue is available, use one T-225 flask for every 75 ml of original adipose tissue. The final volume of medium should be 45–50 ml per cell pellet.
4. Multilocular lipid droplets will begin to appear in the positive control around day 7 of differentiation and continue to accumulate over the next week.
5. The cells can be stored in fixative at 4°C overnight if the plate is wrapped in plastic wrap.

References

1. Guilak F, Awad HA, Fermor B et al (2004) Adipose-derived adult stem cells for cartilage tissue engineering. Biorheology 41:389–399
2. Guilak F, Lott KE, Awad HA et al (2006) Clonal analysis of the differentiation potential of human adipose-derived adult stem cells. J Cell Physiol 206:229–237
3. Gimble J, Guilak F (2003) Adipose-derived adult stem cells: isolation, characterization, and differentiation potential. Cytotherapy 5: 362–369
4. Gimble JM, Grayson W, Guilak F et al (2011) Adipose tissue as a stem cell source for musculo-skeletal regeneration. Front Biosci (Schol Ed) 3:69–81
5. De Toni F, Poglio S, Youcef AB et al (2011) Human adipose-derived stromal cells efficiently support hematopoiesis in vitro and in vivo: a key step for therapeutic studies. Stem Cells Dev 20:2127–2138
6. Kilroy GE, Foster SJ, Wu X et al (2007) Cytokine profile of human adipose-derived stem cells: expression of angiogenic, hematopoietic, and pro-inflammatory factors. J Cell Physiol 212: 702–709
7. Lea-Currie YR, Duffin DJ, Buehrer BM (2011) Use of adipose-derived stem cells in high-throughput screening to identify modulators of lipogenesis. Methods Mol Biol 702: 359–368

Chapter 2

Isolation of Smooth Muscle Cells from Bladder for Generation of Engineered Urologic Organs

Darell W. McCoy

Abstract

The isolation of smooth muscle cells from bladder tissue is a valuable technique used in cell biology research and tissue engineering. Smooth muscle cells can be used for analysis in many areas including, but not limited to, cell function and genotype experimentation. Smooth muscle cells can also be used in tissue engineering applications for research and/or regenerative medicine. Replacement tissue or tissue for augmentation can be created to stem or remediate problems in the urologic system.

Key words Smooth muscle cells, Autologous, Enzymatic digestion, Bladder, Tissue engineering, Primary cell, Explantation

1 Introduction

Generation of tissue for bladder augmentation and replacement involves the use of a bioresorbable scaffold material which gives the cells a base for attachment, function, and regeneration (1). In addition to bladder replacement/augmentation, the scaffolding and cell combination can be formed into a urinary conduit for patients who have had their bladders removed due to cancer (2). These urinary conduits can be used in lieu of small intestine, which causes a host of postsurgical and secondary issues. Smooth muscle cells can be isolated from bladder tissue using two different procedures. The "explantation method" uses mechanical dissociation of the tissue followed by placing the small pieces onto a petri dish and allowing enough time for the tissue to adhere prior to adding medium to the culture. The "enzymatic method" also uses mechanical dissociation; however, the tissue is further broken down using collagenase IV. The enzymatically dissociated suspension is then more conventionally cultured in standard T flasks. The subsequent subculturing of the cells then follows standard cell culture practices. The enzymatic isolation method has an advantage over the explantation method

Joydeep Basu and John W. Ludlow (eds.), *Organ Regeneration: Methods and Protocols*, Methods in Molecular Biology, vol. 1001, DOI 10.1007/978-1-62703-363-3_2, © Springer Science+Business Media New York 2013

due to the fact that your culture is more advanced to start with. The explantation method offers the end user a more affordable method than the enzymatic method since high-quality sterile enzymes generally add cost to the process.

2 Materials

2.1 Explantation Method

1. SMC growth medium:
   (a) (Dulbecco's Modified Eagle Media) DMEM High Glucose.
   (b) 10% Fetal bovine serum (FBS).
   (c) 5 μg/ml gentamicin.
2. Tissue washing medium:
   (a) Phosphate buffered saline (PBS).
   (b) 5 μg/ml gentamicin.
3. 0.25% Trypsin.
4. Bladder tissue—(2.5 cm×2.5 cm) full thickness section from living patient or from fresh cadaveric tissue.
5. 6-well tissue culture dish.
6. 100 mm tissue culture dish.
7. Serological pipettes—10 ml.
8. Forceps (small)—sterile.
9. Microsurgical scissors—sterile.
10. Surgical scalpel—sterile.
11. Pipet-Aid—Drummond Pipet-Aid.
12. Cell culture incubator—humidified, 37°C, 5% CO_2/95% air.
13. Biological safety cabinet.
14. 37°C water bath.

2.2 Enzymatic Digestion Method

1. Collagenase type IV.
2. Dispase II.
3. 500 mM calcium chloride.
4. 0.25% trypsin.
5. SMC growth medium:
   (a) DMEM high glucose.
   (b) 10% FBS.
   (c) 5 μg/ml gentamicin.
6. Tissue washing medium:
   (a) PBS.
   (b) 5 μg/ml gentamicin.

7. Bladder tissue—(2.5 cm × 2.5 cm) full thickness section from living patient or from fresh cadaveric tissue.
8. 6-well tissue culture dish.
9. 100 mm tissue culture dish.
10. 100 μm Steriflip (Millipore Cat# SCNY00100).
11. Serological pipettes—10 ml.
12. Forceps (small)—sterile.
13. Microsurgical scissors—sterile.
14. Surgical scalpel—sterile.
15. Pipet-Aid—Drummond Pipet-Aid.
16. Cell culture incubator—humidified, 37°C, 5% CO_2/95% air.
17. Biological safety cabinet.
18. 37°C water bath.

3 Methods

Carry out all procedures inside a biological safety cabinet unless otherwise specified.

3.1 Explantation Method

3.1.1 Biopsy/Tissue Handling and Preparation

1. Pre-fill the wells of a sterile 6-well plate with 10 ml of pre-warmed tissue washing medium in each well.
2. Using sterile forceps, gently remove the biopsy tissue from the transport container and place in the first well of the 6-well washing plate.
3. Gently agitate the tissue in the well using the sterile forceps for 5–10 s.
4. Carefully lift the tissue from the first well and place in the second well and repeat the agitation procedure above.
5. Continue the successive washing of the tissue through each unused well until all six wells have been used.
6. After washing, carefully move the tissue into a sterile 100 mm tissue culture dish for dissection.

3.1.2 Tissue Dissection

1. Using sterile forceps, orient the tissue so the urothelial cell layer is facing upwards.
2. Using forceps, carefully lift the urothelial layer while cutting it away from the smooth muscle layer with the scissors, and discard.
3. Repeat the lift and cut technique above until all the urothelial layer is removed.
4. Examine the remaining tissue and dissect away any remaining connective tissue, urothelial tissue, fat, or vascular tissue, and discard.

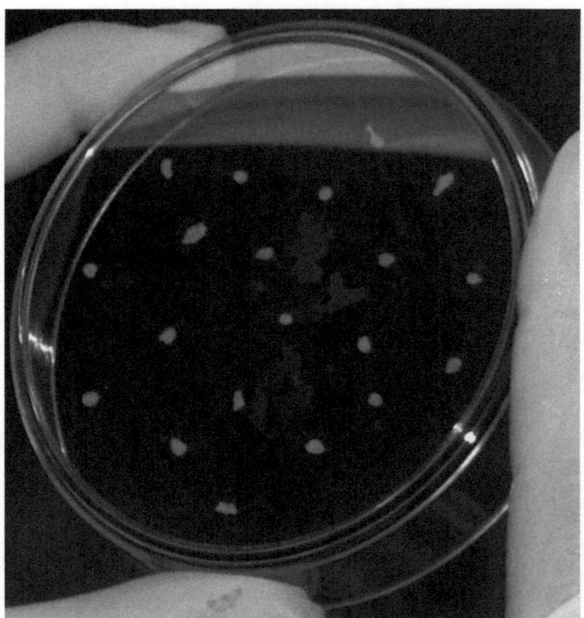

Fig. 1 A 100 mm tissue culture petri dish showing tissue explants adherent to the plastic surface

5. Using sterile forceps and either small sterile scissors or a sterile scalpel, cut small 1 mm diameter pieces of smooth muscle tissue.
6. Using forceps, carefully place each explant onto a 100 mm tissue culture petri dish (see Fig. 1).
7. Continue with the steps above until the dish is evenly covered with 20–25 tissue explants.
8. Allow the plate to sit open inside the biological safety cabinet for 10–15 min to allow the tissue explants time to adhere to the dish (see Fig. 1).
9. Gently add 10 ml SMC growth medium to moisten and submerge tissue fragments without dislodging them from the dish.
10. Place dish into humidified 37°C incubator in 5% CO_2 undisturbed for 3 days.

3.1.3 Smooth Muscle Cell Subculturing

1. After 3 days gently move the explant plates to the biological safety cabinet.
2. Gently aspirate the spent medium from each plate, and replace with 10 ml of fresh SMC growth medium being careful not to dislodge the explants, although some loss is expected.
3. After 5–7 days carefully aspirate any remaining tissue explants from the plate without disturbing the surrounding cell colonies (see Fig. 2).
4. Continue subculturing until the cell colonies around each tissue explant are large and confluent by visual observation (see Fig. 3).

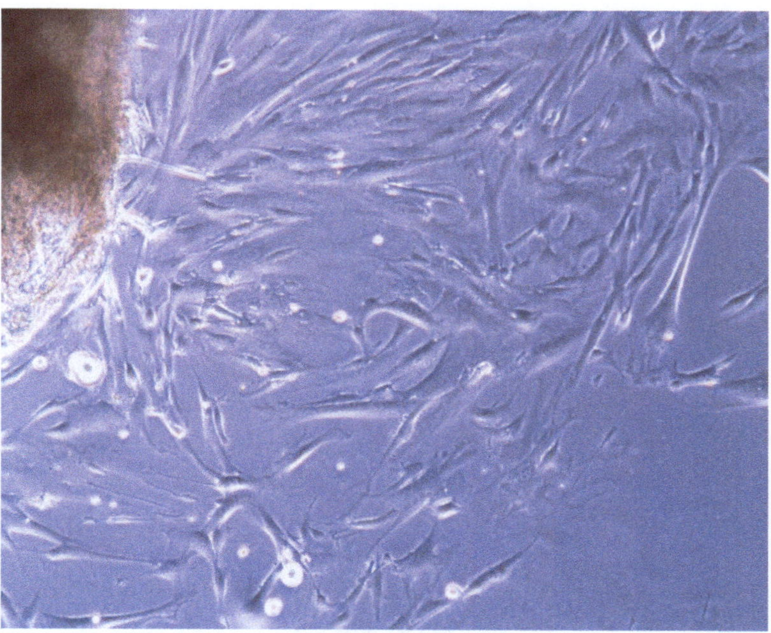

Fig. 2 Phase contrast image of smooth muscle cells migrating away from tissue explant

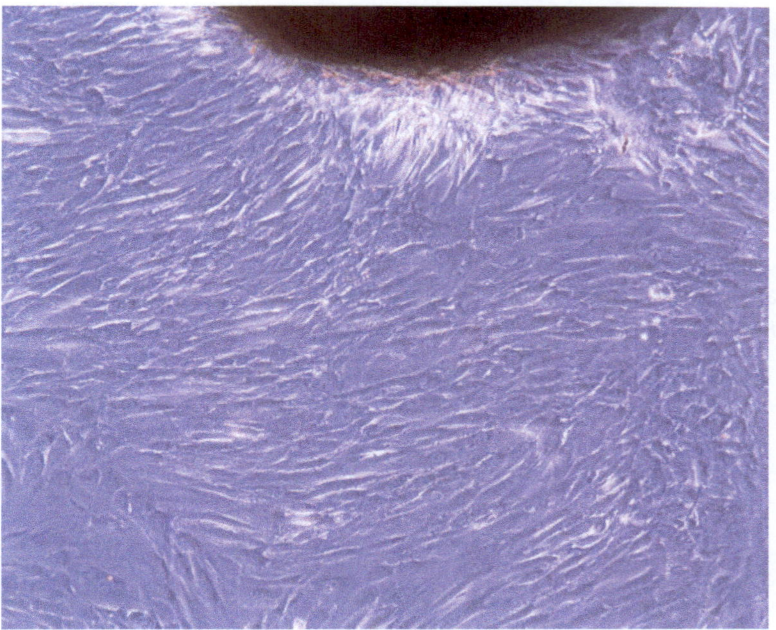

Fig. 3 Phase contrast image of smooth muscle cells surrounding a tissue explant. Proliferation of smooth muscle cells subsequent to migration from tissue explant has resulted in a confluency

5. Upon confluence of the colonies aspirate the spent medium and wash with 5–10 ml of 1× DPBS.
6. Aspirate the DPBS, and then add 5 mL of 0.25% trypsin.
7. Monitor the cell detachment visually.
8. When most of the cells have detached, add 5 ml of SMC growth medium to neutralize the trypsin and detach any lightly attached cells.
9. Transfer the trypsin/medium mixture to a 15 ml centrifuge tube and centrifuge at 300 × *g* for 5 min to pellet the cells.
10. After centrifugation carefully aspirate the supernatant making sure not to disturb the cell pellet.
11. Resuspend the cell pellet in 5 ml of SMC growth medium and count the cells using trypan blue to obtain a total cell number.
12. Subculture using the prior principles and seed continuing cultures at a rate of 2,000–4,000 cells/cm^2 (See Additional Notes 1, 2, and 3).

3.2 Enzymatic Isolation Method

Carry out all procedures inside a biological safety cabinet unless otherwise specified.

3.2.1 Enzyme Preparation

1. Thaw the dispase in a 37°C water bath (20–30 min).
2. Sterilely open the dispase and collagenase in a BSC.
3. Dissolve the collagenase in enough dispase II to make at least 225 ml of solution containing ~450 U/ml (±5) of collagenase IV (e.g., ~450,000 U in 100 ml).
4. Add enough $CaCl_2$ solution to achieve a working concentration of 5 mM.
5. Place the enzyme solution in a 37°C water bath to warm until use.

3.2.2 Biopsy/Tissue Handling and Preparation

1. Pre-fill the wells of a sterile 6-well plate with 10 mL of pre-warmed tissue washing medium in each well.
2. Using sterile forceps, gently remove the biopsy tissue from the transport container and place in the first well of the 6-well washing plate.
3. Gently agitate the tissue in the well using the sterile forceps for 5–10 s.
4. Carefully lift the tissue from the first well and place in the second well and repeat the agitation procedure above.
5. Continue the successive washing of the tissue through each unused well until all six wells have been used.
6. After washing, carefully move the tissue into a sterile 100 mm tissue culture dish for dissection.

3.2.3 Tissue Dissection

1. Using sterile forceps, orient the tissue so the urothelial cell layer is facing upwards.
2. Using forceps, carefully lift the urothelial layer while cutting it away from the smooth muscle layer with the scissors, and discard
3. Repeat the lift and cut technique above until all the urothelial layer is removed.
4. Examine the remaining tissue and dissect away any remaining connective tissue, urothelial tissue, fat, or vascular tissue, and discard.
5. Mince tissue and transfer ~1 g to a 50 ml tube. Repeat for the remainder of the tissue.
6. Add 40 ml of digestion solution, cap tightly, and place on a rocker at 37°C for 1 h (See Note 4).
7. Add 10 ml of SMC growth medium to neutralize the enzyme.
8. Connect the digestion tube to a 100 μm Steriflip, apply a vacuum line, and filter out any remaining connective tissue or large particles.
9. Disconnect the new 50 ml tube from the Steriflip, cap tightly, and centrifuge at 300 × *g* for 5 min.
10. Aspirate the supernatant and resuspend the cell pellet in 40 ml SMC growth medium.
11. Pipet the cell suspension into 4T-75 flasks or 2T-175 flasks.
12. Place dish into humidified 37°C incubator in 5% CO_2 undisturbed for 2–3 days.

3.2.4 Smooth Muscle Cell Subculturing

1. After 3 days remove the flasks from the incubator and evaluate using a light inverted microscope.
2. Place the flasks inside the biological safety cabinet.
3. Using sterile cell culture technique, change the medium on each flask.
4. Continue subculturing until the cells have either generated a confluent monolayer or have developed larger confluent colonies.
5. Upon confluence, aspirate the spent medium and wash with 5–10 ml of 1× DPBS.
6. Aspirate the DPBS, and then add 10–15 ml of 0.25% trypsin.
7. Monitor the cell detachment visually.
8. When most of the cells have detached, add 5 ml of SMC growth medium to neutralize the trypsin and detach any lightly attached cells.

9. Transfer the trypsin/medium mixture to a 50 ml centrifuge tube and centrifuge at 300 × *g* for 5 min to pellet the cells.
10. After centrifugation carefully aspirate the supernatant making sure not to disturb the cell pellet.
11. Resuspend the cell pellet in 40 ml of SMC growth medium and count the cells using trypan blue to obtain a total cell number.
12. Subculture using the prior principles and seed continuing cultures at a rate of 2,000–4,000 cells/cm^2 (See Additional Notes 1, 2, and 3).

4 Notes

1. The smooth muscle cells will begin to enter senescence after about four passages and may begin to decline in growth kinetics as well as morphology.
2. Smooth muscle cells may be cryopreserved effectively in a mixture of 90% smooth muscle cell growth medium/10% DMSO, pipetted into 1 ml cryo-vials and frozen in a Nalgene "Mr. Frosty."
3. Smooth muscle cells may be cryopreserved at any useful concentration, but usually perform their best between 10×10^6 and 50×10^6 cells/ml.
4. Any unused enzyme digestion solution can be stored at −20 to −80°C up to 1 year.

References

1. Jayo MJ, Jain D, Wagner BJ, Bertram TA (2008) Early cellular and stromal responses in regeneration versus repair of a mammalian bladder using autologous cell and biodegradable scaffold technologies. J Urol 180:392–397
2. Basu J, Ludlow JW (2010) Platform technologies for tubular organ regeneration. Trends Biotechnol 28:526–533

Chapter 3

Isolation of Urothelial Cells from Bladder Tissue

Namrata Sangha

Abstract

Presented below is a methodology for the isolation, expansion, and maintenance of urothelial cells derived from human bladder. Such bladder-derived urothelial cells, taken together with bladder or alternately sourced smooth muscle cells, may be complexed with an appropriately shaped biodegradable scaffold to create regenerative constructs capable of seeding formation of new bladder or bladder-like neo-organs upon implantation in human cystectomy patients (1).

Key words Urothelial cells, Bladder, Primary cell culture, Tissue engineering, Regenerative medicine

1 Introduction

Tissue engineering and regenerative medicine approaches offer an alternative, potentially superior methodology to the use of bowel tissue (the current standard of care) for urinary diversion or replacement. In this methodology, the patient's own cells can be sourced from a bladder biopsy and applied to a degradable biomaterial scaffold to create a neo-organ or organ-like construct that, upon implantation within the patient and anastamosis to native components of the urinary system, leads to regeneration of functional, urinary-like neo-tissue capable of storing urine and mediating voiding of urine as needed in response to appropriate neuronal signaling. Such a cell/biomaterial construct catalyzes the regeneration of urinary-like neo-tissue recapitulating native, laminarly organized bladder wall histo-architecture composed of a luminal urothelial layer and multiple smooth muscle layers, appropriately vascularized and innervated. Regeneration of urinary-like neo-tissue is accompanied by progressive degradation of the biomaterial, such that a seamless transition is achieved between the degrading biomaterial and the regenerating urinary-like neo-tissue.

In preliminary experiments to demonstrate the formation of tissue engineered urothelial-like structures in the rabbit, bladder-derived

Joydeep Basu and John W. Ludlow (eds.), *Organ Regeneration: Methods and Protocols*, Methods in Molecular Biology, vol. 1001, DOI 10.1007/978-1-62703-363-3_3, © Springer Science+Business Media New York 2013

urothelial cells were used to seed meshes of nonwoven polyglycolic acid (PGA), which were subsequently implanted within the peritoneal cavity of athymic mice. Upon recovery, structures composed of degrading biopolymer lined with urothelial cells were observed (2). In follow-up studies, combinations of bladder-derived smooth muscle cells and urothelial cells were used to seed tubular-like structures composed of nonwoven PGA mesh. Implantation of these constructs subcutaneously within athymic rabbits led to regeneration of urinary-like tubular organoids composed of urothelial cells lining a central lumen and surrounded by layers of smooth muscle cells, as observed within native bladder tissue. Evidence of neo-vascularization was also noted (3). These studies provided preliminary proof of concept to support the potential for in vivo regeneration of urinary-like neo-organs through implantation of cell-seeded, synthetic biopolymeric scaffolds.

Although current strategies for creation of bladder-like neo-organs leverage principally non-bladder cell sources, initial work on neo-bladder tissue engineering was dependent on patient-derived bladder biopsies as a source of urothelial and smooth muscle cells. For this approach to be commercially viable, the expansion dynamics of cellular growth for both biopsy-derived urothelial and smooth muscle cell populations must be established. Although smooth muscle cells could be reliably expanded from small bladder biopsies without difficulty, the demonstration that a single biopsy-derived source of bladder urothelial cells could also be expanded to the numbers required for effective seeding of urinary neo-organs was critical for establishing the preliminary bioprocess potential of this methodology (4). The alternative would involve multiple biopsy sampling to generate sufficient cell numbers for urinary neo-organ seeding, greatly decreasing the attractiveness of this technology for practical application in the clinic.

Analysis of the dynamics of neo-bladder regeneration in subtotal cystectomized canines serves to further illustrate the dichotomy in outcomes between implantation of acellular and cellularized scaffolds. In another such study, bladder-shaped scaffolds composed of woven PGA felt or PLGA (poly-lactic-*co*-glycolic acid) seeded with autologously sourced bladder-derived urothelial cells and smooth muscle cells facilitated a regenerative response within 1 month post-implantation, as characterized by induction of an extensively vascularized, smooth muscle-like parenchyma. In contrast, acellular PGA/PLGA scaffolds triggered a principally fibrotic, reparative outcome featuring disorganized collagen fibers with minimal vascularization. Baseline urodynamics were reconstituted within 4 months post-implantation with cell-seeded scaffold, whereas urodynamic profiles of animals implanted with acellular scaffolds remained abnormal throughout the 9-month study (5).

In a related cystectomized canine study, native-like trilaminar bladder wall tissue architecture was observed at 3 months post-implantation with a bladder-shaped nonwoven PGA felt scaffold seeded with 1.5×10^8 each of autologously sourced bladder-derived urothelial cells and smooth muscle cells, and normal compliance characteristics of a urinary bladder had developed by 12 months. Regenerated bladders in animals receiving these cell-seeded scaffolds have shown functional and structural stability for up to 2 years post-implantation. Importantly, although the volume of the cell-seeded scaffold was held constant in this particular study, implantation of the construct within dogs of different sizes that had gained varying amounts of weight over the course of the study yielded organs that, as measured by the ratio of bladder capacity to body weight, adapted to the individual recipient animal's size, demonstrating that the regenerated neo-organ was capable of responding to homeostatic mechanisms regulating organ volume (6).

These studies in canines established proof of concept for the application of cell-seeded biodegradable polymeric scaffolds for regeneration of functional, native-like neo-bladders in a large animal cystectomy model. Additional data suggested that smooth muscle cells sourced from diseased bladder could potentially be applied successfully to regenerate neo-urinary tissue (7). These data laid the groundwork for initiation of a proof of concept clinical trial in man. In this seminal study, seven pediatric patients presenting with myelomeningocele (a form of spina bifida) were recruited to receive the first ever human neo-organ implants. As previously described in canines, both urothelial and smooth muscle cells were isolated and expanded from autologously sourced bladder biopsies. Up to five cell passages over 7–8 weeks was required to generate enough cells to seed the neo-bladder scaffold. Using a sterile pipette, the scaffold exterior was seeded with bladder-derived smooth muscle cells at a seeding density of 5×10^7 cells/cm^3. After a 48-h recovery period, the luminal surface of the scaffold was seeded with urothelial cells at a density of 5×10^7 cells/cm^3. The construct was matured in a tissue culture incubator at 37°C for 3–4 days, prior to implantation. Subsequent to implantation, the engineered neo-bladder was cycled (i.e., subjected to serial volume expansion and contraction) as part of regular postoperative care for up to 3 weeks post-implantation; the mechanical forces induced across the neo-bladder during cycling have been found to augment regenerative outcomes. Engineered neo-bladders were found to functionally rescue urologic dynamics and were associated with trilaminar bladder wall architectures upon histological examination of bladder biopsies recovered at 31 months post-implantation (8).

Below, we present a detailed methodology for the isolation, expansion, and maintenance of urothelial cells for application in bladder tissue engineering.

2 Materials

2.1 Tissue Culture

1. Sterile 100 mm and 150 mm tissue culture dishes for working with the bladder biopsy for human study applications. Human bladder typically sourced from NDRI (National Disease Research Interchange, www.ndriresource.org).
2. Sterile 6-well tissue culture plates, 150 mm tissue culture dishes, T75 and T175 tissue culture flasks.
3. 150–500 ml sterile disposable plastic bottles.
4. 15 ml and 50 ml disposable sterile centrifuge tubes.
5. Sterile pipettes: 1, 5, 10, and 25 ml.
6. Sterile pipette tips: 200 μl.
7. Absorbent wipes.
8. Hemocytometer with cover slips.
9. 1.5 ml microcentrifuge tubes.
10. Biopsy transport medium: Dulbecco's Modified Eagle's Media (DMEM) + 50 μg/ml gentamicin.
11. Biopsy Wash Medium 1: DMEM + 50 μg/ml gentamicin.
12. Biopsy Wash Medium 2: DMEM + 5 μg/ml gentamicin.
13. 30% OptiPrep solution: 50% of stock 60% (w/v) Iodixanol + 50% of Keratinocyte Serum Free Medium (K-SFM), (Invitrogen).
14. Urothelial cell (UC) growth medium: K-SFM + 5 ng/ml EGF (epidermal growth factor) + 50 μg/ml bovine pituitary extract (BPE) + gentamicin (5 μg/ml).
15. UC cryopreservation solution: UC growth medium, 10% DMSO, 10% FBS.

2.2 Reagents

1. 70% ethanol.
2. Sterile DPBS (phosphate buffered saline).
3. Stock 60% (w/v) iodixanol (OptiPrep™ Sigma cat # D1556).
4. 0.4% trypan blue.

2.3 Other Equipment

1. Pipettor—Drummond Pipetman.
2. Pipettors—P20, P100.
3. Sterile forceps.
4. Sterile scalpel and sterile disposable blades.
5. Hemocytometer with cover slips.
6. Inverted microscope for cell culture analysis.
7. Incubator—humidified, 37°C, 5% CO_2, and air.
8. -80°C freezer.
9. Liquid nitrogen freezer (long-term storage).

3 Methods

3.1 UC Cell Extraction

1. Clean the biosafety cabinet (BSC) with 70% ethanol (EtOH) before handling tissue (Table 1).
2. Place sterile instruments and 150 mm tissue culture dishes in the clean BSC to start processing biopsy for UC isolation.
3. From the surgical site, the biopsy tissue size of 1 × 1 cm is collected into a 50 ml centrifuge tube containing biopsy transport medium.
4. Once tissue is collected, it can be stored at 4°C until ready to process for cell isolation. Overnight hold at 4°C is preferred to achieve best cell yield.
5. Biopsies can be held up to 96 h in biopsy transport medium at 4°C prior to handling for UC isolation.
6. Spray down the 50 ml centrifuge tube with 70% EtOH and place inside the BSC.
7. Collect the spent biopsy transport medium for sampling of bio-burden and sterility testing (if needed, will not be discussed further here).
8. Place biopsy tissue from the 50 ml centrifuge tube into a clean 150 mm tissue culture dish or sterile disposable bottle.
9. Wash the biopsy by adding fresh wash medium (1) containing (50 μg/ml gentamicin) to the dish or the bottle at room temperature (RT) in the BSC and incubate the wash for 20 min. Aspirate spent wash medium.
10. Repeat above step with fresh wash medium (2) containing (5 μg/ml gentamicin). Aspirate as before.
11. Using clean instruments, trim fat or mucosal tissue off of the biopsy.

Table 1
Medium for cell isolation

Biopsy transport medium	• DMEM with Gentamicin (50 μg/ml)	Invitrogen-Gibco
Wash medium	• DMEM with Gentamicin (50 μg/ml) • DMEM with Gentamicin (5 μg/ml)	Invitrogen-Gibco
OptiPrep solution	• 50% of stock 60%(w/v) Iodixanol • 50% of K-SFM	Sigma Invitrogen-Gibco
Urothelial cell (UC) growth medium	• K-SFM • 5 ng/ml EGF • 50 μg/ml of bovine pituitary extract • Gentamicin (5 μg/ml)	Invitrogen-Gibco[a]

[a]This information better placed in Subheading 3

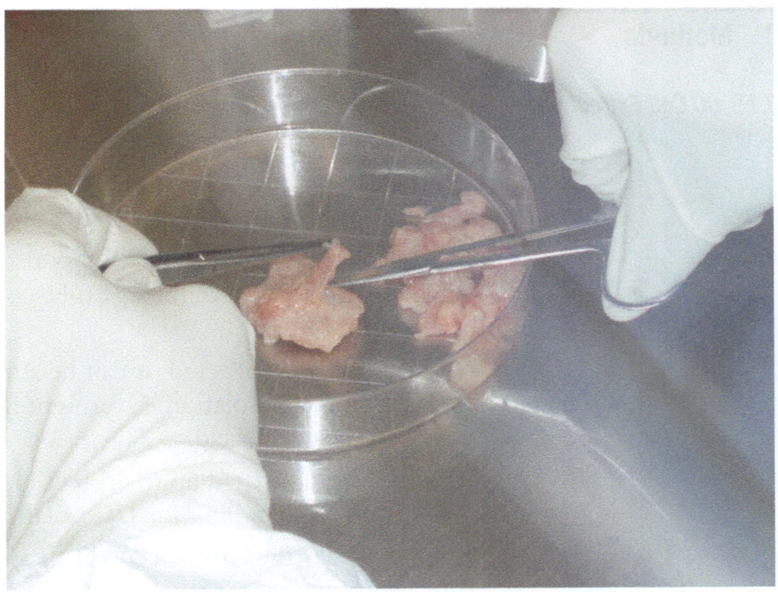

Fig. 1 Detachment of UC layer from the smooth muscle layer

12. Detach the urothelial layer of the biopsy using scissors and forceps (see Note 1, Fig. 1).
13. Transfer the detached UC layer into a sterile 100 mm dish containing 3.5 ml of UC growth medium (K-SFM + 5 ng/ml EGF, 500 μg/ml BPE, and 5 μg/ml Gentamicin sulfate) (see Note 2).
14. Leave the smooth muscle layer (minus the urothelium) tissue in the 150 mm dish for extracting smooth muscle cells or discard tissue.
15. Detach UC by gentle scraping of the lumenal side of the biopsy into the UC growth medium with a sterile scalpel blade. Continue scraping until tissue discoloration is observed and the medium gets cloudy with the suspended cells. Place the tethered, scraped UC layer to one side of the dish (see Note 3, Fig. 2).
16. Suspend the detached UC in 3.5 ml of UC growth medium in a 100 mm tissue culture dish with a 5 ml pipette. Dispense the cells up and down against the edge of the dish to dislodge any clumps for even mixing and dispersion of the cells (see Note 4).
17. Collect all the scraped UC into a 15 ml disposable centrifuge tube and bring up the volume to 5 ml of re-suspended cell suspension with fresh UC growth medium.
18. Prepare a 30% OptiPrep solution from the stock solution.
19. Add 5 ml of 30% OptiPrep solution to a clean 15 ml disposable centrifuge tube.

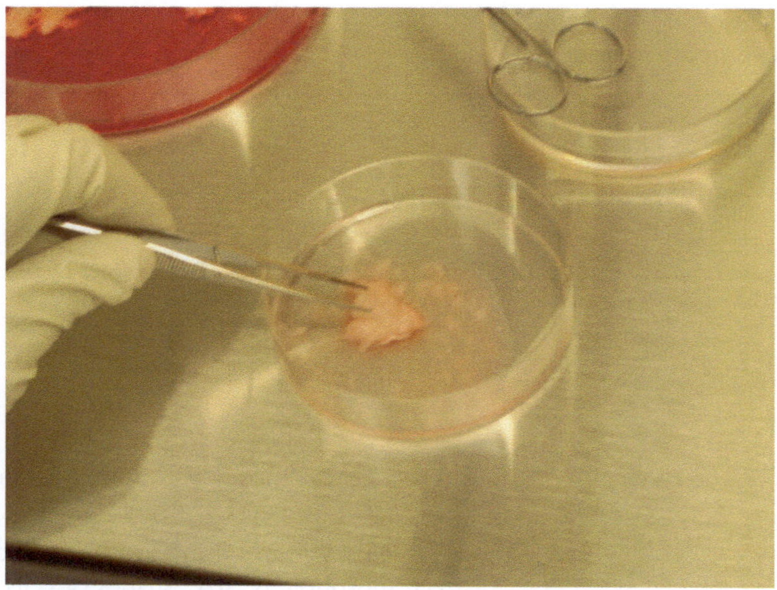

Fig. 2 UC cells in medium post scraping

20. To the tube above, add 5 ml of re-suspended UC suspension from step 17, thoroughly mix the solution by inverting the tube.
21. Overlay the OptiPrep solution/UC mixture with 1 ml of K-SFM, ensuring that the two phases do not mix.
22. Centrifuge the tube at 800×*g* for 15 min at 4°C with no brakes.
23. Take the centrifuged tubes out gently; upon inspection, the UC will be sedimented in a band at the phase interface.
24. Collect the banded UC with a sterile pipette into a 50 ml disposable centrifuge tube.
25. Dilute the collected cell suspension with UC growth medium into a 50 ml disposable centrifuge tube and mix thoroughly so that any residual OptiPrep can be washed out.
26. Centrifuge 300×*g* for 15 min.
27. Count cells using the hemocytometer.
28. Calculate number of cells needed to achieve a plating density of 10,000–50,000 viable cells/cm^2 (see Note 5).
29. Target the cells to be seeded onto a 6 well tissue culture plate/ or other similar vessel to accommodate the cell population captured.
30. Suspend cells in appropriate volume of UC growth medium; dispense into appropriate volume/vessel of choice (see Note 5).
31. Place tissue culture vessels into a humidified 37°C incubator in 5% CO_2.

Table 2
Cell growth medium

UC wash medium	• DPBS with 5 μg/ml Gentamicin	Invitrogen-Gibco
Urothelial cell (UC) growth medium	• K-SFM • 5 ng/ml EGF • 50 μg/ml of bovine pituitary extract • Gentamicin (5 μg/ml)	Invitrogen-Gibco[a]

[a]This information better suited to Subheading 2

3.2 UC Feeding

1. Day 1 (next morning), remove UC culture plates containing extracted human UC (Table 2).
2. Visualize cultures using inverted cell culture microscope. Note cell morphology, size, shapes of cells, and presence of floaters (Fig. 3).
3. Place the dishes in the BSC. Aspirate the medium from the UC culture vessels.
4. Wash cells by adding appropriate amount of DPBS containing 5 μg/ml gentamicin and gently swirl the culture vessel. Aspirate off the wash medium.
5. View the washed UC culture vessel under the microscope; no further washing is necessary if the dish appears clean of cellular debris.
6. Add appropriate amount of UC growth medium to vessel of choice (see Note 5).
7. Return culture plates to a humidified 37°C incubator in 5% CO_2.
8. Monitor cultures on daily basis, notating cell morphology, cell size and shape, cell debris, floaters, and confluency (see Note 6, Fig. 4).
9. Feed cultures every 2–4 days with UC growth medium.

3.3 UC Passaging

1. *Either* at day 7–12 after initial seeding, *or* if the P0 UC cultures reach a confluency of ≥70% or greater, cultures will need to be passaged (see Notes 7–10) (Fig. 5 and Table 3).
2. Remove UC culture dish(s), containing extracted human UC, from the incubator.
3. Visualize UC using the inverted microscope, checking for morphology and confluence.
4. Place the culture vessels in the BSC.
5. Aspirate the medium from the culture vessels containing UC.

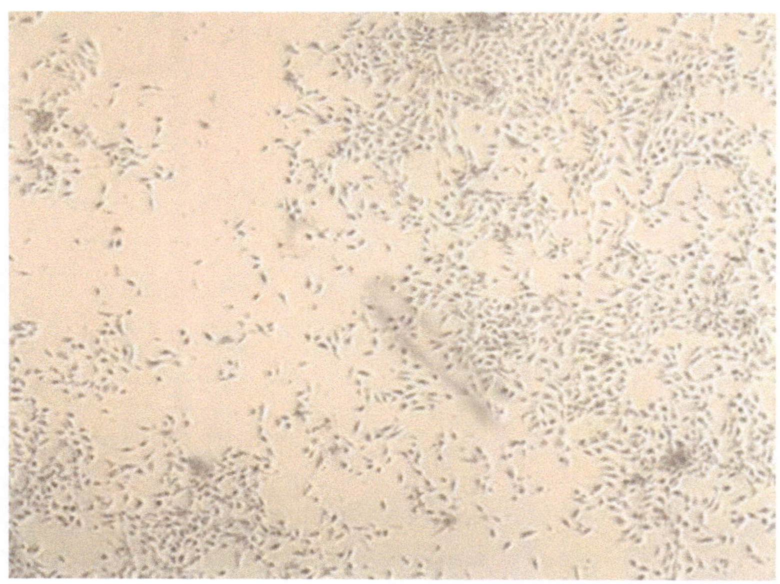

Fig. 3 UC cells—small round, elongated shapes, and some cell debris

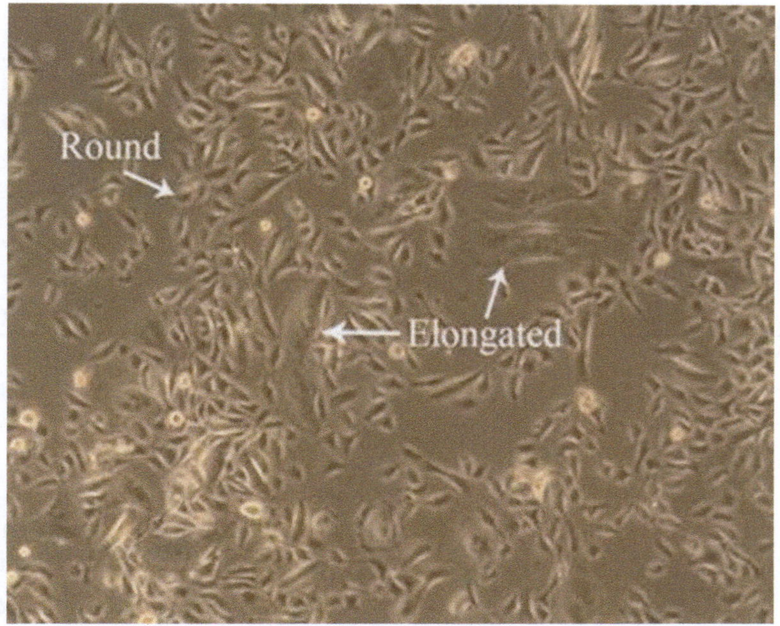

Fig. 4 UC cells—size/shape of cells, confluency

6. Wash cells by adding appropriate amount of DPBS and gently swirl the culture vessel.
7. Aspirate DPBS.
8. Add appropriate amount (see Note 5) of 0.25% Trypsin/ EDTA per culture vessel.

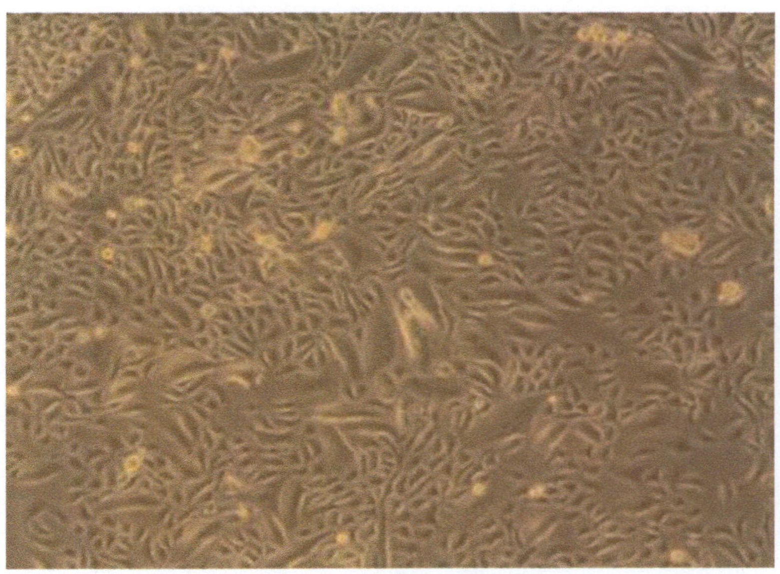

Fig. 5 UC culture ready to be harvested

Table 3
Cell harvesting medium

Wash medium	• DPBS	Invitrogen-Gibco
Trypsinization solution	• 0.25% Trypsin/EDTA	Invitrogen-Gibco
Neutralization solution	• DPBS • 5% FBS	Invitrogen-Gibco
Urothelial cell (UC) growth medium	• K-SFM • 5 ng/ml EGF • 50 μg/ml of bovine pituitary extract	Invitrogen-Gibco[a]

[a]Move to Subheading 2

9. Incubate the UC culture vessels in the BSC for 2–3 min, monitoring detachment under the microscope.
10. Gently tap the dishes/flasks and immediately start washing the vessel surface with a pipette to collect all detached cells.
11. Collect and transfer the trypsinized UC cells to a disposable centrifuge tube containing the 5% fetal bovine serum (FBS) in DPBS to neutralize the trypsinized cells.
12. Centrifuge at 300 × g for 5 min to pellet the cells.
13. Aspirate the supernatant; resuspend the cells in appropriate amount of UC growth medium.

Table 4
Cell freezing medium

UC wash solution	• DPBS	Invitrogen-Gibco
Trypsin neutralization solution	• DPBS • 5% FBS	Invitrogen-Gibco
Cryopreservation solution (UC)	• UC growth medium • 10% DMSO • 10% FBS	Invitrogen-Gibco Sigma-Aldrich[a]

[a]Move to Subheading 3

14. Count and plate culture vessels at 7,000 cells/cm^2.
15. Add appropriate amount of UC growth medium to the re-suspended cells to plate into the culture vessel of choice (see Note 5).
16. Place culture vessel to a humidified 37°C incubator in 5% CO_2.
17. Cultures are monitored on daily basis (Figs. 2 and 3).
18. Feed UC cultures every 2–4 days.
19. UC can be expanded for 4–7 passages (see Notes 7–10).

3.4 UC Cryopreservation

1. UC cultures to be harvested need to be at least 70% or greater in confluency (Table 4).
2. Handle UC passage as noted under Subheading 3.3, UC passaging.
3. After centrifugation, resuspend the UC cell pellet in a small volume of UC growth medium.
4. Count the cells using trypan exclusion with a hemocytometer.
5. Centrifuge the re-suspended cells at $300 \times g$ for 5 min.
6. Determine the cell number/ml; we suggest $5–50 \times 10^6$/vial for freezing.
7. Obtain and label the number of vials necessary to freeze the amount of cells at hand.
8. Aspirate after centrifugation and resuspend the pellet in a drop-like fashion with UC cryopreservation solution.
9. Once pellet is fully re-suspended in UC cryopreservation solution, aliquot 1 ml/vial at a defined UC concentration.
10. Place labeled vials into a cryogenic container for freezing at −80°C freezer.

11. Post overnight storage at −80°C, move the cryo-vials to liquid nitrogen for long-term storage.

4 Notes

1. If the two layers (UC layer and SMC layer) are not cleanly dissected/separated, a heterogenous cell population is likely. However, the continual culture feeding with UC growth medium (containing no serum) will select for the UC population.
2. Gentamicin was added to the medium at P0 to reduce the risk of contamination. No adverse effect on cell expansion was noted with this addition of antibiotic into the medium.
3. Gentle and deliberate scraping is necessary to obtain the UC. Once the medium appears cloudy, and the tissue looks tethered, it is an indication that the tissue can be discarded and the medium with the scraped cells can be collected to proceed to the next UC processing step.
4. If the cells look clumpy, they need to be mixed gently with a sterile pipette to break up the clumps and diluted with more UC growth medium before further processing.
5. Guidelines for UC seeding density, cell numbers, type of vessel, volume of trypsin/vessel, and targeting volumes for media/vessel were based on recommendations from the following websites: www.protocol-online.org/prot/Protocols/Recommended-working-medium--trypsin-volume-and-cell-inoculation-density-49.html *and* bts.ucsf.edu/desai/protocols/Working%20Volumes%20for%20Tissue-Culture%20Vessels.xls.
6. Properly dispersed UC attach readily to primaria dishes at initial seeding and then NUNC flasks/plates with continual passaging. At initial seeding, there is an unusual amount of cell/tissue debris which lessens with every passage of the cultures.
7. Properly handled cultures can be maintained for 4–10 passages before cells senesce.
8. All cultures have a growth phase of 5–7 days with the exception of initial cultures which can vary from 7 to 10 days before harvesting. Much of this is dependent on biopsy variability, seeding density, and passage number.
9. UC cultures should be carried out at >70% confluency to ensure proper growth phase.
10. UC populations maintain healthy growth kinetics, function, and visual morphology when following these criteria.

References

1. Basu J, Ludlow JW (2010) Platform technologies for tubular organ regeneration. Trends Biotechnol 28:526–533
2. Atala A et al (1992) Formation of urothelial structures in vivo from dissociated cells attached to biodegradable polymer scaffolds in vitro. J Urol 148:658–662
3. Atala A et al (1993) Implantation in vivo and retrieval of artificial structures consisting of rabbit and human urothelium and human bladder muscle. J Urol 150:608–612
4. Cilento BG et al (1994) Phenotypic and cytogenetic characterization of human bladder urothelia expanded in vitro. J Urol 152: 665–670
5. Jayo MJ et al (2008) Early cellular and stromal responses in the regeneration of a functional mammalian bladder. J Urol 180:392–397
6. Jayo MJ et al (2008) Long-term durability, tissue regeneration and neo-organ growth during skeletal maturation with a neo-bladder augmentation construct. Regen Med 3:671–682
7. Lai JY et al (2002) Phenotypic and functional characterization of in vivo tissue engineered smooth muscle from normal and pathological bladders. J Urol 168:1853–1858
8. Atala A et al (2006) Tissue engineered autologous bladders for patients needing cystoplasty. Lancet 306:1241–1246

Chapter 4

Isolation of Pulsatile Cell Bodies from Esophageal Tissue

John W. Ludlow, Joydeep Basu, Christopher W. Genheimer, Kelly I. Guthrie, and Namrata Sangha

Abstract

Pulsatile cell bodies, three-dimensional cell clusters with satellite streaming cells, can be isolated from esophageal tissue. One of the key features of these clusters is that they pulsate at rhythmic rates and demonstrate contractility under several in vitro conditions. Their ability to pulsate appears to be due to the presence of interstitial cells of Cajal (ICC), which mediate signal transmission from nerve to muscle cells. As predicted, the cells comprising these clusters express phenotypic and genotypic markers characteristic of smooth and skeletal muscle, neuronal, and epithelial cells. Because of the critical role of ICC in gastrointestinal tract motility, loss of function in these cells can result in a variety of pathologies. Cultures of pulsatile cell bodies may have utility as an in vitro model to study tissue engineering and regenerative medicine approaches to treating defects in gastrointestinal rhythmicity due to disease or injury.

Key words Interstitial cells of Cajal, ICC, Pulsatile cell bodies, Peristalsis, Rhythmic contraction, Cell clusters, Organoids, 3-D cultures

1 Introduction

The interstitial cells of Cajal (ICC) are specialized gastrointestinal cells responsible for the regulation of intestinal rhythmic motility (reviewed in ref. 1). The evidence to date supports that ICC converts chemical signals from myenteric nerves into electrical potentials which are then propagated to smooth muscle cells through gap junctions (2, 3). Cellular distribution and density of these cells appear to be important for normal gut motility, as a reduced number of these cells correlates with abnormal peristalsis (2, 4).

Isolated cell clusters resulting from incomplete enzymatic digestion of gastrointestinal (GI) tissue exhibit rhythmic rates of contraction and relaxation under certain culture conditions (5). These cell clusters contain multiple cell types, including ICC, epithelial, smooth muscle, and neuronal (6). As such, these clusters recapitulate, to a degree, the structure and function of the intact

Joydeep Basu and John W. Ludlow (eds.), *Organ Regeneration: Methods and Protocols*, Methods in Molecular Biology, vol. 1001, DOI 10.1007/978-1-62703-363-3_4, © Springer Science+Business Media New York 2013

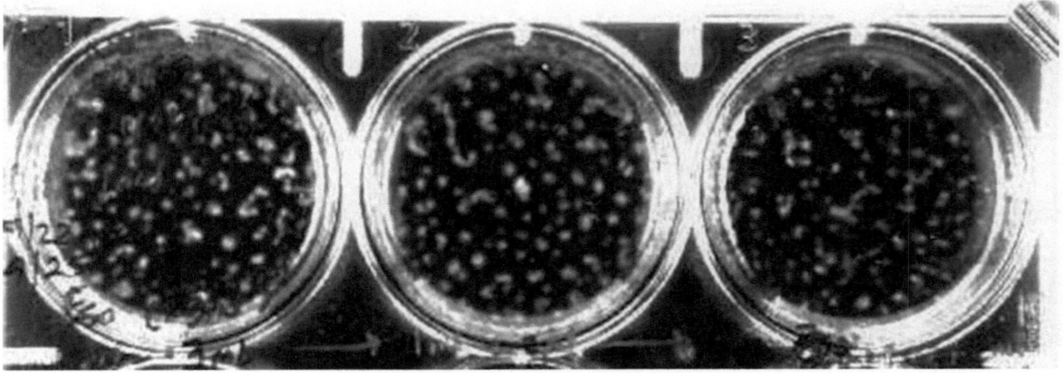

Fig. 1 Pulsatile bodies formed on Matrigel-coated wells of a 24-well dish after 6 weeks in culture

tissue of the organ from which they were isolated. Sometimes referred to as "organoids," the advantage of using these types of clusters in culture is that they allow for a three-dimensional (3-D) model of cells to be tested. Such 3-D models have a proven utility in bridging the gap between traditional two-dimensional (2-D) cell cultures and in vivo animal studies. These 3-D models are designed to emulate certain salient features found in vivo, such as different cell types being in close contact and interacting with one another, thus taking advantage of well-defined aggregation geometries while concurrently leveraging the same tools used to study cells in traditional cell cultures. Traditional 2-D cultures do not allow researchers to test questions pertinent to an integrated cell response as a 3-D culture does, particularly those addressing questions relevant to developing tissues and organs.

Pulsatile cell bodies are defined here as 3-D cell clusters with satellite streaming cells exhibiting rhythmic contraction and relaxation. Figure 1 illustrates cluster formation observed at a gross level through a light microscope. Higher magnification reveals the 3-D structure and higher order geometry of the aggregation (Fig. 2). Connection of pulsatile bodies with one another via structures reminiscent of neuronal axons (Fig. 3) enables coordinated and sequential contraction and relaxation visualized as a "wave" of pulsation from one body to the other.

The method below describes conditions for isolating and culturing from rat esophagus. The same method may be applied to other species, as well as other GI tissues, such as stomach and small intestine. Pulsatile cell body cultures provide an in vitro tool for addressing questions around organs and tissues which have disruptions in their electropotential. In so doing, these cultures facilitate research and development in tissue engineering and regenerative medicine approaches to treating defects in gastrointestinal rhythmicity due to disease or injury.

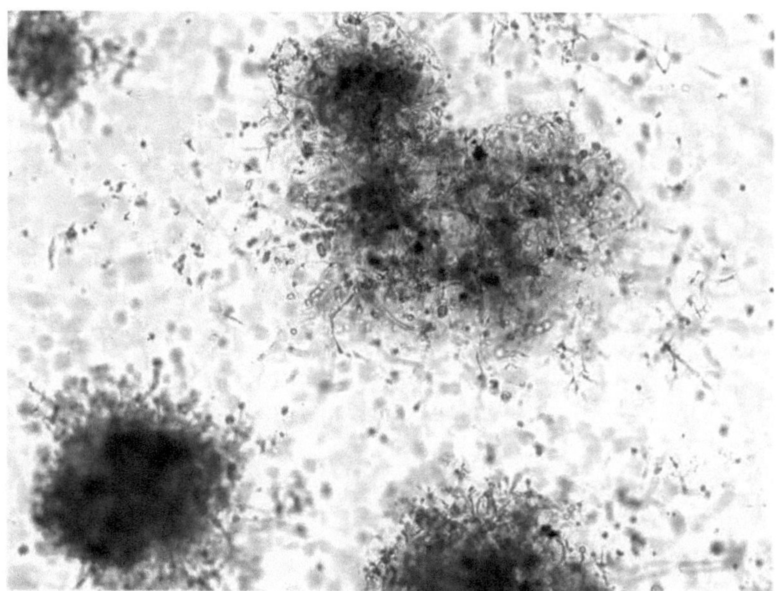

Fig. 2 Higher magnification (×20) of pulsatile bodies. Note higher order geometry of the aggregations and satellite streaming cells at the outer edges of the bodies

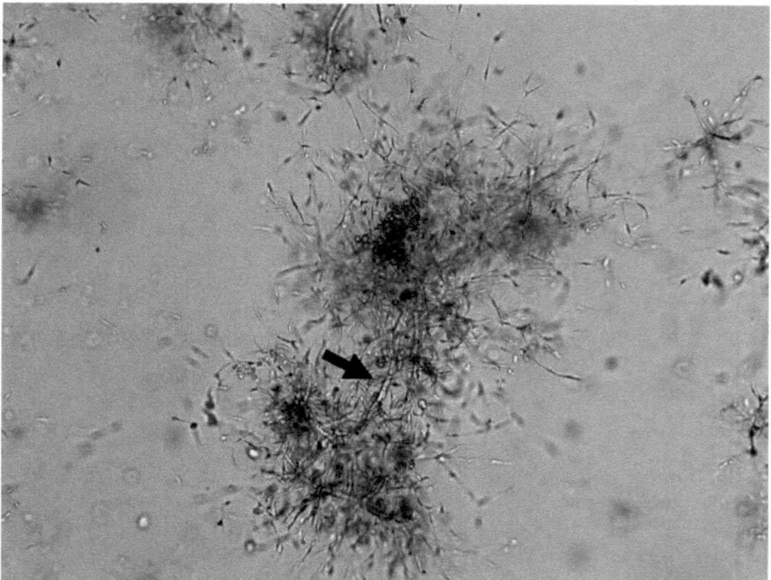

Fig. 3 Connection of pulastile bodies with one another via structures reminiscent of neuronal axons (see *arrow* at the center between bodies) enables coordinated and sequential contraction and relaxation, visualized as a "wave" of pulsation from one body to the other (×20 magnification)

2 Materials

2.1 Adult Rat Esophagus

1. Lewis Strain.
2. Charles River Laboratories International Inc., Wilmington, MA, USA).

2.2 Culture Plasticware

1. Sterile Petri dish, 100 mm.
2. Pipettes, 5, 10, 25 mL volumes.
3. Sterile 24-well tissue culture dish.
4. Centrifuge tube, 50 mL.
5. Filter sterilization unit, 0.22 μm.

2.3 Stainless Steel Instruments and Surgical Material

1. Scissors.
2. Scalpel.
3. Forceps.
4. Hemostats.
5. Betadine surgical scrub solution.
6. Sterile gauze pads, 12-ply, 4×4 in.

2.4 Equipment

1. Class II biosafety cabinet.
2. Centrifuge with swinging-bucket rotor.
3. Variable speed test tube mixer.

2.5 Culture Media Components and Supplements

Storage conditions, shelf life, and expiration dates for all media, components, and supplements are provided by the manufacturer.

1. Dulbecco's Modified Eagle Medium, high glucose (DMEM-HG), containing sodium pyruvate (Invitrogen, Life Technologies Corp., Grand island, NY, USA).
2. Kerotinocyte Serum-Free Medium (KSFM) containing L-glutamine and supplied with prequalified human recombinant epidermal growth factor 1–53 (EGF 1–53) and bovine pituitary extract (BPE) in separate packaging (Invitrogen).
3. Fetal bovine serum (Invitrogen).
4. Insulin–Transferrin–Selenium G solution (ITS) 100× (Invitrogen).
5. Antibiotic/Antimycotic, 100× (Invitrogen).
6. Sorbitol (Sigma-Aldrich, St. Louis, MO, USA).
7. Gentamicin, 50 mg/mL concentration (Invitrogen).
8. Collagenase Type XI (Sigma).
9. Dispase, 5 mg/mL concentration, in Hank's Buffered Saline Solution (STEMCELL Technologies, Vancouver, BC, Canada).
10. Matrigel solution (BD Biosciences, San Jose, CA, USA).

2.6 Formulations

All manipulations take place in a biosafety cabinet to reduce the risk of microbial contamination.

1. Digestion solution—300 U/mL collagenase XI in appropriate volume of Dispase solution. Sterilize through a 0.22 μm filter. Make fresh right before use. Excess may be stored at −20°C for up to 1 month, thawed once, and used immediately.
2. Growth medium—1:1 mixture DMEM-HG: KSFM supplemented with 10% FBS, kit supplements (EGF 1–53 and BPE), 1× ITS, and 0.1% antibiotic/antimycotic. Store at 4°C, shelf life of 1 month.
3. Wash medium—DMEM-HG containing 50 μg/mL gentamicin and 0.1% antibiotic/antimycotic. Store at 4°C, shelf life of 1 month.
4. Neutralization medium—DMEM-HG containing 2% FBS, and 0.1% antibiotic/antimycotic. Store at 4°C, shelf life of 1 month.

3 Methods

3.1 Preparation of Matrigel Plates

Matrigel is a reconstituted basement membrane preparation comprising approximately 60% laminin, 30% collagen type IV, and 8% enactin (7).

1. Thaw the Matrigel in an ice-water bath (see Note 1).
2. Dilute Matrigel to 1 mg/mL final concentration in serum-free, ice-cold DMEM-HG.
3. Add 1 mL of 1 mg/mL Matrigel solution per well of a 5-well plate (see Note 2).
4. Gently rotate the plate to ensure that the entire bottom surface of each well is coated with the Matrigel solution (see Note 3).
5. Allow the plates to gel at room temperature for 1 h or overnight at 4°C.
6. Wash each well with 1 mL growth medium before plating with cells.

3.2 Esophageal Organoid Isolation and Pulsatile Body Culturing

The procedure for isolating esophageal organoids, from which pulsatile bodies are cultured, was modified from Evans et al. (8). Unless otherwise stated, all manipulations take place in a biosafety cabinet to reduce the risk of microbial contamination.

1. Sacrifice rat according to IACUC-approved methods (see Note 4).
2. With the rat firmly secured on its back, flood the thoracic area with the antiseptic solution betadine.
3. Blot excess betatine with sterile gauze pads.

4. Using scissors, trim fur to as close to the skin as possible.
5. Repeat steps 2 and 3.
6. With a scalpel, make a vertical incision running the entire length of the thoracic area.
7. Carefully dissect away skin, underlying fascia, and muscle tissue to expose the esophagus.
8. Secure esophagus with forceps or hemostat and cut a section of esophagus with scissors (see Note 5).
9. Excised rat esophageal tissue (approximately 5 mm in length, weighing 0.2–1.0 g) is placed in a sterile 100 mm Petri dish and a longitudinal incision is made along the entire length of the organ (see Note 6).
10. Wash the tissue by adding 10 mL of wash medium to the Petri dish and gently swirl for 10 s (see Note 7).
11. Aspirate the medium.
12. Repeat steps 2 and 3 an additional two more times, for a sum total of three washes (see Note 8).
13. Transfer the washed tissue to a new clean, sterile Petri dish (see Note 9).
14. Using a scalpel, mince the tissue into pieces having a size <1 mm (see Note 10).
15. Transfer minced tissue to a 50 mL centrifuge tube.
16. Add 20 mL of the digestion solution to the centrifuge tube.
17. Using a test tube platform mixer, with an angle of 48° and a setting of 15 rpm, mix the tissue suspended in the enzyme for 25 min on the bench top at room temperature (see Note 11).
18. Take the centrifuge tube back into the biosafety cabinet and vigorously pipette the suspension 150 times with either a 25 or a 50 mL pipette (see Note 12).
19. Sediment by gravity (1 min).
20. Aspirate supernatant and discard.
21. Repeat steps 13–17 an additional two more times for a sum total of three digests (see Note 13).
22. At the end of the final digest, sediment by gravity for 1 min and collect supernatant in a clean tube.
23. Add an equal volume of neutralization medium and mix.
24. Centrifuge at $50 \times g$ for 2–3 min.
25. Aspirate supernatant and discard.
26. Repeat steps 20–22 an additional two more times for a sum total of three times.
27. Resuspend pellet in growth medium and plate onto Matrigel.

28. Incubate at 37°C in a humidified, 5% CO_2-containing atmosphere.
29. Change medium every 2–3 days as needed, until pulsatile bodies are observed under a light microscope (see Note 14).

4 Notes

1. This may take overnight. To ensure that the solution does not get above 10°C, at which point it will begin to gel, place the ice-water bath in a 4°C refrigerator or cold room. Alternatively, you may warm the solution more rapidly by hand or in tepid water, being careful that there is some frozen portion still remaining, and then place into the ice-water bath.
2. Make sure that all pipettes and pipette tips are cooled to 4°C before transfer, as Matrigel mixture may begin to gel if they are at room temperature. Thickness of the Matrigel layer may be adjusted by decreasing (thinner layer) or increasing (thicker layer) the volume pipetted into the well. Gel thickness (μL/cm^2) is calculated by dividing the growth area (cm^2) of the culture vessel by the volume (μL) of Matrigel used. For example, the 1 mL volume per well of a 6-well dish (growth area = 9.6 cm^2) will give an approximate thickness of 104 μL/cm^2. BD Biosciences defines a thin gel as 50 μL/cm^2, whereas a thick gel is defined at 150–200 μL/cm^2 (7).
3. This is especially important to ensure complete coverage of the well bottom if a thinner layer is targeted.
4. The method we used was anesthetization by continuous inhalation of isoflurane until the animal was unconscious, followed by cervical dislocation.
5. Securely hold esophagus during cutting since the tissue has a tendency to snap back into the neck region or lower thoracic area, making retrieval most difficult.
6. The purpose is to open the esophagus, thus changing the structure of the tissue from a tube to a flat sheet to gain better access to the luminal epithelial layer. Forceps and scissors are preferred over the use of a scalpel.
7. The lumen of the esophagus is exposed to the outside environment, and is therefore not sterile. As such, it is important to remove any loosely adhered material from the inner lumen to reduce the chance of contaminating the tissue culture.
8. Successive washes with these volumes will give a theoretical 1,000-fold dilution of any microbes loosely adhering to the tissue. Combined with the presence of gentamicin and anti/anti, resulting cultures should be at minimal risk for microbial contamination.

9. Transferring to a new clean, sterile Petri dish further reduces the risk of carrying over microbial contamination to subsequent tissue manipulation steps.
10. Digestion efficiency for minced tissue greater than 1 mm in size is greatly reduced at the time and temperature used, thereby resulting in fewer pulsatile bodies isolated.
11. Increasing the rocking time or the temperature will increase the efficiency of the digestion, which can result in fewer pulsatile bodies isolated.
12. The purpose here is to free the partially digested tissue clusters from the larger minced tissue. Depending upon the final size of the minced tissue, it may not pass through the bore of a 25 mL pipette, so a 50 mL pipette may be required to efficiently disrupt the tissue.
13. A series of partial digestions are better for isolating pulastile bodies compared to a single more complete digestion.
14. Pulsatile bodies start to be observed between 10 and 15 days after culturing. Pulsatile function can take up to 6 weeks before being observed.

References

1. Takaki M (2003) Gut pacemaker cells: interstitial cells of Cajal (ICC). J Smooth Muscle Res 39:137–161
2. Kilic A, Luketich JD, Landreneau RJ et al (2008) Alterations in the density of interstitial cells of Cajal in achalasia. Dig Dis Sci 53:1488–1492
3. Sanders KM, Ordog T, Koh SD et al (2000) A novel pacemaker mechanism drives gastrointestinal rhythmicity. News Physiol Sci 15:291–298
4. Midrio P, Alaggio R, Strojna A et al (2010) Reduction if interstitial cells of Cajal in esophageal atresia. J Pediatr Gastroenterol Nutr 51: 610–617
5. Ito-Dufros Y, Funakoshi Y, Uehara A et al (2007) In vitro development of gut-like tissue demonstrating rhythmic contractions from embryonic mouse intestinal cells. Neurogastroenterol Motil 19:288–300
6. Epperson A, Hatton WJ, Callaghan B et al (2000) Molecular markers expressed in cultured and freshly isolated interstitial cells of Cajal. Am J Physiol Cell Physiol 279:C529–C539
7. BD Matrigel frequently Asked Questions. http://www.bdbiosciences.com/documents/BD_CellCulture_Matrigel_FAQ.pdf. Accessed December 15, 2011
8. Evans GS, Flint N, Somers AS et al (1992) The development of a method for the preparation of rat intestinal epithelial cell primary cultures. J Cell Sci 101:219–231

Chapter 5

Cell Isolation Through Whole-Liver Perfusion and Preparation of Hepatocytes for Cytochrome P450 Analysis

Sonya O. Meheux

Abstract

Activity of cytochrome p450 (CYP450) enzyme is used to measure the ability of hepatocytes to metabolize pharmaceutical compounds. When determining functionality of hepatocytes, the cells may be induced in order to determine metabolic activity after drug induction. Hepatocytes, whether in suspension or plated, are used in the pharmaceutical industry as a surrogate to assess in vivo drug metabolism. Within this chapter, isolation of hepatocytes from whole-liver tissue and subsequent preparation for CYP450 isotype 3A4 activity is discussed.

Key words Cytochrome p450, Human hepatocyte isolation, HPLC, Rifampicin induction

1 Introduction

Cytochrome P450 (abbreviated CYP) is a superfamily of enzymes involved in the oxidation of a variety of organic substances. One particular member of this superfamily, cytochrome 3A4 (CYP3A4), catalyzes many reactions which are involved in drug metabolism. Most drugs become deactivated by the activity of CYP3A4, either directly by alteration of their physical or chemical state or indirectly by having their excretion from the body facilitated by this enzyme. CYP3A4 activity can also be induced by ligand binding, which results in increased transcription and expression of this gene. While located in the inner membrane of the mitochondria or in the endoplasmic reticulum of many cell types, CYP3A4 activity in the liver is of particular interest to pharmaceutical companies, as this organ is the primary site of drug metabolism in the body.

Joydeep Basu and John W. Ludlow (eds.), *Organ Regeneration: Methods and Protocols*, Methods in Molecular Biology, vol. 1001, DOI 10.1007/978-1-62703-363-3_5, © Springer Science+Business Media New York 2013

The ideal source of human hepatocytes used for drug metabolism studies is adult livers. This is due to the fact that adult hepatocytes have greater overall levels of cytochrome p450 activity, including CYP3A4. Measuring the ability of adult hepatocytes to metabolize drugs is a key to determining the in vivo effectiveness of these therapeutic compounds being developed by pharmaceutical companies.

This chapter covers hepatocyte isolation from the whole organ and preparation of these cells for CYP3A4 activity assays. In brief, the liver is enzymatically digested to a single-cell suspension, followed by cellular selection and concentration steps using Percoll density gradients. The recovered hepatocytes are then assayed by high-performance liquid chromatography (HPLC) for CYP3A4 activity, utilizing 7-ethoxycoumarin and 7-hydroxycoumarin as substrates, and rifampin as an inducer (1, 2).

2 Materials

2.1 Chemicals

1. Sodium chloride.
2. Potassium chloride.
3. Sodium phosphate, monobasic.
4. D-Glucose.
5. Selenious acid.
6. Acetonitrile (271004; Sigma-Aldrich, St. Louis, MA, USA).
7. Glacial acetic acid.
8. 4-(2-Hydroxyethyl)-1-piperazineethanesulfonic acid (HEPES).
9. Percoll (P4937, Sigma Aldrich).
10. 10× concentration of phosphate-buffered saline (10× PBS).
11. Dexamethasone.
12. Ethylene glycol tetraacetic acid (EGTA).

2.2 Culture Media and Supplements

1. Fetal bovine serum.
2. Insulin (bovine pancreatic or human recombinant).
3. GlutaMAX (35050-061, Gibco, Grand Island, NY).
4. SPITE (S5666; Sigma-Aldrich, St. Louis, MO).
5. Type 1A Collagenase from *Clostridium histolyticum* (C9891, Sigma-Aldrich).
6. Williams E medium (W1878, Sigma-Aldrich).
7. L-Glutamine (G8540, Sigma-Aldrich).
8. Rifampin (R8883, Sigma-Aldrich).
9. Williams E Medium + rifampin: 200 μM final concentration of rifampin.

10. Williams E Medium + L-glutamine: 0.292 g/l L-glutamine in Williams E medium.
11. ECOD media: Williams E + 1× GlutaMax + 1× SPITE + 10 μM dexamethasone.
12. 7-Ethoxycoumarin (7EC) stock: 150 mM 7EC (E1379; Sigma-Aldrich) made in Williams E medium.
13. 7-Hydroxycoumarin (7HC) stock: 150 mM 7HC (MET104A; Sigma-Aldrich) made in Williams E medium.

2.3 Buffers and Solutions

1. Chelation buffer: 10 mM HEPES, 130 mM sodium chloride, 0.3 mM potassium chloride, 1 mM sodium phosphate monobasic, 10 mM D-glucose, 0.5 mM EGTA, 5 μg/ml insulin (bovine pancreatic or human recombinant), 3×10^{-11} M selenious acid, pH 7.4. Shelf life is 6 months when stored at 4°C.
2. Digestion buffer: 10 mM HEPES, 130 mM sodium chloride, 0.3 mM potassium chloride, 1 mM sodium phosphate monobasic, 10 mM D-glucose, 1 mM calcium chloride, 5 μg/ml insulin (bovine pancreatic or human recombinant), 3×10^{-11} M selenious acid, pH 7.4. Shelf life is 6 months when stored at 4°C.
3. Digestion buffer with collagenase: Digestion buffer supplemented with 300 mg/l of collagenase and filter sterilized just before use.
4. Collection buffer: RPMI 1640 with L-glutamine, without phenol red, 5 μg/ml insulin (bovine pancreatic or human recombinant), 3×10^{-11} M selenious acid, pH 7.4. Shelf life is 12 months when stored at 4°C.
5. Collection buffer with FBS: Collection buffer is supplemented with FBS to a final concentration of 10% just before use.
6. Isotonic Percoll: 9:1 Percoll: 10× PBS.
7. 50:1 Acetonitrile/acetic acid solution: 50:1 acetonitrile:glacial acetic acid.

2.4 Tissue Culture Materials

1. 12-well plastic plate.
2. Sterile plastic pipettes.

2.5 Hardware

1. 1,000-μm filter (CPN-1000-D, Small Parts, Logansport, IN).
2. 800-μm filter (B0044KN6EO, Small Parts).
3. 500-μm filter (CMN-0500-D, Small Parts).
4. Tubing (see Fig. 1)—Make all tubing 46″ in length (Cole Parmer EW-96410-24) where the inner diameter fits around the cannulae (T1), the large tubing contains a male quick connect plug connects to the port on the basin (T2), and the tubing that is leading from the digestion buffer has a quick connect female plug that connects to the male plug connected to the basin (T3).

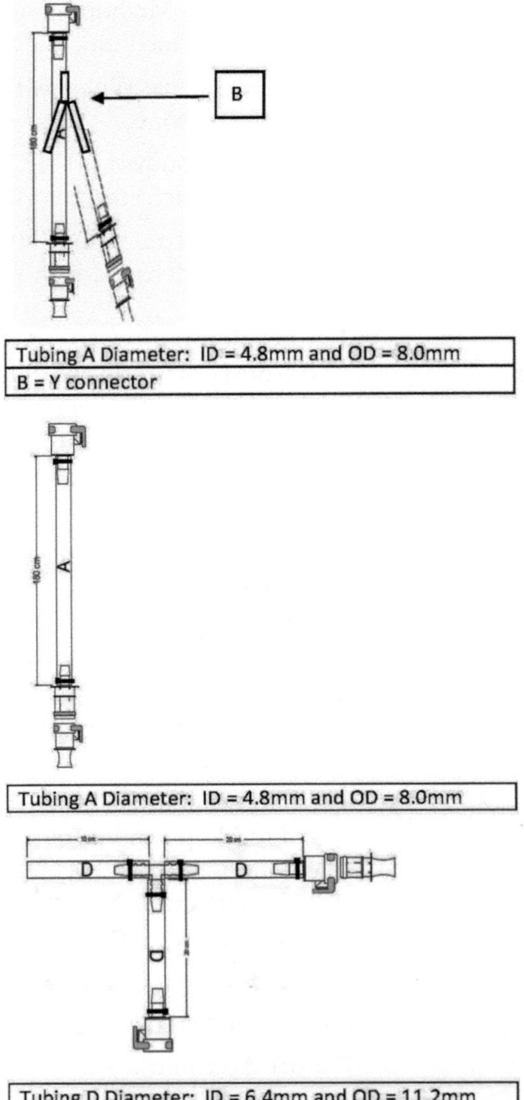

Fig. 1 Schematic showing tubing connections

5. 4 l beakers.
6. Sterile basin with lid.
7. Hemoclips with applicator (Cardinal, Dublin, OH W523400 applicator = RICA 523105).
8. Scalpels.
9. Various sized barbed cannulae (Cole Parmer, Vernon Hills, IL EW-06361-10 or EW-06362-40).
10. One 60 cc syringe.
11. One perfusion basin (3LER9; contains a port for recirculation of buffer, Grainger, Raleigh, NC).

2.6 Surgical Instruments

1. Scissors.
2. One curved needle holder.
3. One straight needle holder.
4. 2–4 Tissue forceps (with teeth).
5. One surgical forceps (smooth edged).
6. Two suture (2–0 silk).
7. Three umbilical clamps (HL9423, Cardinal).

2.7 Equipment

1. Two peristaltic pumps where one is double headed and the other is single headed (Watson Marlow, Wilmington, MA Bredel Pumps 323).
2. Digital scale.
3. Heated working surface/apparatus (maintains 37°C).
4. Heating panels.
5. Heated Hepa filtered box.
6. Biosafety cabinet (BSC).
7. Heating element within the BSC.

3 Methods

3.1 Preparation (See Fig. 2)

1. Place the 3 l chelation and 3 l digestion buffers in the water bath at 37°C and allow to reach temp (~30 min to an hour).
2. Place the peristaltic pumps atop each other in the BSC where the single-headed pump is atop the double-headed pump (see Note 1).
3. Set up a heating element within the hood in order to maintain the temperature of 37°C during perfusion and digestion. Place this as close to the perfusion basin as possible.

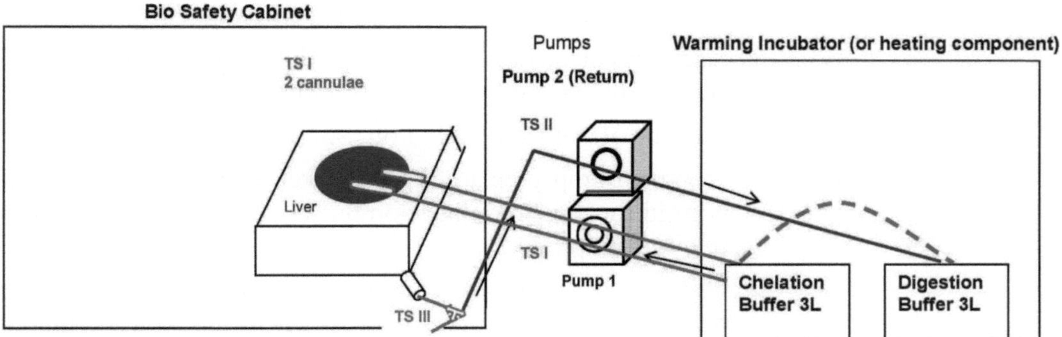

Fig. 2 Schematic showing tubing system setup for liver perfusion

4. Place T1 through the double-headed pump such that the ends are inside the BSC that are to be connected to the cannulae. One end of each tube will be within the 4 l beaker and the other end in the perfusion basin.
5. Connect T2 to T3 to connect the basin to the digestion buffer recirculation using the quick connections (see Note 2).
6. Place T3 through the top single pump head.

3.2 Whole-Liver Perfusion

This process is adapted from that reported previously by Cho et al. (2).

1. In the BSC, place the liver into a perfusion basin with a port on the side. It is not necessary for the port to be on the basin as the recirculation may be performed by placing tubing into the bowl.
2. Prepare the liver within the basin by removing the gallbladder, diaphragm, and extra fat from the liver.
3. If the liver is whole, place a cannulae into the hepatic artery and portal vein. Secure each with suture by tying around the barbs so that they are not easily removed from each vessel. If working with a resection, place two cannulae into the vessels that allow the liver to balloon when buffer is pushed into the vessel.
4. Pour the 3 l chelation buffer into the beaker.
5. Start the bottom pump to allow chelation buffer to flow through the liver.
6. Maintain a speed to allow the buffer to flow at a steady rate where the liver is ballooning and blanching.
7. Continue perfusion with the chelation buffer for 20 min and/or until all chelation buffer has been utilized but do not allow the lines to contain air.
8. During the 20-min chelation, prepare 300 mg/l of collagenase for the digestion buffer. The amount may be preweighed, but do not add it to the buffer until right before use in order to maintain a high level of enzymatic activity.
9. Add the collagenase to 30 ml of digestion buffer and shake vigorously in order to reconstitute the collagenase.
10. Add the 30 ml to the 3 l of digestion buffer.
11. Stop the pumps, replace the 4 l beaker with another 4 l beaker, pour in the prepared digestion buffer + collagenase, and restart the bottom pump.
12. Once the beaker has ~1 l remaining, start the top pump in the reverse direction so that buffer that is in the basin is running back into the beaker.

13. If the basin does not have a port, place the line that would have been connected to the basin directly into the basin and ensure that it is fully submerged in the digestion buffer within the basin. Do not block the flow out of the basin.
14. Allow digestion buffer to perfuse for 30 min or less if the cells appear to pull away from the capsule.
15. During the 30-min digestion period, add a final concentration of 10% FBS to the collection buffer.
16. Stop the bottom pumps at 29 min and continue to reverse the digestion buffer into the beaker until the basin is empty.
17. Place the prepared collection buffer in a 4 l beaker and run the bottom pump for 5 min.
18. Remove the liver from the perfusion basin and transfer it into a secondary basin lined with a 1,000 μm mesh.
19. Using a scalpel, slice the capsule of the liver in long parallel cuts. Then, carefully remove the capsule and collect the cells. Rinse the mesh with 1 l of 10% FBS + collection buffer and remove the mesh (see Note 3) along with the non-digested tissue.
20. Place a large funnel over a beaker (see Note 4) and place an 800 μm mesh atop the funnel. Pour the cells from the basin over the funnel and allow the cells to filter into the beaker.
21. Place a large funnel over a beaker and place a 500 μm mesh atop the funnel. Pour the cells from the basin over the funnel and allow the cells to filter into the beaker (see Note 5).
22. Pour the volume of cells into 500 ml conical tubes where there is 200 ml of cells and add the remaining volume of 300 ml of collection buffer with 10% FBS. Centrifuge at $200 \times g$ for 5 min. This is a concentration step.

3.3 Percoll Gradient

1. Once the cells have been concentrated as per step 22 in Subheading 3.2 above, count the hepatocytes using trypan blue exclusion.
2. Make a Percoll/cell mix with 25% of isotonic Percoll and 75% of the cells where the cells are at a final concentration of 5×10^6/ml (e.g., 200 ml centrifugation spin uses 1×10^9 total cells).
3. Centrifuge at $100 \times g$ for 5 min.
4. Remove the supernatant and resuspend the pellet in 300 ml of collection buffer with 10% FBS to wash the cells at $200 \times g$ for 5 min.
5. Count the cells to determine the concentration.

3.4 ECOD Assay Preparation (Suspension)

1. Remove 5×10^6 cells from cell suspension and centrifuge in ECOD media at $200 \times g$ for 5 min. Remove the supernatant.

2. Resuspend the cells in 5 ml of the ECOD media.
3. Add 0.5 ml of cell suspension each to 4 wells of a 12-well plastic plate.

   If induction is to occur, the 200 μM of rifampin will be added; otherwise do not induce and proceed to the next step.
4. Pre-incubate at 37°C for 30 min.
5. Prepare the dosing solutions:
   (a) 6 ml ECOD media + 8 μl of 150 mM 7EC stock solution.
   (b) 6 ml ECOD media + 8 μl of 150 mM 7HC stock solution.
6. Warm the dosing solutions in a 37°C water bath.
7. Remove the plate from the incubator and add the dosing solutions.
   (a) Add 0.5 ml of 7EC solution to 2 wells.
   (b) Add 0.5 ml of 7HC solution to 2 wells.
8. Incubate the plate for 60 min at 37°C.
9. Remove 0.5 ml sample from each well and place in 1.5 ml tube on ice.
   (a) Add 150 μl of cold acetonitrile/acetic acid (50:1) to each tube to stop the reaction.
   (b) Vortex the tubes and place at −80°C until ready for HPLC analysis (2).

4 Notes

1. Silicone tubing is the best option for use in the peristaltic pumps.
2. During the perfusion, if the basin for recirculation does not have a port in which to connect the recirculation tubing, place the tubing in the basin and ensure that it is submerged in the digestion buffer so that the beaker does not run out of digestion buffer.
3. If the filter 24″ × 12″ is too narrow, Small Parts also sells rolls of mesh that can be cut to the desired size.
4. If the height of the beaker with the funnel is an issue when filtering, use a shorter beaker and filter smaller amounts of cells at a time.
5. If the filter is overly clogged, gently pass back and forth with a cell scraper (Fisher 08-100-242) which will aid the cells to pass through the filter.

References

1. Miltry RR, Hugher RD, Dhawan A (2002) Progress in human hepatocytes: Isolation, culture & cryopreservation. Cell Dev Biol 13:463–467
2. Cho J-J, Joseph B, Sappal BS et al (2004) Analysis of the functional integrity of cryoperserved human liver cells including xenografting in immunodeficient mice to address suitability for clinical applications. Liver Int 24:361–370

Chapter 6

Ex Vivo Culture and Separation of Functional Renal Cells

Andrew T. Bruce, Kelly I. Guthrie, and Rusty Kelley

Abstract

The following methods outline the procedures for isolating primary renal cells from kidney tissue via enzymatic digestion, followed by their culture, harvest, and then fractionation of renal subpopulations from primary culture. The current methods describe procedures to sub-fractionate biologically active cells that have been used to treat and stabilize renal function in models of chronic kidney disease (Kelley et al. Am J Physiol Renal Physiol 299(5):F1026–F1039, 2010).

Key word Renal cell therapy

1 Introduction

The incidence and prevalence of CKD are rising worldwide and especially in the United States. New treatment approaches are required to improve quality of life among CKD and ESRD populations who currently have limited healthcare options. Previous methods have been successful in isolating kidney epithelial cells for regenerative medicine (1). Evaluation of a recently identified and novel renal cell-based therapy (4) may predict the potential impact of these therapies in CKD patients. Recent reports using fate mapping strategies identified the resident renal tubular epithelia, not a specialized progenitor cell (2), as the primary cell source responsible for repairing the kidney (3). Taking advantage of this fundamental learning of kidney repair and regeneration, we recently reported a novel cell-based treatment approach demonstrating the reproducible isolation and expansion ex vivo of a selected population of renal cells enriched for tubular epithelia (4). Herein we describe a method for the selection of a population of renal cells established to have reparative and regeneration ability to augment renal tissue function in animal models of CKD.

Joydeep Basu and John W. Ludlow (eds.), *Organ Regeneration: Methods and Protocols*, Methods in Molecular Biology, vol. 1001, DOI 10.1007/978-1-62703-363-3_6, © Springer Science+Business Media New York 2013

2 Materials, Reagents, and Equipment

2.1 Materials

1. Freshly harvested kidneys collected not more than 1 day prior; perfused with phosphate-buffered saline (PBS) at the time of harvest and stored at 2–8°C in Viaspan.
2. Batch record or notebook.
3. Freezer box for sample storage and analysis.
4. Pipette tips capable of dispensing volumes ranging from 2–1000 μl.
5. Serological pipettes: 2, 5, 10, 25, 50 ml; aspirating pipettes.
6. Transfer pipettes.
7. Sterile 1.5 ml microcentrifuge tubes for RNA and Western Blot collection.
8. Non-sterile 1.5 ml microcentrifuge tubes for preparing cells for counting.
9. 15 and 50 ml conical bottom polypropylene tubes.
10. Pre-weighed 50 ml conical tubes for tissue digestion.
11. 1.5, 15, and 50 ml tube racks.
12. Sterile 150 cm dishes (or comparable) to hold instruments and for mincing kidneys.
13. Sterile 150 cm dishes pre-weighed for minced tissue.
14. 100 μm Steriflip® filters (Millipore, SCNY00100).
15. Fine-tipped forceps (curved or straight) (sterile).
16. Scalpel handle with appropriate blade or disposable scalpel (sterile).
17. Alcohol wipes.
18. Bench Wipes (paper towels).
19. Hemocytometer.
20. Bench Counter (for counting cells on hemocytometer).
21. Sterile Nunc T500 flasks (Item #132913).
22. 70% ethanol spray bottle.
23. Bacdown spray bottle.

2.2 Reagents

1. Wash buffer: DPBS calcium and magnesium free (Invitrogen Gibco 14190-235).
2. 50:50 growth medium 50% DMEM high glucose (4.5 g/l) (GIBCO 11995), 50% Keratinocyte-SFM (GIBCO 17005) containing human recombinant epidermal growth factor 1–53 (EGF 1–53), Bovine Pituitary Extract, 5% FBS (GIBCO 16000), 1× Anti–Anti (GIBCO 15240), and 1× Insulin Transferrin Selenium (GIBCO 41400).
3. Trypan Blue 0.4% (GIBCO 15250).

4. Digestion buffer: Dispase I 4 U/ml (StemCell Technologies 07913), 5 mM final calcium chloride (Sigma C2661), Collagenase type IV (Worthington 4212) reconstituted with dispase to a final concentration of 300 U/ml. Always check units on collagenase bottle b/c it changes with every lot (see Note 1).
5. Calcium chloride stock at 500 mM (100×). For 58 ml of final volume add 5.8 ml.
6. Hold digestion buffer at 37°C until ready for use.
7. 40 ml of digestion buffer is needed for each gram of tissue being digested.
8. Neutralization buffer: 5% FBS in DMEM (or comparable medium such as KGM) for neutralizing digestion buffer.
9. 30% Optiprep gradient medium (50% Sigma Iodixanol stk 60%w/v: 50% KSFM).
10. Keratinocyte-SFM.
11. 1× PBS (GIBCO 14190).
12. Liquid nitrogen for snap freezing.
13. 4% Paraformaldehyde for cell fixation (if needed for flow cytometry).
14. Freezing media (80% HTS-FRS, 10% DMSO, 10% FBS).
15. Sterile deionized (Di) H_2O for cleaning instruments.

2.3 Equipment

1. Biological safety cabinet (BSC)/tissue culture hood.
2. 37°C, 5% CO_2, humidified incubator.
3. Water bath 37°C.
4. Upright light microscope.
5. Centrifuge capable of 800 × *g* with appropriate bucket adapters for various tube sizes.
6. Tube rocker.
7. –80°C freezer.
8. Digital scale for weighing tubes and tissue.
9. Sonicator and milk bath for cleaning surgical instruments.
10. Pipettors (Ranin) capable of dispensing volumes ranging from 2–1000 μl.
11. Serological pipet controller (Drummond).

3 Methods

3.1 Renal Cell Isolation

1. Prior to kidney processing, clear the BSC and clean all surfaces of the cabinet with Bacdown (or sufficient substitute) followed by 70% ethanol or isopropyl alcohol.

2. Note catalog numbers and lot numbers of all reagents being used on batch record.
3. Prepare digestion buffer (see Subheading 2.2 for recipe) and hold at 37°C until use. Use estimate of tissue weight to determine the amount to prepare or wait until tissue weight is determined. 40 ml of digestion buffer required per gram of tissue to be digested.
4. Prepare hoods with necessary tools needed for procedure (surgical instruments (forceps, scalpels, and blades), plates, alcohol wipes, markers, tubes, pipettes, pipette tips, wash buffer (PBS w/o Ca^+, Mg^+), sterile microcentrifuge tubes).
5. Pre-weigh several 50 ml conical polypropylene tubes and record weight on tube. One tube will be required per gram of tissue to be digested. These weights will be used to check that 1 ± 0.1 g is added to each tube.
6. Pre-weigh two P150 dishes for weighing intact kidneys as well as kidney tissue prepared for digestion.
7. Prepare tube rocker by sanitizing with 70% ethanol and place in incubator to allow equilibration to 37°C.
8. Retrieve pre-flushed and cooled kidneys from Viaspan or other suitable storage solution. Verify that sample is cold (not frozen or warm) and verify integrity of packaging (no leaks, holes, etc.). Verify shipping sheet and sample contents match. On batch record, note the unique sample IDs and condition of sample (cold, warm, etc.), and indicate shipping/storage media (Viaspan) strain, and age of sample if applicable.
9. Spray primary container with 70% ethanol and wipe down.
10. Remove parafilm from around the mouth of the tube (if applicable), spray down with 70% ethanol, and wipe. Place tube in the BSC.
11. Aspirate the remaining shipping medium and wash the kidney twice with CMF-DPBS.
12. Transfer the kidney to pre-weighed P150 and weigh. Record the weight of pre-weighed dish, kidneys + dish, and determine the weight of the kidney tissue alone.
13. After washing, aspirate buffer, place tissue in dish or plate, and then remove and discard connective tissue, calyx, and capsule around the kidney using sterile forceps and scalpel.
14. Manually mince together the remaining kidneys using scalpel blade making finely chopped slurry of tissue. Kidney tissue pieces should be minced finely.
15. Add ~1.0 g (±0.1 g) kidney tissue to each pre-weighed 50 ml conical tube. Check weights of tube to confirm that weight of tissue within each tube is approximately 1 g (±0.1 g). Adjust tissue amounts if needed.

16. Add 20 ml of premade digestion buffer to each 50 ml conical tube containing ~1.0 g kidney tissue.
17. Digestion 1: Incubate digestion tubes at 37°C on a tube rocker located in the incubator for 20 min.
18. After 20 min, remove tubes from incubator/rocker and place in BSC.
19. Allow undigested tissue to settle to the bottom of the tube.
20. Aspirate supernatant from the top of each tube and discard (leaving only undigested tissue in the bottom of the tube, less than 5 ml). This helps to remove red blood cells, connective tissue, and other cell debris from sample.
21. Add an additional 20 ml of warmed digestion buffer to each tube containing undigested tissue.
22. Digestion 2: Incubate digestion tubes at 37°C on a tube rocker located in the incubator for 40 min.
23. After 40 min, remove tubes from rocker/incubator, combine the contents of two tubes to yield approximately 40 ml per tube, and run each sample through a 100 μm SteriFlip® filter using an aspirator.
24. Flow through will now be in a new, sterile 50 ml tube. This is an *unfractionated* sample that should be reserved.
25. Add 5 ml of neutralization buffer to each tube and mix well. Centrifuge at 300 × *g* for 5 min.
26. Aspirate supernatant and resuspend pellet in KSFM. Utilize sufficient volume such that a count can be made from this cell suspension. Typically a minimum of 10 ml per gram of tissue digested is required.
27. Remove small cell sample for counting using a hemocytometer (see Note 2). It is recommended to use a manual hemocytometer in order to obtain an accurate count during this step (see Note 3).
28. Record live cell, dead cell, cell viability, cells per ml, and total number of cells on batch record. Prepare samples of unfractionated cells for initial analysis (see Note 4).
29. For the portion of UNFX (unfractionated) cells that are to be cultured immediately, proceed as described below.
30. Based on the total remaining cell number in sample, calculate how many gradient tubes (15 ml tubes) are needed to load 60–75e6 cells/tube.
31. Pellet cells as described previously, 300 × *g* for 5 min, and wash once more with KSFM.
32. After wash resuspend cells to a final concentration of 10–15 × 10e6 in KSFM (see Note 5).

33. Establish 15% mixing gradient: Add 5 ml of 30% Optiprep to each 15 ml tube required for gradients.
34. Aliquot 5 ml of resuspended cells into each 15 ml tube.
35. Mix tubes by inversion six times.
36. Carefully layer 1 ml of PBS on top of Optiprep/cell mixture. (This will form a liquid barrier between the cell mixture and outside the air and will protect cells after centrifugation.) If 50 ml tubes are used, layer 5 ml.
37. Centrifuge at 800 × *g* for 15 min without brake (ensure that there is a balance tube in the centrifuge, if needed).
38. Once the centrifugation is complete, carefully remove tube(s) and make sure that there is a cell band located around the 9–10 ml mark and a pellet at the bottom of the tube.
39. Collect and combine all cell bands by aspirating via transfer pipette and transfer band contents into a new 50 ml conical tube. (Collecting 1 ml above or below the cell band is acceptable for accuracy.)
40. Aspirate out the remaining density medium and discard via vacuum aspiration (leaving the pellet in the tube).
41. Resuspend all pellets with KSFM.
42. Combine collected bands and pellets together adding enough KSFM to mixture to reach a minimum of 2:1 ratio KSFM to cell/Optiprep mixture. Sample may need to be split into multiple 50 ml conical tubes to achieve the 2:1 ratio.
43. Mix tubes by inversion six times (see Note 6).
44. Centrifuge at 300 × *g* for 5 min.
45. Aspirate supernatant via vacuum aspiration and discard (leaving pellet in tube).
46. Resuspend cell pellet with KSFM to desired volume. This is the initial 15% band + pellet (B + P) sample.
47. To establish cultures for scheduled use: Rat kidney cells are plated at 50,000 cells/cm^2 at isolation, typically in T500 with 150 ml of kidney growth medium per flask at 37° C/5% CO_2 under standard tissue culture (TC) oxygen. If however, the culture period is to be 4 days for scheduling purposes (3 days 21% O_2/1 day 2% O_2), cells are plated at 30,000 cells/cm^2.
48. To prepare cells for cryopreservation for future culture, harvest, and implantation or analysis, continue as described below.
49. Centrifuge collected volume containing the cell # desired to cryopreserve in 50 ml tubes, centrifuge at 300 × *g* for 5 min, and then wash pellet with unsupplemented KSFM. Centrifuge at 300 × *g* for 5 min.

3.2 Renal Cell Cryopreservation

1. Prepare cryopreservation media as described previously (FRS 80% vol, FBS 10% vol. and DMSO 10% vol). 1 ml of cryopreservation media is required per vial of cells to be frozen. Cells may be frozen at up to 50 × 10e6/ml.
2. After final wash is complete, aspirate supernatant. Tap tube containing pelleted cells at this point to resuspend the cells.
3. Add previously prepared cryopreservation media to the pellet in a dropwise manner until volume is equivalent to that required for freezing.
4. Transfer 1 ml of cells in cryopreservation media to each prelabeled cryovial. Tubes should be labeled with sample ID, cell number/vial, and date frozen.
5. Transfer labeled and filled vials to the rate-controlled freezer to freeze.
6. Once frozen, transfer to liquid nitrogen storage.
7. Retrieve all snap-frozen samples from liquid nitrogen and place in a labeled freezer box. Place box in −80°C freezer for analytical interpretation.
8. All cells should be distributed at this point to immediate culture, cryopreservation, RNA, Western Blot, or other analytic methods.

3.3 Renal Cell Culture and Harvest Procedure

1. Upon isolation of primary kidney cells, verify cell number.
2. Place 50:50 culture medium, trypsin, and KSFM in 37°C water bath to pre-warm.
3. For cultures to be harvested 3 days later, plate 50,000 cells/ cm^2 in desired TC-treated vessel using complete 50:50 media at a volume that is recommended by vessel manufacturer (see Note 7).
4. Incubate cells for 48 h without disturbing in 150 ml total volume.
5. After 48 h (day 2 postseed) change medium to remove unattached cells and debris.
6. Replenish flasks with 100 ml of fresh medium.
7. Switch all culture vessels to 37°C, 5% CO_2, low O_2 (2%), humidified incubator.
8. Maintain cultures at low oxygen O_2 (2%) for 18–24 h.
9. On the day of harvest (day 3 postseed) monitor confluency of cells via light microscopy and image to document morphology and confluency and record on batch record or notebook. Cells should be approximately 60–80% confluent at this point (Fig. 1).

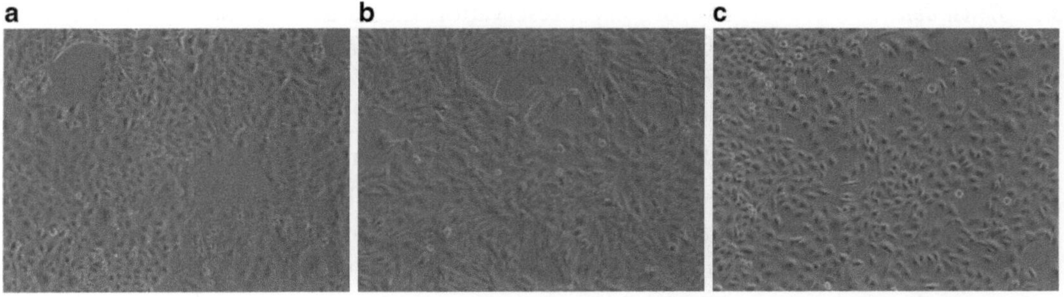

Fig. 1 Primary kidney cell culture morphology from (**a**) rat, (**b**) canine, (**c**) human kidney prior to harvest and passaging (×100)

10. Remove culture medium by pouring into a waste bottle (ensuring to maintain sterile technique) (see Note 8).
11. Wash each culture vessels with 25 ml sterile PBS (for T500). Add enough PBS to remove all traces of culture media (especially serum) and cellular debris. Volume will vary depending on the culture vessel.
12. Remove PBS wash via pouring into a waste bottle.
13. Add appropriate volume of pre-warmed trypsin to cell monolayer at a volume that is recommended by culture vessel manufacturer (for a T500 add 25 ml of trypsin).
14. Monitor cell layer via light microscopy until cells have detached. This should take typically 5–7 min but less than 10 min.
15. Prepare a sterile collection vessel for collection of the trypsinized cells. Add a volume of 5% FBS DMEM or other comparable medium to the vessel such that the neutralization medium is equivalent to 10–20% of the final collected volume (see Note 9).
16. When the cells have detached, gently tap flasks to detach all cells. Pour the trypsinized contents into the collection vessel.
17. Continue process until all of the cells have been harvested. Stagger process if possible so that cells are not exposed to trypsin for extended times.
18. Transfer neutralized cell suspension to conical tubes for centrifugation (300 × *g* for 5 min).
19. Aspirate supernatant and after tapping to loosen pellet, resuspend in KSFM in sufficient volume for counting. Variation of buffers used in the wash/prep process can affect the banding of cells on the density gradient and should be avoided.
20. Remove 18 μl cell sample for counting (using either a hemacytometer or an automatic cell counter) (see Note 10).
21. Record live cell, dead cell, cell viability, cells per ml, and total number of cells on batch record. Prepare sample of unfractionated (pre-gradient) cells for analysis.

22. Determine the distribution of cells that will be fractionated or banded for injection/analysis, cryopreserved, designated for flow cytometry, WB, or other functional/characterization-based needs.
23. If applicable, aliquot the portion of cells that are to be cryopreserved into 50 ml conical tube(s) previously described in Subheading 3.2.
24. Aliquot the portion of cells that are to be fractionated/banded into 50 ml conical tube(s).
25. Centrifuge all tubes at 300 × *g* for 5 min if using 50 ml conical tubes.
26. Carefully aspirate supernatant via vacuum aspiration and discard, leaving only cell pellet at the bottom of the tube.
27. Tap tube to loosen pellet and resuspend cells.
28. Resuspend pellet such that the concentration of cells is equivalent to 30–37.5e6 cells per ml. These cells will be used for subfractionation using a discontinuous Optiprep step gradient in the next section.
29. At this point the cells are ready to be fractionated/banded and cryopreserved for later culture.

3.4 Renal Cell Fractionation

1. Determine the amount of subcultured primary cells that have been attained.
2. Each density step gradient requires a cell number ranging from 60 to 75 million cells per tube. Once the number of cells has been determined, calculate the number of gradient tubes to be generated.
3. Prepare 7 and 16% Optiprep as previously described in sufficient volumes to prepare needed gradients. Typically no less than 40 ml is prepared to minimize pipetting error associated with smaller volumes.
4. Make a discontinuous density step gradient(s) by first pipetting 4 ml of 16% Optiprep in KSFM into each of the required 15 ml conical centrifuge tube(s) to form the bottom layer of the two-step gradient.
5. Carefully layer 4 ml of 7% Optiprep in KSFM onto the bottom layer of the density gradient by tilting the tube at a 45° angle and letting the medium slowly run down the side of the tube. This will minimize mixing at the interface between the two different densities.
6. Once the top of the two-step density gradient has been layered, carefully pipette 2 ml of cell suspension containing between 60 and 75 million cells in KSFM medium on top of the step gradient using the above layering method. Continue until all of the gradients have been loaded with cell suspension.

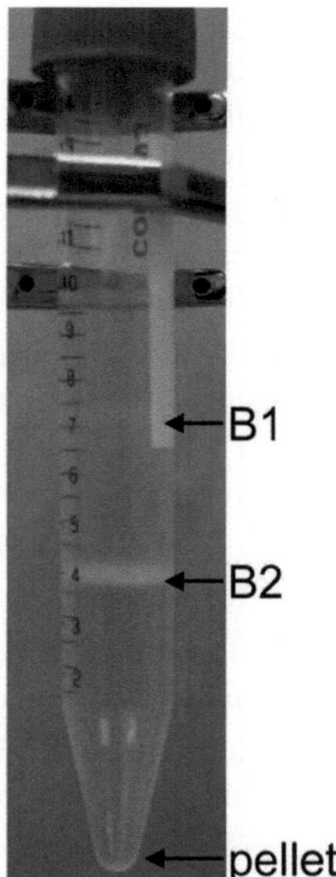

Fig. 2 Discontinuous OptiPrep 2—step gradient. 7% (w/v) layered on top of the 16% (w/v) provide an interface for the biologically active cells to settle. The upper Band 1 (*B1*) cells consist of the larger tubular epithelial cells as well as collecting duct cells while the Band 2 (*B2*) cells comprise a majority of proximal tubular cells. The pellet composition is mainly dead cell and debris

7. Once the cells have been loaded, carefully place tubes into the centrifuge and spin at 800 × *g* for 20 min without brake.
8. After centrifugation, collect tubes and visually inspect gradient bands to verify banding pattern. Between the 7–8 ml mark there should be a thin band of cells referred to as Band 1. A heavy second band should be visible near the 4 ml mark and is referred to as Band 2. A faint pellet of cells will also be present (Fig. 2).
9. Collect gradient bands by aspirating each band using either sterile bulb or 5 ml pipette.
10. Place each cell band in a separate tube. Separately combine all Band 1's, Band 2's, and then pellets.
11. Once all bands have been collected and combined, dilute at least twofold with KSFM medium and mix well by inversion. Resuspend the residual pellet(s) with equal volume of KSFM.
12. Wash out residual Optiprep by centrifuging at 300 × *g* for 5 min.
13. Resuspend the pellets from all of the collected bands in 50:50 medium and perform cell count.

14. After cell count has been determined, collect samples for cell analysis or cryopreserve the remaining cells using previously described methods for later use.
15. The biologically active Band 2 cells from rodent and canine have been successfully used to treat model chronic kidney disease.

4 Notes

1. Example: Stock collagenase = 17,440U. 17,440U/300U = 58 ml of dispase (at 4 U/ml) needed to add to collagenase to make collagenase 300 U final.
2. Initial unfractionated cells are very "tube-like" in appearance and it can be difficult to count individual cells.
3. Example of trypan blue exclusion using hemocytometer: To make a count using a 1.1× dilution, take 18 μl cell sample and 2 μl of 0.4% trypan blue, load 10 μl of cell count mixture, and view through upright light microscope. Count at least two squares of the grid and calculate the number of viable cells for instance:
   (a) For a 1.1 (dilution factor) × 10,000 × (no. of live cells/no. of squares counted) = total live cells/ml.
   (b) Multiply total live cells/ml by total volume of neutralized collection volume = total no. of live cells.
4. Testing initial unfractionated cells will provide information on baseline expression levels for RNA (or Western Blot) as described below. Samples can be taken at this point for additional testing as needed for FACS analysis using Guava or for functional analysis such as GGT/LAP. Make sure that all samples are correctly labeled.
   (a) Remove 1e6 cells for RNA isolation (and 1e6 for Western Blot if applicable) and place in a sterile 1.5 ml microcentrifuge tube.
   (b) Add DBPS to microcentrifuge tube(s) (for wash step) and centrifuge at 300 × *g* for 5 min. Aspirate contents down to pellet and snap freeze pellet in liquid nitrogen. (Label tube with assigned RNA number from RNA folder indicating sample designation and record in batch record.)
   (c) Cells taken for flow analysis are washed in DPBS, fixed in 2% paraformaldehyde for 30 min, then washed, and stored at 4°C.
5. It is possible to load 225e6 cells in 50 ml conical tubes.
6. If too little KSFM is added and/or if tubes are not mixed well, the Optiprep mixture will re-band the cells.

7. Recommended seeding for T500 Nunc flask: 25e6 cells/flask, 150 ml of 50:50 media. This sample will be passage 0 (p0) once seeded. If cultures will be harvested 4 days later, plate at 30,000 cells/cm^2. It is recommended to vent flask caps, even if flasks have filter caps, to promote adequate gas exchange across cell monolayer.
8. Contamination mat arises if the collection procedure was not performed using aseptic technique. If contamination is observed, immediately remove all culture vessels from incubator. Save contaminated medium for microbiological analysis. Dispose of contaminated material properly with bleach and thoroughly clean BSC and incubators with Bacdown and 70% EtOH (or autoclave contaminated materials if applicable).
9. If 10×T500 flasks are to be trypsinized, the total trypsinized volume will be 250 ml. Add 25–50 ml of neutralization buffer to the collection vessel.
10. The harvested cultured kidney cells are usually in a fairly uniform single-cell suspension, so it is possible for an automated counter to get an accurate count at this point such as a Cedex or an equivalent.

References

1. Aboushwareb T, Egydio F, Straker L, Gyabaah K, Atala A, Yoo JJ (2008) Erythropoietin producing cells for potential cell therapy. World J Urol 26(4):295–300
2. Humphreys BD, Czerniak S, DiRocco DP, Hasnain W, Cheema R, Bonventre JV (2011) Repair of injured proximal tubule does not involve specialized progenitors. Proc Natl Acad Sci U S A 108(22):9226–9231
3. Humphreys BD, Valerius MT, Kobayashi A, Mugford JW, Soeung S, Duffield JS, McMahon AP, Bonventre JV (2008) Intrinsic epithelial cells repair the kidney after injury. Cell Stem Cell 2(3):284–291
4. Kelley R, Werdin ES, Bruce AT, Choudhury S, Wallace SM, Ilagan RM, Cox BR, Tatsumi-Ficht P, Rivera EA, Spencer T, Scott Rapoport H, Wagner BJ, Guthrie K, Jayo MJ, Bertram TA, Presnell SC (2010) Tubular cell-enriched subpopulation of primary renal cells improves survival and augments kidney function in rodent model of chronic kidney disease. Am J Physiol Renal Physiol 299(5):F1026–F1039

Chapter 7

Isolation and Myogenic Differentiation of Mesenchymal Stem Cells for Urologic Tissue Engineering

Rongpei Wu, Guihua Liu, Shantaram Bharadwaj, and Yuanyuan Zhang

Abstract

Cell-based tissue engineering is one of the most promising areas in biotechnology for restoring tissues and organ function in the urinary tract. Current strategies for bladder tissue engineering require a competent biological scaffold that is seeded in vitro with the patient's own bladder cells. This use of autologous cells avoids graft rejection and the long-term use of immunosuppressive medications usually required after allogeneic transplantation. However, suitable bladder cells from the patient are sometimes limited or unobtainable. When suitable cells are unavailable for seeding due to bladder exstrophy, malignancy, or other reasons, the use of other cell types originating from the patient may be an alternative. A suitable alternative to autologous bladder cells could be mesenchymal stem cells (MSC). MSC reside primarily in the bone marrow, although they exist in other sites as well, including adipose tissue, peripheral and cord blood, liver tissue, and fetal tissues. Bone marrow-derived stromal cell populations contain few MSC (one MSC in 10^4–5×10^7 marrow cells), with the exact number depending on the age of the patient. Despite their limited numbers, MSC possess both the ability to self-renew for extended periods of time and the potential to differentiate into several different specialized cell types under the appropriate conditions. MSC are capable of expansion and tissue-specific differentiation in vitro based on external signals and/or the environment. There are different methodologies for induction and maintenance of a differentiated cell phenotype from MSC. For example, MSC can differentiate into a smooth muscle cell (SMC) phenotype in vitro when exposed to stimuli such as conditioned medium derived from SMC cultures or specific myogenic growth factors (PDGF-BB, HGF, TGF-β). These differential cells can migrate to a scaffold for differentiation into smooth muscle-like cells in vivo. Furthermore, stem cell-seeded scaffolds that are implanted into the bladders repopulate and reorganize the tissue rapidly, thus reducing fibrosis and restoring appropriate neural functionality.

In this chapter, we describe the methods we use for the isolation of human bone marrow mesenchymal stem cells (BMSC), and demonstrate evidence of their myogenic differentiation capacity for potential use in urologic tissue engineering.

Key words Mesenchymal stem cells, Bone marrow, Myogenic differentiation, Tissue engineering, Urology

Joydeep Basu and John W. Ludlow (eds.), *Organ Regeneration: Methods and Protocols*, Methods in Molecular Biology, vol. 1001, DOI 10.1007/978-1-62703-363-3_7,

1 Introduction

Bladder cancer is the fourth most commonly diagnosed malignancy in men and the eighth most common malignancy in women in the United States. It is estimated that more than 60,000 new cases of bladder cancer are diagnosed in the United States per year (1). Cystectomy and bladder augmentation/reconstruction are options for people with advanced-stage bladder cancer. These surgical approaches restore bladder capacity and compliance and prevent or improve hydronephrosis and renal failure. Currently, augmentation cystoplasty is usually accomplished by placing a stomach patch or detubularized segment of intestine onto the bladder to replace the disease segment of tissue. Although these techniques are functional, complications can arise from using gastric flaps or intestinal segments for urinary reconstruction. Several serious side effects associated with the use of stomach and bowel tissue include electrolyte abnormalities, infection, intestinal obstruction, mucus production, perforation, and carcinogenicity. Cell-based tissue engineering provides an alternative for patients with bladder cancer or other end-stage bladder diseases (2). Autologous bladder-derived cells provide a cell source for urological tissue engineering because they can be used without rejection. However, suitable bladder cells might not be available in patients with bladder cancer or other advanced conditions; thus, an appropriate alternative to utilizing bladder cells becomes necessary. Stem cells derived from skin (3), adipose tissue (4, 5), skeletal muscle (6), bone marrow (7–9), and hair follicles (10) are potential cell sources for this purpose and many of these cell types have been applied to tissue engineering technology for ultimate use in the urinary tract.

Mesenchymal stem cells (MSC), which can be isolated from bone marrow, skeletal muscle, and adipose tissue, possess the capacity to differentiate into cells of connective tissue lineages, including muscle. Information about the isolation and characterization of MSC, and the control of their myogenic differentiation, that has been derived from both preclinical and clinical studies has suggested their potential for use in urological regenerative medicine and tissue engineering. Currently, the most comprehensively characterized types of mesenchymal stem cells are from bone marrow. Mesenchymal stem cells derived from bone marrow have been shown to differentiate into a number of specialized cell types, including hepatocytes (11–13), neural cells (14–17), and many mesodermal derivatives such as bone, cartilage, cardiac muscle, skeletal muscle, and fat. Bone marrow mesenchymal stem cells (BMSC) are a good cell source for cell-based therapy because these cells possess plasticity and the ability to self-renew. BMSC are also probably safe for bladder cancer patients who need autologous tissue-engineered bladder augmentation because bone and/or bone marrow are not usually targets for bladder carcinoma metastasis. If BMSC are placed on a proper biodegradable scaffold and implanted, they can act as anti-fibrotic, angiogenic, anti-apoptotic, and mitotic agents. Recently, BMSC were evaluated as an alternative cell type for use in replacement of bladder SMC when native bladder muscle tissue is unavailable. The potential of BMSC to differentiate into cells with bladder SMC characteristics was assessed in vitro (18) and in different animal models (18–24).

Recently, we have investigated the impacts of soluble growth factors, bladder extracellular matrix (ECM), and 3D dynamic culture on cell proliferation and differentiation of human BMSC into smooth muscle cells (18, 19). Myogenic growth factors (PDGF-BB and TGF-β1) alone or combined either with bladder ECM or dynamic cultures induced BMSC to express smooth muscle-specific genes and proteins. Either ECM or the dynamic culture alone promoted cell proliferation but did not induce myogenic differentiation of BMSC. A highly porous poly-l-lactic acid (PLLA) scaffold provided a 3D structure for maximizing the cell-matrix penetration, maintained myogenic differentiation of the induced BMSC, and promoted tissue remolding with rich capillary formation in vivo (19). These data suggest that myogenic-differentiated BMSC seeded on a nanofibrous PLLA scaffold can be used for cell-based tissue engineering for bladder augmentation or replacement in bladder cancer patients requiring cystoplasty.

2 Materials

2.1 Cells Isolation and Culture

Unless otherwise stated, all cell isolation and culture medium, supplements, and solutions materials were obtained from Invitrogen, Life Technologies Corp., Grand Island, NY, USA (see Note 1).

1. Minimum essential medium, alpha modified (α-MEM) with L-glutamine, without ribonucleosides or deoxyribonucleosides.
2. Fetal bovine serum (FBS) (Atlanta Biologicals, Lawrenceville, GA, USA).
3. L-Glutamine, 200 mM.
4. Penicillin G (10,000 U/ml), and streptomycin sulfate (10,000 μg/ml) in a solution of 0.85% NaCl.
5. Complete culture medium (CCM): 440 ml α-MEM, 50 ml FBS (final conc. 10%), 5 ml L-glutamine (final conc. 2 mM), and 5 ml Penicillin G and streptomycin sulfate (final conc. 100 U/ml penicillin and 100 μg/ml streptomycin).
6. Dulbecco's modified eagle medium (DMEM, low glucose).
7. Hank's balanced salt solution (HBSS) w/o Ca^{2+} and Mg^{2+}.

8. Ficoll-Paque (Histopaque-1077, Sigma-Aldrich, St. Louis, MO, USA).
9. 10 ml syringe, 20 ml syringe (Becton Dickinson, Franklin lakes, NJ, USA).
10. Bone marrow biopsy and aspiration system with 15 G puncture needle (OnControl™ Aspiration System, Vidacare, Shavano Park, TX, USA).
11. 10 ml Vacutainer blood collecting tube, sodium heparin spray coated (Becton Dickinson).
12. Dulbecco's phosphate buffered saline (DPBS), without Ca^{2+} or Mg^{2+}.
13. Trypsin (0.25%)–EDTA.
14. Trypan blue, 0.4%.
15. Sterile 500 ml filter units, 0.2 μm pores.
16. Biological Safety cabinet Class II plugged into a vacuum system.
17. Water bath set at 37°C.
18. Water jacketed CO_2 incubator with HEPA filter system that can maintain a humidified atmosphere at 37°C and 5% CO_2.
19. Bench centrifuge with swinging bucket rotor and brake ON/OFF option.
20. Inverted phase microscope.
21. Hemocytometer.
22. 2-, 5-, 10-, and 25-ml individually wrapped sterile cell culture plastic pipettes.
23. Pipet-Aid, or other electric or manual pipette filler/dispenser.
24. 15- and 50-ml sterile conical centrifuge tubes.
25. 75 and 175 cm^2 tissue culture flasks.
26. 100- and 150-mm tissue culture treated culture dishes.
27. 10–1,000 μl single channel pipettors.
28. 10-, 20-, 200-, and 1,000-μl sterile aerosol barrier pipette tips.
29. Collagen-IV solution: 5 mg/ml in 10 mM acetic acid. Filter sterilize through a 0.20 μm cellulose acetate filter.

2.2 Myogenic Differentiation of Human BMSC

1. Recombinant human TGF-β1 (240-B-010/CF, R&D Systems, Inc., Minneapolis, MN, USA).
2. Recombinant human PDGF-BB (220-BB-050, R&D Systems, Inc., Minneapolis, MN, USA).
3. 100 mm Costar transwell plate with 0.4 μm barrier membrane 75 mm insert.

4. Orbital shaker.
5. Poly-L-lactic acid (PLLA) scaffold, stored at 4°C.

2.3 Flow Cytometry

1. FACS Calibur flow cytometer (Becton Dickinson) or equivalent.
2. CellQuest Pro software or equivalent.
3. CD14-FITC antibody (BD-555397, BD Biosciences, San Jose, CA, USA).
4. CD29-PE antibody (303004, Biolegend, San Diego, CA, USA).
5. CD34-PE antibody (BD-550761, BD Biosciences, San Jose, CA, USA).
6. CD45-FITC antibody (BD-347463, BD Biosciences, San Jose, CA, USA).
7. CD105 Alexa Fluorl 488 antibody (323210, Biolegend, San Diego, CA, USA).
8. CD166-PE antibody (BD-559263, BD Biosciences, San Jose, CA, USA).
9. IgG2a-FITC Isotype Control (BD-555573, BD Biosciences, San Jose, CA, USA).
10. IgG1-PE Isotype Control (400112, Biolegend, San Diego, CA, USA).
11. IgG1-PE Isotype Control (BD-555749, BD Biosciences, San Jose, CA, USA).
12. IgG1-FITC Isotype Control (BD-340755, BD Biosciences, San Jose, CA, USA).
13. IgG1-Alexa Fluorl 488 Isotype Control (400129, Biolegend, San Diego, CA, USA).
14. FACS buffer: 1% bovine serum albumin (100M1781V, Sigma-Aldrich, St. Louis, MO, USA) in PBS.

2.4 Immunofluorescence Staining

All antibodies were diluted in antibody dilution solution obtained from Dako, Carpinteria, CA, USA.

1. 1:100 dilution of αSMA monoclonal antibody (Sigma-Aldrich).
2. 1:10,000 dilution of calponin monoclonal antibody (Sigma-Aldrich).
3. 1:50 dilution of desmin monoclonal antibody (Santa Cruz Biotechnology Inc., Santa Cruz, CA, USA).
4. 1:50 dilution of myosin heavy chain (MHC) monoclonal antibody (Santa Cruz).
5. 1:2,000 dilution of fluorescein isothiocyanate (FITC)-conjugated goat anti-mouse IgG (Vector Laboratories).

Table 1
Sequence information of primers used for RT-PCR and their expected product size

Target gene	Forward primer (5′–3′)	Reverse primer (5′–3′)	Amplicon Size (bp)
MHC	GGACGACCTGGTTGTTGATT	GTAGCTGCTTGATGGCTTCC	656
αSMA	ACCCACAATGTCCCCATCTA	TGATCCACATCTGCTGGAAG	595
Calponin	ATGTCCTCTGCTCACTTCA	TTTCCGCTCCTGCTTCTCT	453
Desmin	CCATCGCGGCTAAGAACATT	TCGGAAGTTGAGGGCAGAGTA	440
GAPDH	CGGATTTGGTCGTATTGG	TCAAAGGTGGAGGAGTGG	861

6. Vectashield Mounting Medium for Fluorescence with 4′,6-diamidine-2-phenylindole (DAPI) kit or propidium iodide (PI) kit (Vector Laboratories, Burlingame, CA, USA).
7. Serum-free protein block solution (X090930, Dako).
8. 4% paraformaldehyde solution.
9. Acetone.
10. 8-well chamber slides.

2.5 PCR

1. TRIzol reagent (15596-026, Invitrogen).
2. SuperScript™ II Reverse-Transcriptase reagents (11917-010, Invitrogen).
3. The PCR primer sequences required are listed in Table 1.
4. Thermal cycler (Techne TC-4000, Bibby Scientific Limited, Staffordshire, UK) or equivalent.
5. Gel electrophoresis system.
6. 1.5% agarose gel incorporating SYBR-safe DNA gel stain (S33102, Invitrogen).
7. Human glyceraldehyde-3-phosphate dehydrogenase (402869, Applied Biosystems, Foster City, CA, USA).
8. Kodak Gel Logic 200 Imaging System or equivalent.

2.6 Western Blotting

1. Tris-buffered saline (TBS) buffer: Dilute in deionized water from 10× TBS (Boston BioProducts, St. Ashland, MA, USA) containing 0.5 M Tris-HCl and 1.5 M NaCl (pH 7.4).
2. TBS/Tween-20 buffer (TBST): Dilute from 10× TBST buffer (Boston BioProducts) containing TBS buffer with 0.5% Tween-20.
3. Blocking buffer: 5% nonfat dry milk in TBST buffer stored at 4°C.
4. Antibody dilution buffer: Dilute blocking buffer to 1% nonfat dry milk with TBST buffer.

5. 1:1,000 dilution of αSMA monoclonal antibody (ABcam, Cambridge, MA, USA), freshly prepared before use.
6. 1:1,000 dilution of calponin monoclonal antibody (Sigma-Aldrich), freshly prepared before use.
7. 1:500 dilution of smooth muscle myosin heavy chain (MHC) monoclonal antibody (Sigma-Aldrich), freshly prepared before use.
8. 1:200 dilution of desmin monoclonal antibody (Santa Cruz), freshly prepared before use.
9. 1:2,000 dilution of mouse anti-human β-actin monoclonal antibody (Santa Cruz), freshly prepared before use.
10. 1:2,000 dilution of horseradish peroxidase (HRP)-conjugated goat anti-mouse IgG (Santa Cruz), freshly prepared before use.
11. Polyvinylidene difluoride membranes.
12. Restore™ Western Blot Stripping Buffer (Pierce Chemical Co., Rockford, IL, USA).
13. RIPA buffer with Triton: 50 mM Tris–HCl (pH 7.4), 150 mM NaCl, 1% Triton X-100, 0.5% sodium deoxycholate, and 0.1% SDS (Boston BioProducts).
14. Protease inhibitor cocktail tablet (05892970001, Roche, IN, USA), each tablet is sufficient for a volume of 10 ml extraction solution (RIPA buffer with Triton).
15. Super Signal West Femto reagent (Pierce).
16. Fuji film imaging system LAS 3000 or equivalent.

3 Methods

3.1 Isolation and Culture of Bone Marrow-Derived MSC

1. Under local anesthetic, bone marrow aspirates should be collected from the health human donor's iliac crest and placed in 10 ml sodium heparin spray-coated Vacutainer tubes prefilled with 3 ml αMEM. The tubes and samples should be kept on ice until transported to the laboratory and processed.
2. Transfer each aspirate into a 50 ml conical tube and dilute to 15 ml with HBSS. Rinse aspirate tubes twice with 5 ml of HBSS and combine with the diluted aspirate (25 ml total volume).
3. For each aspirate, place 10 ml of pre-warmed (37°C) Ficoll-Paque into a separate 50 ml conical tube (see Note 2).
4. Gently overlay each aspirate onto the Ficoll-Paque. Take care to angle the tube containing Ficoll-Paque and very slowly pipette the diluted aspirate over the border of the Ficoll-Paque meniscus. Once done, gently replace the tube in a vertical position.

5. Centrifuge tubes at 400 × *g* for 30 min at room temperature in a swinging bucket rotor with the brake OFF (see Note 3).
6. After centrifugation, carefully collect the buffy coat, located at the Ficoll-Paque/HBSS interface, with a sterile Pasteur transfer pipette and place the cells into a clean 50 ml conical tube.
7. Dilute each sample to 25 ml with HBSS and invert the tube 3–5 times to mix.
8. Centrifuge tubes at 250 × *g* for 10 min in a swinging bucket rotor with the brake ON.
9. Remove the supernatant by vacuum aspiration and resuspend the cells with 30 ml of prewarmed complete culture medium (CCM).
10. Count viable cells with a hemocytometer using trypan blue and plate at a cell density of 50–100 cells/cm^2 in 175 cm^2 flasks or 150 mm dishes.
11. Incubate the cells at 37°C with 5% humidified CO_2 for 24 h to allow adherent cells to attach.
12. After 24 h, remove the medium and nonadherent cells (see Note 4).
13. Add 10 ml of prewarmed PBS to the culture, rock gently to cover the entire surface area, and aspirate. Repeat the wash two additional times.
14. Add 30 ml of fresh CCM to each flask and return flasks to the incubator.
15. Examine cultures daily with phase contrast microscopy.
16. Every 3 days, remove the medium and rinse the flask with 10 ml of pre-warmed PBS. Aspirate the wash and feed cultures with 30 ml of fresh CCM. Continue until the cells reach 70–80% confluence.
17. To harvest cultures, remove the medium and rinse the flask with 30 ml PBS and aspirate.
18. Add 10 ml of prewarmed Trypsin–EDTA solution to the flask. Distribute the trypsin across the surface area of the flask. Incubate the flask for 2–5 min at 37°C. Examine the cells with phase contrast microscopy.
19. After 80–90% of the cells have rounded up or become detached, gently tap the sides of the flask to dislodge any remaining attached cells.
20. Add 10 ml CCM to the flask. Rock the flask back and forth to swirl the medium around the flask and transfer the entire cell suspension into a clean 50-ml conical tube.
21. Rinse the flask with 30 ml of 1× PBS and combine with the cell suspension.

22. Centrifuge at 400 ×*g* for 5 min in a swinging bucket rotor with the brake ON.
23. Remove the supernatant and resuspend the cells in 1–2 ml of pre-warmed PBS.
24. Count the cells with a hemocytometer and trypan blue.
25. Reseed harvested cells at a density of 50–100 viable cells/cm^2 in an appropriate culture vessel (see Note 5). The resultant BMSC cultures can usually be successfully expanded through passage 3 or 4 without significant loss of the stem cell phenotype.

3.2 Flow Cytometry

1. Obtain harvested BMSC and resuspend cells at 10^6 cells per 100 μl in FACS buffer.
2. Stain 100 μl aliquots of cells with conjugated anti-human IgGs (20 μl of CD14-FITC, CD34-PE, CD45-FITC, CD166-PE, respectively. 10 μl of CD29-PE, 5 μl of CD105 Alexa Fluorl 488) or properly matched isotype IgG controls at 4°C for 1 h in the dark.
3. Wash the cells with 5 ml FACS buffer, spin 5 min at 300 ×*g*.
4. Resuspend the cells in 300 μl FACS buffer.
5. Perform quantitative flow cytometry on a FACS Calibur flow cytometer according to the manufacturer's instructions.
6. Flow cytometry should confirm that the isolated BMSC express the stem cell markers CD29, CD105, and CD166, but do not express the hematopoietic stem cell markers CD14, CD34, and CD45 (see Note 6).

3.3 Differentiation of Human BMSC In Vitro: Myogenic Growth Factors

1. Culture BMSC with DMEM (low glucose) plus 10% FBS to sub-confluence (see Note 7).
2. Remove medium and wash cells twice with PBS. Add enough Trypsin–EDTA to cover the cells while gently rocking the flask back and forth.
3. Incubate cells at 37°C for 2–3 min. Monitor the progress of detachment visually under the microscope. If cells have not detached, return the flask to the incubator for an additional 2 min.
4. As soon as cells have detached, stop the reaction by adding 5–10 ml DMEM (low glucose) with 10% FBS and homogenize the cell suspension by repetitive pipetting with a 5 or 10 ml serological pipette.
5. Determine the concentration of cells by removing a small aliquot, mixing the aliquot 1:1 with trypan blue, and counting viable cells in a hemocytometer. The number of viable cells should be greater than 98%.
6. Adjust the concentration of cells to 30,000 cells/ml through the addition of culture medium.

7. Seed cells in multi-well plates or chamber slides at a concentration of approximately 1,000 cells/cm^2 in growth medium.
8. For immunostaining experiments, 8-well chamber slides should be used and 10 cm dishes should be used for protein extraction experiments. Cell morphology should be recorded before and after growth factor addition for up to 14 days.
9. For myogenic differentiation, DMEM containing 2.5 ng/ml TGF-β1 (see Note 8) and 5.0 ng/ml of PDGF-BB (see Note 9) is needed (see Note 10). The final serum concentration should be maintained at 10% in the myogenic differentiation medium.
10. Place reseeded cultures back in the incubator and allow cells to expand until they reach 80–90% confluence.
11. Replace differentiation medium every third day.
12. To study the effects of 3D dynamic culture on human BMSC growth, seed BMSC on culture plates and allow to attach for 6 h. After this, load the plates on an orbital shaker and culture at 40 rpm for up to 6 days.
13. Harvest cells after 7–14 days of treatment for SMC differentiation analysis.

3.4 Differentiation of Human BMSC In Vitro: Cocultured with Bladder SMC or SMC Conditioned Medium

1. Bladder SMC conditioned medium can be derived from bladder cells by collecting the medium from cultured human bladder SMC at 70–90% confluence every 12 h.
2. The collected bladder SMC-conditioned medium must be centrifuged at 400 × *g* for 5 min to remove cells, filtered (0.2 μm), and diluted with an equal volume of DMEM with 10% FBS before use (see Note 11).
3. Culture BMSC with DMEM (low glucose) plus 10% FBS to sub-confluence.
4. Remove medium and wash cells twice with PBS, then add enough Trypsin–EDTA to cover the cells while gently rocking the flask back and forth.
5. Incubate cells at 37°C for 2–3 min. Monitor the progress of detachment visually under the microscope. If cells have not detached, return the flask to the incubator for an additional 2 min.
6. As soon as cells have detached, stop the reaction by adding 5–10 ml DMEM (low glucose) with 10% FBS and homogenize the cell suspension by repetitive pipetting with a 5 or 10 ml serological pipette.
7. Determine the concentration of cells by removing a small aliquot. Mix the aliquot 1:1 with trypan blue and count viable cells in a hemocytometer. The number of viable cells should be greater than 98%.
8. Coculture 2,500–3,000 cells/cm^2 BMSC with bladder cells in transwell plates or with bladder cell-derived conditioned medium for 14 days.

9. Bladder cells (top chamber) should be indirectly cocultured with BMSC (bottom chamber) in a 100 mm transwell plate with a barrier membrane (0.4 μm) between the cell types.
10. Care should be taken during culture medium replenishments (every third day) to avoid cell or medium leakage to the bottom chamber.

3.5 BMSC Proliferation on Bladder ECM

3.5.1 Preparation of the Bladder Tissue-Specific ECM

1. To obtain tissue-specific ECM, use the bladder tissue harvested from adult swine as reported previously (25).
2. To prepare the decellularized bladder ECM, the frozen muscle tissue should be thinly sliced using a razor blade and decellularized through exposure to a series of solutions with continuous agitation at 4°C as described below.
3. Wash the tissue for 2 days in deionized water to induce cell lysis, and follow this by decellularization via exposure to 1% Triton X-100 for 5 days, with daily solution changes.
4. Complete the process with 2 days of rinsing in deionized water followed by a 1 day rinse in PBS.
5. Assess completion of decellularization by sampling the tissue piece (3 mm × 5 mm × 10 mm) using histology and DNA analysis.
6. To prepare decellularized ECM for lyophilization, treat pieces with citrate buffer (pH 4.3) for 48 h at 4°C with constant shaking to cause swelling of the tissue pieces.
7. Remove the buffer and freeze the tissue pieces for 1 day at −80°C prior to lyophilization.
8. The lyophilized tissue should then be powdered using a micro grinder for 15 min and stored at −80°C prior to dissolution in 2 M urea for 3 days at 4°C with constant shaking.
9. Remove the undissolved material via centrifugation at 5,000 × *g* for 20 min.
10. Pass the solution through a 40 μm filter into dialysis tubing.
11. Dialysis against distilled water for 2–3 days.
12. Finally, lyophilize and use the lyophilized dialysate as the matrix powder following total protein determination.

3.5.2 BMSC Proliferate and Myogenic Differentiation on Bladder ECM

1. Coat culture dishes with bladder ECM solution created in the steps outlined above at a final concentration of 0.1 mg/ml by overnight incubation on a rocker at 4°C.
2. The next day, aspirate excess matrix solution out of the plates and rinse the plates gently with PBS solution.
3. Sterilize the coated dishes by γ-irradiation (10^6 rad). These dishes can be used immediately or stored at 4°C for up to 2 weeks.

4. BMSC should be cultured on 12-well plates coated with bladder ECM (1,000 cells/well).
5. Perform cell counts using a standard hemocytometer for six wells at each specific time point (days 1, 3, 5, and 7).
6. Coat control plates with collagen-IV via incubation at room temperature for 1 h.
7. After 1 h coating time, aspirate excess solution and wash the plates gently with PBS (two times). The coated plate should be allowed to dry in the culture hood before use. All coated plates must be hydrated using PBS before culturing cells.

3.6 In Vivo Myogenic Differentiation

1. BMSC induced to differentiate into smooth muscle cells using appropriate growth factors can be seeded onto a nanoporous matrix. We have used a highly porous poly-L-lactic acid (PLLA) scaffold that was created in the laboratory of Professor Peter X. Ma (19).
2. The PLLA matrix should be pre-wet with absolute alcohol for 30 min, wash three times with PBS (30 min each), and incubate twice in serum-containing medium (2 h each).
3. Resuspend the BMSC (1×10^6) in 50–100 ml of culture medium and slowly load onto the matrix.
4. Leave the matrix undisturbed for 6–12 h before culture medium being added.
5. Implant the matrix subcutaneously into nude mice and analyze 1 month after implantation.
6. Process the implants for immunocytochemistry with smooth muscle-specific antibodies, desmin and myosin.
7. For quantitation of immunohistochemical images, count the cells in four random fields (400×). Count a total number of 300–400 cells for each of the experimental condition. The stained cells can be expressed as a percentage of the total cell number in that particular field per cm^2.

3.7 Immunofluorescence Staining of SMC Markers

1. Perform immunofluorescence staining for SMC markers.
2. Seed cells in 8-well chamber slides and allow to adhere overnight.
3. Wash cells twice with PBS, fix with 4% paraformaldehyde, and permeabilize with ice cold acetone.
4. Block each slide using serum-free protein block solution and incubate overnight at 4°C with monoclonal primary antibodies to αSMA, calponin, desmin, and myosin heavy chain (MHC), respectively.
5. Remove unbound primary antibodies by washing twice in PBS and followed by application of FITC-conjugated anti-mouse secondary antibody.

6. Incubate at room temperature for 45 min.
7. Following incubation, remove unbound secondary antibody with several PBS washes. Mount the slides in Vectashield Mounting Medium for Fluorescence with propidium iodide (PI) or with 4′,6-diamidino-2-phenylindole (DAPI) for visualization by fluorescence microscopy.
8. Cells cultured using the 3D dynamic culture system should also be analyzed for smooth muscle-specific protein expression after 14 days.

3.8 PCR for SMC-Specific Gene Expression

1. Isolate total cellular RNA from cell cultures using TRIzol reagent.
2. Synthesize complementary DNA (cDNA) from this RNA using the SuperScript™ II Reverse-Transcriptase reagents according to the manufacturer's instructions.
3. A touch down PCR is used for the detection of SMC-specific transcripts.
4. The PCR primer sequences required are listed in Table 1.
5. The initial denaturing step is at 94°C for 6 min, followed by 94°C for 30 s, 61°C for 30 s, and 72°C for 1 min for three cycles.
6. Carry out the subsequent cycling steps by decreasing the annealing temperature by 2°C for every three cycles, until a final temperature of 53°C (30 cycles) is reached.
7. Analyze amplified products by electrophoresis (15 μl/well) using a 1.5% agarose gel incorporating SYBR-safe DNA gel stain for visualization, visualize with Kodak Gel Logic 200 Imaging System or equivalent.
8. Human glyceraldehyde-3-phosphate dehydrogenase (GAPDH) can be used as a housekeeping gene.

3.9 Western Blot for SMC Markers

3.9.1 BMSC Cell Detachment and Cell Lysate Preparation

1. Wash a 75 cm^2 flask containing BMSC twice with calcium–magnesium-free DPBS and add 1.5 ml Trypsin–EDTA.
2. Incubate at 37°C for no more than 5 min.
3. Shake the plate gently to detach the cells from the surface.
4. Inactivate the trypsin by adding 4.5 ml basal culture medium containing 10% FBS to the flask.
5. Harvest the detached cells in a 15 ml conical tube.
6. To determine cell number, ensure cell solution is thoroughly mixed and then mix 20 μl of cell suspension with 80 μl (1:5 dilution) or 180 μl (1:10 dilution) 0.4% trypan blue in a 1 ml Eppendorf tube. Load approximately 10 μl of the mixture to each side of a hemocytometer and determine cell concentration.
7. Wash cells twice with DPBS and remove the supernatant.

8. Resuspend the cell pellet in RIPA buffer (approximately five times of the volume of the cell pellet) and protease inhibitor cocktail and thoroughly homogenize by repetitive pipetting and vortexing on ice.
9. After homogenization, centrifuge the cell lysate at 14,000 × *g* at 4°C for 30 min.
10. Transfer the supernatant to a clean Eppendorf test tube and remove 25 μl of lysate for protein quantification.

3.9.2 Western Blot Analysis of SMC-Specific Markers

1. After measurement of total protein concentration in each sample using the Lowry method, load protein samples (25–50 μg/well) on to a 10% SDS-polyacrylamide gel after denaturating the samples by boiling for 5 min.
2. Transfer the separated proteins onto a nitrocellulose membrane following gel electrophoresis.
3. Block nonspecific binding sites on the membrane by incubating it in TBST containing 5% nonfat dry milk at room temperature with constant shaking for 1 h.
4. Following blocking, probe the membrane with appropriate antibodies to smooth muscle-specific proteins such as αSMA, calponin, smooth muscle myosin heavy chain (MHC), and desmin, respectively.
5. The housekeeping protein, β-actin, should be used as a loading control.
6. Incubate the primary antibodies on the membrane overnight at 4°C (see Note 12).
7. Wash the membrane three times with TBST and incubate with the secondary antibody (anti-mouse-HRP) at room temperature for 1 h.
8. Detection of protein hybridization can be accomplished by using the chemiluminescent reagent, Super Signal West Femto reagent, and images captured using a Fuji film imaging system or equivalent.
9. For quantification of band intensities, we use the Multi Gauge V 3.0 software from Fujifilm and the values are normalized to the β-actin band in that particular lane.

4 Notes

1. The cell culture procedure should be performed in a tissue culture hood using aseptic techniques. All materials to be used in the biological safety cabinet should be wiped down with 70% ethanol before bringing it in to the biological safety

cabinet. Lab coats and gloves should be worn. Culture media and buffers should be pre-warmed to 37°C before use.

2. If the Ficoll-Paque and HBSS-cell suspension layers are mixed, the mononuclear cells will not completely and efficiently separate during centrifugation.

3. The brake is left off to allow a slow deceleration that helps to avoid disturbance of the Ficoll–HBSS-cell suspension interface.

4. If the nonadherent cells are not removed, hematopoietic cells may become attached and contaminate the BMSC culture.

5. It is important to allow the chambers containing CCM to equilibrate to 37°C and 5% CO_2 before use.

6. When flow cytometry is performed, the variation in values for log (%G/%T) should be established against samples containing 0.5 or 1 million BMSC/ml when the following parameters are varied: (a) the flow rate is 250, 500, or 900 cells/s; (b) the forward scatter (FS) is assayed with 67 or 122 V and a gain of 2 or with 353 V and a gain of 1; and (c) the peak for FS of the 20 mm bead is set at 550, 650, or 750; and (d) the peak for SS for the 7 mm bead is set at 350, 450, or 550.

7. Cell culture confluence is important, and cells should be harvested when they are less than 80% confluent.

8. Preparation of growth factors: Stock solution for TGFβ1 was prepared by dissolving 2 μg of TGFβ1 in 400 μl dilution buffer (1% BSA in PBS, store at 4°C), resulting in a final concentration of 5 μg/ml. Aliquot 100 μl/tube and store at −70°C.

9. PDGF-BB stock solution (10 μg/ml) was made by dissolving 50 μg PDGF-BB in 5 ml dilution buffer. Aliquot 500 μl/tube and store at −70°C.

10. Preparation of growth factors: Dilute the various growth factors in 1% BSA in PBS to minimize adsorption of growth factors onto the wall of plastic Eppendorf tubes.

11. We optimized the differentiation medium for SMC differentiation. We used myogenic medium consisting of SMC-derived conditioned medium and fresh DMEM medium with 10% FBS at different ratios. When the ratio of SMC-derived conditioned medium and fresh medium reached 1:1, this ratio was better in terms of cell differentiation and proliferation than those at other ratios (such as 1:5, 1:2.5, 2.5:1, and 5:1).

12. Diluted primary SMA and calponin antibodies can be recycled by collecting the solution in 15 ml conical tube and stored at 4°C for up to 1 week.

References

1. Ac S (2008) Cancer facts and figures. American Cancer Society, Atlanta, GA
2. Atala A, Bauer SB, Soker S et al (2006) Tissue-engineered autologous bladders for patients needing cystoplasty. Lancet 367(9518): 1241–1246
3. Brehmer B, Rohrmann D, Rau G et al (2006) Bladder wall replacement by tissue engineering and autologous keratinocytes in minipigs. BJU Int 97(4):829–836
4. Jack GS, Almeida FG, Zhang R et al (2005) Processed lipoaspirate cells for tissue engineering of the lower urinary tract: implications for the treatment of stress urinary incontinence and bladder reconstruction. J Urol 174(5): 2041–2045
5. Jack GS, Zhang R, Lee M et al (2009) Urinary bladder smooth muscle engineered from adipose stem cells and a three dimensional synthetic composite. Biomaterials 30(19): 3259–3270
6. Usas A, Huard J (2007) Muscle-derived stem cells for tissue engineering and regenerative therapy. Biomaterials 28(36):5401–5406
7. Shukla D, Box GN, Edwards RA et al (2008) Bone marrow stem cells for urologic tissue engineering. World J Urol 26(4):341–349
8. Zhang Y, Lin HK, Frimberger D et al (2005) Growth of bone marrow stromal cells on small intestinal submucosa: an alternative cell source for tissue engineered bladder. BJU Int 96(7):1120–1125
9. Zhang Y, Bharadwaj S, Liu Y, Atala A (2008) Differentiation of human bone marrow stem cells into smooth muscle for urinary tract reconstruction. American Academy of Pediatrics, Joint Pediatric Urology Meeting [Abstract book]
10. Drewa T (2008) Using hair-follicle stem cells for urinary bladder-wall regeneration. Regen Med 3(6):939–944
11. Yang Y, Qu B, Huo JH et al (2009) Serum from radiofrequency-injured livers induces differentiation of bone marrow stem cells into hepatocyte-like cells. J Surg Res 155(1):18–24
12. Saulnier N, Lattanzi W, Puglisi MA et al (2009) Mesenchymal stromal cells multipotency and plasticity: induction toward the hepatic lineage. Eur Rev Med Pharmacol Sci 13(Suppl 1):71–78
13. Mizuguchi T, Hui T, Palm K et al (2001) Enhanced proliferation and differentiation of rat hepatocytes cultured with bone marrow stromal cells. J Cell Physiol 189(1):106–119
14. Yaghoobi MM, Mowla SJ (2006) Differential gene expression pattern of neurotrophins and their receptors during neuronal differentiation of rat bone marrow stromal cells. Neurosci Lett 397(1–2):149–154
15. Xin H, Li Y, Chen X et al (2006) Bone marrow stromal cells induce BMP2/4 production in oxygen-glucose-deprived astrocytes, which promotes an astrocytic phenotype in adult subventricular progenitor cells. J Neurosci Res 83(8):1485–1493
16. Shichinohe H, Kuroda S, Yano S et al (2006) Improved expression of gamma-aminobutyric acid receptor in mice with cerebral infarct and transplanted bone marrow stromal cells: an autoradiographic and histologic analysis. J Nucl Med Soc Nucl Med 47(3):486–491
17. Scintu F, Reali C, Pillai R et al (2006) Differentiation of human bone marrow stem cells into cells with a neural phenotype: diverse effects of two specific treatments. BMC Neurosci 7:14
18. Tian H, Bharadwaj S, Liu Y et al (2010) Differentiation of human bone marrow mesenchymal stem cells into bladder cells: potential for urological tissue engineering. Tissue Eng Part A 16(5):1769–1779
19. Tian H, Bharadwaj S, Liu Y et al (2010) Myogenic differentiation of human bone marrow mesenchymal stem cells on a 3D nano fibrous scaffold for bladder tissue engineering. Biomaterials 31(5):870–877
20. Chung SY, Krivorov NP, Rausei V et al (2005) Bladder reconstitution with bone marrow derived stem cells seeded on small intestinal submucosa improves morphological and molecular composition. J Urol 174(1):353–359
21. Kanematsu A, Yamamoto S, Iwai-Kanai E et al (2005) Induction of smooth muscle cell-like phenotype in marrow-derived cells among regenerating urinary bladder smooth muscle cells. Am J Pathol 166(2):565–573
22. Sharma AK, Hota PV, Matoka DJ et al (2010) Urinary bladder smooth muscle regeneration utilizing bone marrow derived mesenchymal stem cell seeded elastomeric poly(1,8-octanediol-co-citrate) based thin films. Biomaterials 31(24):6207–6217
23. Sharma AK, Fuller NJ, Sullivan RR et al (2009) Defined populations of bone marrow derived mesenchymal stem and endothelial progenitor cells for bladder regeneration. J Urol 182(4 Suppl):1898–1905
24. Sharma AK, Bury MI, Marks AJ et al (2010) A non-human primate model for urinary bladder regeneration utilizing autologous sources of bone marrow derived mesenchymal stem cells. Stem Cells 29:241–250
25. Liu Y, Bharadwaj S, Lee SJ et al (2009) Optimization of a natural collagen scaffold to aid cell-matrix penetration for urologic tissue engineering. Biomaterials 30(23–24): 3865–3873

Chapter 8

Xeno-Free Adaptation and Culture of Human Pluripotent Stem Cells

Tori Sampsell-Barron

Abstract

Human pluripotent stem cells (hPSCs), including embryonic stem (ES) and induced pluripotent stem (iPS) cells, have been historically cultured in media containing xenogeneic animal components. As hPSC-derived cells and tissues are being developed for human therapies, the application of culture systems to reduce potential immunoreactivity and improve reproducibility becomes increasingly vital. Methods for directly adapting hPSCs to a commercially available culture system free of nonhuman proteins (xeno-free) are described in this chapter.

Key words Human embryonic stem cell, Induced pluripotent stem cell, Human pluripotent stem cell, Xeno-free

1 Introduction

Human embryonic stem (ES) and induced pluripotent stem (iPS) cells (hPSCs) can be expanded indefinitely and directly differentiated to potentially any cell type in the human body (1–3). These features promote their application as an unlimited source of material for drug discovery and screening, developmental research, and regenerative therapy. However, the culture and maintenance of hPSCs are challenging and historically fraught with inconsistent results mostly due to variability in media, matrix, and culture conditions (4). As hPSC-derived cells and tissues enter the clinic, it becomes more crucial to culture them in a defined and reproducible manner emphasizing the removal of xenogeneic animal components from culture conditions (5).

Since the first formulation in 1998 (1), ES cell culture systems have advanced from undefined, serum-rich medium and a mouse embryonic fibroblast (MEF) feeder matrix to more defined systems with media containing KnockOut™ serum replacer (KSR) in combination with feeder-free matrices (6–9). More recently, culture media developed with human serum albumin (10) or chemically

Joydeep Basu and John W. Ludlow (eds.), *Organ Regeneration: Methods and Protocols*, Methods in Molecular Biology, vol. 1001, DOI 10.1007/978-1-62703-363-3_8, © Springer Science+Business Media New York 2013

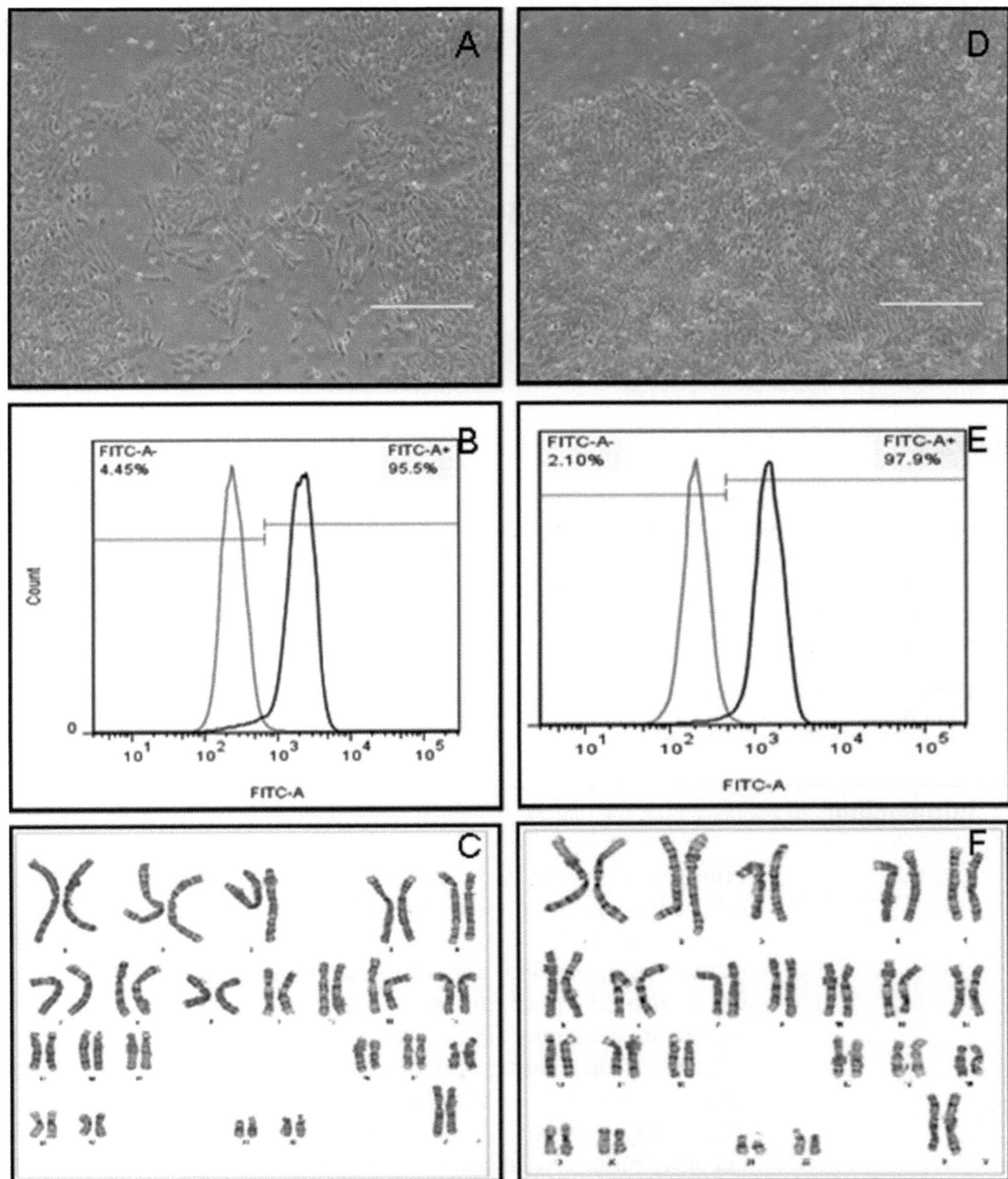

Fig. 1 (**a**) KSR-XF/CELLstart™-adapted WA09p74 (15 passages) ES cells in normoxic conditions, (**b**) flow cytometry results of FITC-conjugated anti-human Nanog pluripotency marker on same cells, and (**c**) Karyotype results 46, *XX* at passage 75. (**d**) KSR-XF/CELLstart™-adapted WA09p74 (15 passages) ES cells in hypoxic conditions (12 passages), (**e**) flow cytometry results of FITC-conjugated anti-human Nanog pluripotency marker on same cells, and (**f**) Karyotype results 46, *XX* at passage 75 (Karyotype Images **c** and **f** courtesy of WiCell Cytogenetics)

defined components (11) have been shown to support the derivation and growth of hPSCs. In addition, researchers have begun deriving and culturing hPSCs in incubators that regulate both oxygen and carbon dioxide levels to more closely approximate those levels found in utero during human fetal development (11, 12).

While culture systems vary across industry and academia, it is widely accepted that a defined and commercially available system free of nonhuman proteins (xeno-free) that mimics the in vivo environment will provide advantages for the development of safer cell therapies, improved drug testing platforms, and the reproducible study of human development. Here, methods are described for the direct adaptation of hPSCs from KSR-based medium on either feeder or feeder-free matrices to a completely xeno-free and feeder-free culture system using KnockOut™ SR Xeno-Free CTS™-based medium and CELLstart™ substrate from Life Technologies. Notably, ES cells cultured in this xeno-free system under both normoxic and hypoxic conditions demonstrate high expression of Nanog (a common pluripotency marker), and maintain genetic stability over twelve passages (Fig. 1), while further characterization and efforts to directly differentiate hPSCs in this xeno-free system are ongoing.

2 Materials

2.1 Culture Components and Media

1. KnockOut™ DMEM/F12 (Life Technologies, Carlsbad, CA, USA).
2. KnockOut™ SR Xeno-Free CTS™ (Life Technologies): Thaw overnight at 4°C; prepare 20 mL aliquots and store at −20°C.
3. GlutaMAX™-I CTS™ (Life Technologies).
4. Basic fibroblast growth factor (bFGF—recombinant human): Rehydrated to 10 μg/mL in 10% KnockOut™ SR Xeno-Free CTS™ solution in KnockOut™ DMEM/F12, aliquot and store at −20°C. Avoid freeze/thaw cycles.
5. KnockOut™ SR GF Cocktail CTS™ (Life Technologies): Thaw overnight at 4°C; prepare 1 mL aliquots and store at −20°C.
6. hPSC xeno-free medium (KSR-XF complete medium): Add 78 mL KnockOut™ DMEM/F12, 20 mL KnockOut™ SR Xeno-Free CTS™, 1 mL Glutamax-I CTS™, 1 mL 100× KnockOut™ SR GF Cocktail CTS™, and 200 μL 10 μg/mL bFGF to a bottle-top vacuum filter and filter sterilize (see Note 1). Aseptically add 2-mercaptoethanol to filter-sterilized medium at 0.1 mM final concentration. Aliquot daily required volume (Table 1) of KSR-XF complete medium to conical tubes for daily feeding or passaging. Equilibrate the aliquot before use to 37°C and incubator gases (5% CO_2/~20% O_2-normoxic: or 5% CO_2, 8% O_2-hypoxic) (see Note 2) by loosening lid and

Table 1
Volumes required for medium exchange and passaging

Culture dish type	Surface area (per well) (cm^2)	Volume of KSR-XF (daily exchange/seeding) (mL)	Volume of KSR-XF (passaging-scraping /rinses) (mL)	Volume of TryPLE™ select CTS™ (mL)
6-well	9.5	2.0–2.5/well	2.0/well	0.5/well
12-well	3.8	0.5/well	0.5/well	0.2/well
6 cm plate	21.0	5.0/dish	3.0/dish	1.0/dish
10 cm plate	55.0	15.0/dish	5.0/dish	3.0/dish

placing in appropriate incubator for 30 min. Store the remaining medium at 2–8°C for up to 1 week.

7. TrypLE™ Select CTS™ (TrypLE™) (Life Technologies).
8. DPBS–/– (*minus* calcium and magnesium).
9. Optional: UltraPure™ distilled water (Life Technologies).
10. *Optional*: Thiazovivin (StemGent, San Diego, CA, USA): Prepare 2 mM 1,000× stock in UltraPure™ distilled water and store at –20°C. Avoid freeze/thaw cycles (see Note 3).
11. *Optional*: Y-27636 ROCK Inhibitor (StemGent): Prepare 10 mM 1,000× stock in Ultrapure™ distilled water and store at –20°C. Avoid freeze/thaw cycles (see Note 3).
12. Dimethyl Sulfoxide (DMSO) Hybri-Max™ (Sigma-Aldrich, St. Louis, MO, USA).
13. 55 mM 2-mercaptoethanol (Life Technologies): Aliquot 15 mL 1,000× working stock to conical, store at 4°C protected from light.
14. 70% Ethanol.
15. DPBS+/+ (*plus* calcium and magnesium): Store at room temperature.
16. Cryopreservation medium A: 9:1 KSR-XF complete medium: KnockOut™ SR Xeno-Free CTS™.
17. Cryopreservation medium B: 8:2 KSR-XF complete medium: DMSO.
18. CELLstart™ CTS™ humanized substrate for stem cell culture (Life Technologies): Store at 4°C protected from light.

2.2 Cultureware

1. Tissue culture dishes (6-well, 10 cm, 6 cm) (Corning® Costar, Corning, NY, USA).
2. StemPro® EZPassage™ Disposable Stem Cell Passaging Tool (EZPassage™) (Life Technologies).
3. Corning® Cell Lifter.

4. 15 and 50 mL conical tubes (Corning®, Corning, NY, USA).
5. 1 and 5 mL glass (borosilicate) pipettes (Thermo Fisher Scientific, Waltham, MA, USA) (see Note 4).
6. 2 mL Cryovials (Nunc, Thermo Fischer Scientific).
7. Flame-pulled Pasteur pipette: UV irradiated or autoclaved (see Note 5).
8. Bottle-top vacuum filter—0.22 μm pore (Corning®).
9. Polystyrene serological pipettes (5, 10, 25, 50 mL sizes) (see Note 4).
10. Parafilm®.

2.3 Equipment

1. Phase contrast microscope 5×, 10×, 20× objectives.
2. Freezing container (Mr. Frosty, Nalgene®, Rochester, NY, USA).
3. Stereoscope/brightfield microscope within biosafety cabinet (EVOS® xl core, Advanced Microscopy Group, Bothell, WA, USA).
4. Incubator: Normoxic (Nu-Aire, Plymouth, MN, USA) NU-4750, 37°C and incubator gases 5% CO_2; Hypoxic (Binder, Bohemia, NY, USA) CB 150-UL, 37°C and incubator gases 5% CO_2, 8% O_2 (see Note 2).
5. Drummond Pipet-Aid® (three speed).

3 Methods

To protect users and prevent culture contamination, all preparations must be performed in a Class II biosafety cabinet under sterile conditions wearing personal protective equipment (disposable gloves, safety glasses, clean laboratory coats, etc.). UV irradiate biosafety cabinet for 15 min prior to use and spray surface with 70% ethanol. Keep all materials sterile, opening all packaging within biosafety cabinet and spraying media bottles with 70% ethanol prior to placing in biosafety cabinet. Spray gloved hands frequently with 70% ethanol while working with hPSCs.

hPSCs require daily medium exchange (feeding) and monitoring to prevent differentiation. While cells can be adapted to xeno-free conditions from either feeder or feeder-free conditions (see Note 6), it is critical to start with a mitotically robust and undifferentiated hPSC culture population. Adaptation to xeno-free conditions can vary by cell type and line. The following protocol describes a direct method of adaptation to xeno-free conditions; however, some cell lines may require a more gradual adaptation process which can be achieved by sequentially increasing the ratio of xeno-free medium to control medium (see Note 7). Prior to xeno-free adaptation, scale and cryopreserve backup cultures in starting conditions and maintain a parallel culture of hPSCs in starting conditions during the adaptation process. Passaging of hPSCs may be performed manually or enzymatically depending on researcher preference, scale of culture, and downstream applications.

3.1 Culture Dish preparation

1. Dilute CELLstart™ 1:50 with DPBS+/+ in conical tube (Table 2). Pipette several times gently to resuspend.
2. Add required volume of CELLstart™ solution (Table 2) to culture dish, swirl to completely cover, and incubate 1–2 h at 37°C under normoxic or hypoxic conditions.
3. Remove from incubator and equilibrate CELLstart™ culture dishes to room temperature in biosafety cabinet for 30–60 min prior to cell seeding.
4. Aspirate CELLstart™ solution from culture dish immediately prior to cell seeding (no rinse required).
5. Parafilm® and store unused CELLstart™ dishes overnight at 4°C for next-day use (see Note 8).

3.2 Negative Selection

1. Observe cells microscopically to identify differentiation (Fig. 2). If differentiation is identified, perform negative selection in spent medium prior to daily medium exchange or passaging.
2. Aliquot and equilibrate daily required volume (see Table 1) of KSR-XF complete medium to conical tube for feeding or passaging.
3. If less than 20% differentiation is observed, transfer culture to microscope platform in biosafety cabinet (see Note 9).
4. Remove differentiation from hPSC culture by gently scraping with a sterile flame-pulled Pasteur pipette in spent medium (Fig. 2).
5. Transfer negatively selected culture dish to cell culture biosafety cabinet and aspirate spent medium.

Table 2
Volumes required for culture dish preparation

Culture dish type	Surface area (per well) (cm^2)	Volume of CELLstart™ 1:50 (mL)	Volume of CELLstart™ 1:50 (μL)	Volume of DPBS+/+ (mL)
6-well	9.5	1.0	20	0.980
12-well	3.8	0.4	8	0.392
6 cm plate	21.0	2.0	40	1.960
10 cm plate	55.0	5.0	100	4.900

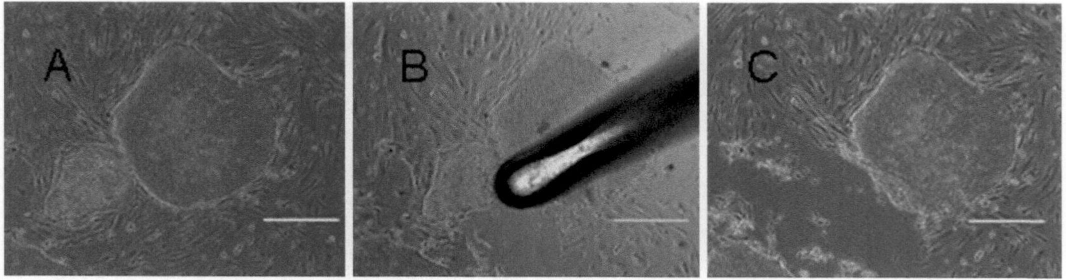

Fig. 2 (**a**) Gibco® hIPS cells in KSR/MEF with area of differentiation (*left*). (**b**) A flame-pulled Pasteur pipette is used to remove differentiation (**c**) prior to passaging (4×)

6. Add pre-equilibrated KSR-XF for daily medium exchange (Table 1) and return to 37°C incubator. Alternately, proceed to enzymatic or manual passaging (Subheading 3.3 or 3.4).

3.3 Passaging hPSCs: Enzymatic

Use enzymatic passaging when adapting hPSC cultures growing on fibroblast feeder substrates (see Note 6). Manual passaging is recommended if starting cultures are on a feeder-free substrate and once cultures have been adapted to CELLstart™-coated dishes. Routine passaging should be performed once cultures are 70–80% confluent (typically every 4–5 days). Maintain backup cultures in both control and adaptation conditions until successful passaging results are confirmed.

1. Observe cells microscopically to confirm they are 70–80% confluent and free of differentiation (<5%). Perform negative selection (Subheading 3.2) if needed prior to passaging (Fig. 2).
2. Determine the number of culture dishes to be passaged and prepare CELLstart™-coated culture dishes (Subheading 3.1) based on a 1:2 passaging ratio.
3. Equilibrate CELLstart™-coated culture dishes to room temperature. If using premade CELLstart™ dishes stored overnight at 2–8°C, equilibrate to room temperature by incubating for 15 min at 37°C or placing in biosafety cabinet for 90 min.
4. Aliquot and equilibrate required volume of KSR-XF complete medium. Calculate medium requirements based on a passaging ratio of 1:2 until cells are completely adapted to KSR-XF complete medium (4–5 passages) (see Note 10).
5. Warm TrypLE™ Select CTS™ (Table 1) in 37°C water bath for 15 min.
6. Rinse cells gently with DPBS-/- and aspirate to remove debris.
7. Add pre-warmed TryPLE™, swirl dish gently to cover, and incubate at 37°C for *2–3 min maximum* (see Note 11).

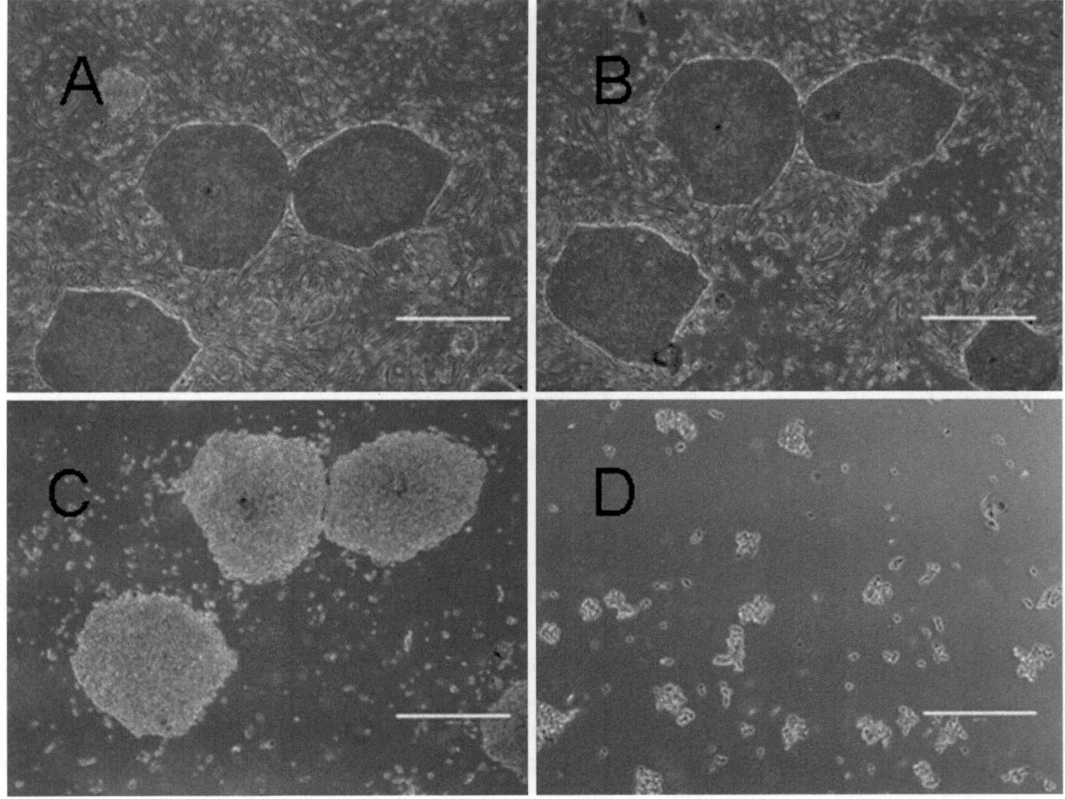

Fig. 3 TryPLE™ passaging of Gibco® hIPS cells from MEFs/KSR to CELLstart™/KSR-XF. (**a**) Passage 7, day 4 cells (4×), (**b**) following negative selection (4×), (**c**) following 2-min TrypLE™ incubation (4×), and (**d**) as seeded cell clumps to CELLstart™-coated dish (10×)

8. Tilt culture dish and aspirate TryPLE™. Rinse culture dish very gently (see Note 11) with DPBS−/− two times (Fig. 3).
9. Add equilibrated KSR-XF complete medium to culture dish and gently scrape cells with a Corning® Cell Lifter. Use a 1 mL or a 5 mL glass pipette to scrape cultures in formats smaller than 6-well plates.
10. Transfer scraped cells to a 15 mL conical tube with a 1 or a 5 mL glass pipette (see Note 12).
11. Rinse culture dish with equilibrated KSR-XF complete medium (Table 1) to obtain any remaining cells (see Note 12). Add rinsed cells to cell solution from step 10 in Subheading 3.3 in 15 mL conical tube.
12. Spin 15 mL conical tube at 200 × *g* for 5 min.
13. Aspirate medium carefully to avoid dislodging cell pellet, and then gently flick cell pellet.
14. Aspirate CELLstart™ solution from room-temperature culture dishes and add one-half seeding volume of equilibrated KSR-XF complete medium (e.g., add 1 mL KSR-XF complete medium to each well of a 6-well plate to be seeded).

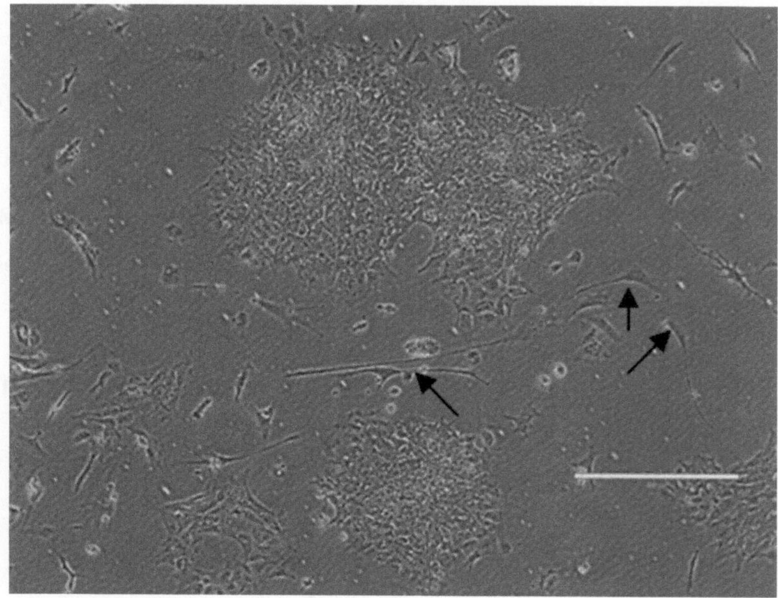

Fig. 4 Typical morphology of Gibco® hIPS cells 24 h following passaging with TryPLE™ directly from MEF/KSR to CELLstart™/KSR-XF. Note cell morphology: Cells are more flattened and loosely packed, colony edges are undefined. Mitotically inactivated MEF cells are still present (*arrows*) and will be eliminated with successive passaging (10×)

15. Add one-half final seeding volume (Table 1) of equilibrated KSR-XF complete medium, with a 5 mL glass pipette, to cell pellet and gently draw up the cell solution (e.g., add 6 mL KSR-XF complete medium for 1:2 passaging ratio from three wells of a 6-well plate to six wells of a 6-well plate). *Do not triturate*!
16. Add cells to CELLstart™-coated culture dish in a slow, dropwise fashion (see Note 13), distributing cells evenly.
17. Transfer cell culture dish to 37°C hypoxic or normoxic incubator and quickly move dish side to side three times, and then front to back three times, with one more side-to-side motion to disperse and evenly distribute floating cells prior to attachment.
18. Twenty-four hours later, monitor the cell culture for attachment with a phase contrast microscope (Fig. 4) (see Note 14).
19. Exchange medium daily with equilibrated KSR-XF complete medium and monitor cell culture for expansion, differentiation (negative selection requirements), and timing of next passaging (approximately every 3–5 days).

3.4 Passaging hPSCs: Manual

1. Observe cells microscopically to confirm they are 70–80% confluent and free of differentiation (<5%). Perform negative selection (step 3.3) (Table 1) if needed prior to passaging.
2. Prepare CELLstart™-coated culture dishes (Subheading 3.1) for a 1:2 passaging ratio and equilibrate to room temperature.

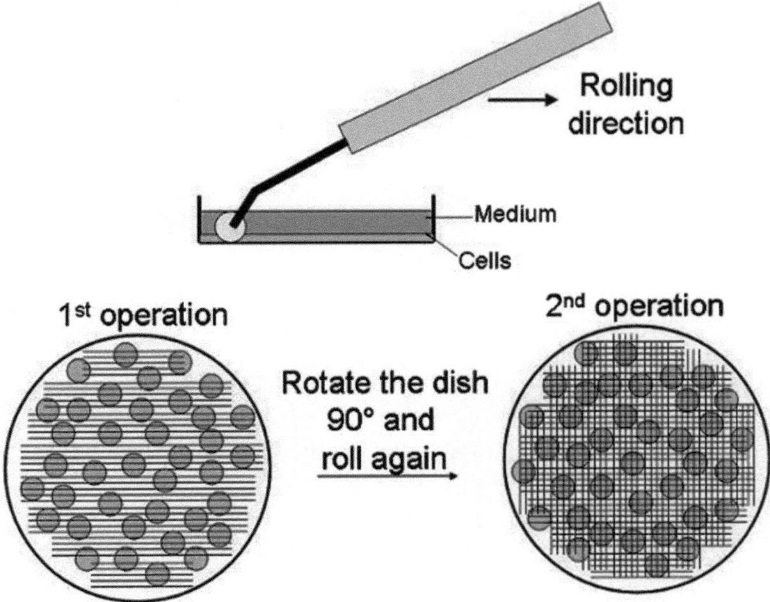

Fig. 5 StemPro® EZPassage™ Disposable Stem Cell Passaging Tool (courtesy of Soojung Shin, Life Technologies)

If using premade CELLstart™ dishes stored overnight at 2–8°C, equilibrate to room temperature by incubating for 15 min at 37°C or placing in biosafety cabinet for 90 min. Do not allow CELLstart™ to evaporate.

3. Aliquot and equilibrate required volume of KSR-XF complete medium. Calculate medium requirements based on a passaging ratio of 1:2 until cells are completely adapted to KSR-XF complete medium (4–5 passages) (see Note 10).
4. Rinse cells gently with DPBS−/− and aspirate to remove debris.
5. Add equilibrated KSR-XF complete medium (Table 1) to culture dish.
6. Open StemPro® EZPassage™ Disposable Stem Cell Passaging Tool (EZPassage™ tool) in biosafety cabinet.
7. Roll the EZPassage™ tool across the entire culture dish in one direction without overlapping rows (Fig. 4). Apply light but consistent pressure when rolling to keep rows parallel (see Note 15).
8. Rotate the culture vessel 90° and repeat step 7 in Subheading 3.4 (Fig. 5).
9. Move culture vessel to the microscope and confirm a grid-like pattern over most of the colonies (Fig. 6).
10. Scrape the culture dish gently with a Corning® Cell Lifter for hPSCs in feeder-free conditions. Scrape the culture dish using

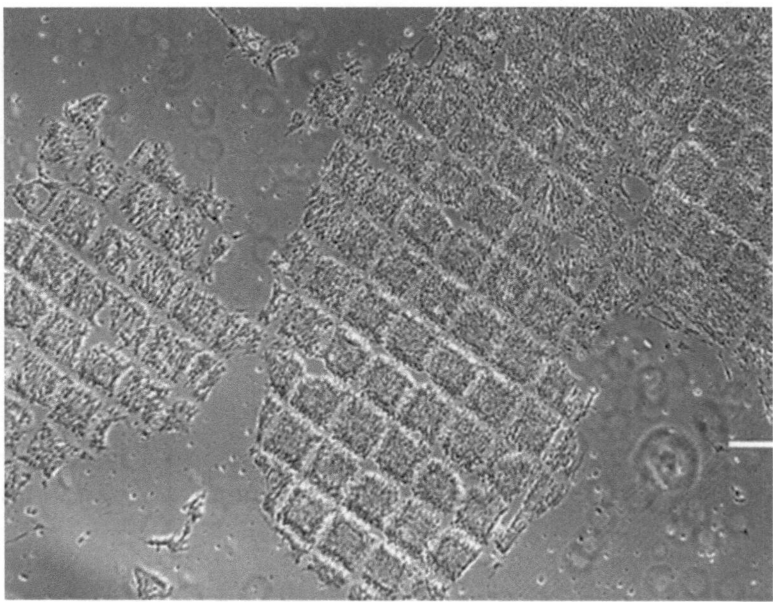

Fig. 6 KSR-XF adapted Gibco® hIPS cells (passage 12, day 5) following cutting with StemPro® EZPassage™ Disposable Stem Cell Passaging Tool (4×)

a 5 mL glass pipette for hPSCs on fibroblast feeders. Use a 1 or a 5 mL glass pipette to scrape cultures in formats smaller than 6-well plates.

11. Transfer scraped cells to a 15 mL conical tube with a 1 or a 5 mL glass pipette (see Note 12).
12. Rinse culture dish with equilibrated KSR-XF complete medium (Table 1) to obtain the remaining cell colonies (see Note 11). Add rinsed cells to cell solution from step 11 in Subheading 3.4 in 15 mL conical tube.
13. Spin 15 mL conical tube at 200 ×*g* for 5 min.
14. Aspirate supernatant carefully to avoid dislodging cell pellet and then gently flick cell pellet.
15. Aspirate CELLstart™ solution from room-temperature culture dishes and add one-half seeding volume of equilibrated KSR-XF complete medium (e.g., add 1 mL KSR-XF complete medium to each well of a 6-well plate).
16. Add one-half final seeding volume (Table 1) of equilibrated KSR-XF complete medium to cell pellet with a 5 mL glass pipette, and *gently* triturate cell solution 3–4 times to create uniformity of colony size (e.g., add 6 mL KSR-XF complete medium for 1:2 passaging ratio from three wells of a 6-well plate to six wells of a 6-well plate).
17. Add hPSCs to CELLstart™-coated culture dish in a slow, dropwise fashion (see Note 13), distributing cells evenly.

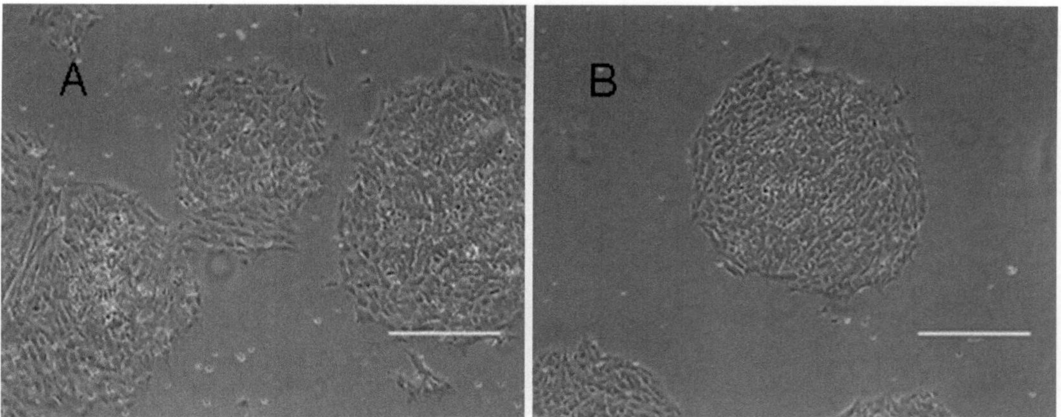

Fig. 7 Typical morphology of KSR-XF/CELLstart™-adapted (seven passages) WA09p65 ES cells 24 h following passaging with StemPro® EZPassage™ Disposable Stem Cell Passaging Tool in (**a**) normoxic conditions and (**b**) hypoxic conditions (three passages). Note cell morphology: Cells are flattened and loosely packed, colony edges are undefined

18. Transfer cell culture dish to 37°C hypoxic or normoxic incubator and quickly move dish side to side three times, and then front to back three times, with one more side-to-side motion to disperse and evenly distribute floating cells prior to attachment.
19. Twenty-four hours later, monitor the cell culture for attachment with a phase contrast microscope (Fig. 7) (see Note 14).
20. Exchange medium daily with equilibrated KSR-XF complete medium and monitor cell culture for expansion, differentiation (negative selection requirements), and timing of next passaging (approximately every 3–5 days).
21. Discard EZPassage™ tool after use. Do not reuse.

3.5 Cryopreservation

1. Observe that hPSCs are 80–90% confluent and free from differentiation prior to cryopreservation. Perform negative selection if needed (Subheading 3.2).
2. hPSCs cultured in KSR-XF complete medium should be cryopreserved in a total volume of 1 mL per cryovial with a 1–2:1:1 ratio for harvest, cryopreservation, and later thawing (e.g., harvest 1–2 wells of a six-well plate and cryopreserve to one cryovial for future thawing to one well of a six-well plate).
3. Chill Mr. Frosty at 4°C for 1–2 h.
4. Warm Cryopreservation medium A (step 16 in Subheading 2.1) in 37°C water bath for 15–30 min. Chill Cryopreservation medium B (step 17 in Subheading 2.1) on ice for 15–30 min.
5. Harvest hPSCs to be cryopreserved enzymatically (Subheading 3.3) or manually (Subheading 3.4).

6. Gently resuspend cell pellet in one-half final volume (for 1 mL per cryovials) with Cryo A medium minimizing colony breakage.
7. Add one-half final volume Cryo B medium slowly drop-wise (see Note 13) (one drop per 5 s) to cells in Cryo A medium and gently flick the tube every two drops to prevent osmotic shock.
8. Add cell solution evenly to cryovials at 1 mL per cryovial and transfer to prechilled Mr. Frosty. Store overnight in −80°C freezer and transfer to liquid nitrogen the next day.

3.6 Thawing

1. Equilibrate required seeding volume of KSR-XF complete medium (Table 1) for thawing to appropriate size CELLstart™-coated culture dish.
2. Equilibrate additional 5 mL KSR-XF complete medium for each cryovial to be thawed.
3. Prepare CELLstart™-coated culture dish (Subheading 3.1).
4. Remove cryovial of hPSCs from liquid nitrogen storage (see Note 16), roll between palms, and immerse bottom two-thirds of vial in 37°C water bath with gentle agitation.
5. Remove cryovial from water bath when small ice crystal remains and spray the entire cryovial generously with 70% ethanol.
6. Move cryovial to biosafety cabinet and transfer the entire contents with 1 mL glass pipette to a 15 mL conical tube.
7. Add 4 mL KSR-XF complete medium slowly drop-wise (see Note 13) (one drop per 5 s) to hPSCs in 15 mL conical tube and gently flick the tube every two drops to prevent osmotic shock.
8. Rinse cryovial with 1 mL KSR-XF complete medium. Transfer 1 mL rinsed hPSCs drop-wise and combine to 15 mL conical tube.
9. Spin 15 mL conical tube at 200 × *g* for 5 min.
10. Aspirate medium carefully to avoid dislodging cell pellet, and then gently flick cell pellet.
11. Aspirate CELLstart™ solution from room-temperature culture dishes and add one-half seeding volume of equilibrated KSR-XF complete medium (e.g., add 1 mL KSR-XF complete medium to one well of a 6-well plate).
12. With a 5 mL glass pipette, add one-half final seeding volume (Table 1) of equilibrated KSR-XF complete medium to cell pellet and *gently* triturate cell solution 3–4 times to create uniformity of colony size (e.g., add 1 mL KSR-XF complete medium when thawing 1 cryovial containing hPSCs harvested and cryopreserved from 1 to 2 wells of 6-well plate for seeding to 1 well of a 6-well plate).

13. Add hPSCs to CELLstart™-coated culture dish in a slow, drop-wise fashion (see Note 13), distributing cells evenly.
14. Transfer cell culture dish to 37°C hypoxic or normoxic incubator and quickly move dish side to side three times, and then front to back three times, with one more side-to-side motion to disperse and evenly distribute floating cells prior to attachment.
15. Twenty-four hours later, monitor the cell culture for attachment (see Note 14).
16. Exchange medium daily with equilibrated KSR-XF complete medium and monitor cell culture for expansion, differentiation (negative selection requirements), and timing of next passaging (approximately every 3–5 days).

4 Notes

1. Alternately, KnockOut™ DMEM may be substituted for KnockOut™ DMEM/F12, KnockOut™ SR Xeno-Free CTS™ may be added at 15% final concentration, and bFGF may be added aseptically following filter sterilization to minimize filter binding of protein.
2. hPSCs have been traditionally cultured under normoxic conditions of ~20–21% O_2 and 5% CO_2. Researchers have recently begun culturing hPSCs under hypoxic conditions of 5% CO_2 and 5–10% O_2 to more closely approximate the in vivo environment (11, 12).
3. If single-cell passaging is desired, increased cell survival and clonal efficiency may be achieved by the addition of a ROCK inhibitor such as Y-27632 (StemGent) at 10 μM final concentration (10, 13) or Thiazovivin (StemGent) at 2 μM final concentration (15) to the KSR-XF complete medium during passaging manipulations and seeding.
4. Use disposable glass (borosilicate) serological pipettes when manipulating hPSCs during thawing, passaging, and cryopreservation to reduce cell loss. Disposable polystyrene serological pipettes can be used for daily medium exchange and medium preparation.
5. To make a flame-pulled Pasteur pipette, hold pipette on both ends over an alcohol burner or a flame source (such as propane torch) so that flame contacts pipette approximately 2 in. from thin end. When pipette heats to orange, simultaneously bend and pull pipette away to create a negative selection tool with a thin closed end and a 15–45° angle according to user preference (14). UV sterilize for 15 min prior to negative selection and dispose after use.
6. For best xeno-free adaptation results, adapt hPSCs from a human or a murine fibroblast matrix cultured in traditional

KSR-based medium. If adapting hPSCs grown on a feeder-free matrix (e.g., Matrigel™ or GelTrex™), it is best to adapt from a traditional MEF-conditioned medium (MEF-CM (6)) versus a defined or KSR-based medium. If adaptation to KSR-XF on human feeders is desired, seed human feeders on CELLstart™-coated dishes and omit bFGF from the KSR-XF medium formulation when culturing hPSCs.

7. Some cell lines may require a more gradual adaptation to xeno-free conditions. If proliferation is decreased or cell death is observed, use the following method of sequential adaptation: Passage 1: 75% control medium + 25% KSR-XF complete medium. Passage 2: 50% control medium + 50% KSR-XF complete medium. Passage 3: 25% control medium + 75% KSR-XF complete medium. Passage 4: 100% KSR-XF complete medium.

8. Do not use CELLstart™-coated dishes older than 24 h for hPSC seeding. Keep culture dishes on level surface to insure complete coverage with CELLstart™ and prevent areas of dehydration.

9. Greater than 20% differentiation may require positive selection in fresh control medium to achieve a pluripotent culture prior to beginning adaptation to xeno-free conditions. If positive selection is needed, use a sterile flame-pulled Pasteur pipette to dissect pluripotent colonies in fresh control medium within a biosafety cabinet. Transfer dissected colonies aseptically to a culture dish format that achieves high confluency.

10. Refer to Table 1 and use the following formulas to calculate volume of KSR-XF complete medium to equilibrate for cell culture work:

    Daily medium exchange

    Volume of KSR-XF complete medium to equilibrate = (# of culture well/dishes to "feed" × corresponding volumes in column 3).

    Enzymatic or manual passaging

    Volume of KSR-XF complete medium to equilibrate = ((# of culture wells/dishes to passage × corresponding volumes in column 4) × 2) + (for passaging ratio 1:2 – 2 × corresponding volumes in column 3).

11. TrypLE™Select CTS™ is a very-fast-acting enzyme that will singularize hPSCs when incubated longer than 3 min. Care should be taken to perform DPBS–/– rinses gently and quickly, keeping hPSCs in clumps for passaging unless singularization is required for specific cell seeding densities (see Note 3). Perform enzymatic passaging of multiple culture dishes sequentially to prevent singularization.

12. To maximize cell retention, avoid generating bubbles during harvests and rinses. To minimize bubbles, do not expel or draw up the entire volume of the medium in the serological pipette.

13. For required "drop-wise" addition of cells or media, use Pipet-Aid® on "slow" setting.
14. If poor attachment is observed, first check that medium and culture dishes were prepared correctly. Repeat passaging on backup culture with one or more of the following modifications:
    - Generate larger colony pieces for passaging by reducing TrypLE™ incubation time and/or resuspend cell pellet more gently.
    - Reduce passaging ratio from 1:2 to 1:1.5.
    - Add a ROCK inhibitor (see Note 3) to KSR-XF complete medium during passaging and seeding steps.
    - Use fresh (same day) CE LLstart™-coated dishes for seeding.
    - Optimize CELLstart™ dilution from 1:50 to 1:100.
    - Some cell lines may require sequential adaptation to xeno-free conditions (see Note 6).
15. Do not remove the culture medium before rolling the plate. For first-time users, use a "practice" well and observe EZPassage™ technique and pattern. If too much pressure is applied with the EZPassage™ tool, cells may begin to lift off; if too little pressure is applied, uncut colonies that are too large for passaging may be generated.
16. The entire thawing procedure should be performed as quickly as possible to improve cell survivability.

Acknowledgments

The author extends many thanks to Marian Piekarczyk for her significant contributions to this work and critical review of the chapter. Thanks to David Piper for his encouragement and critical review of the chapter. Thanks to Justin Wetter and Jun Wang for their assistance with hPSC culture and flow cytometry. Lastly, thanks to my family for their love and support.

References

1. Thomson JA, Itskovitz-Eldor J, Shapiro SS et al (1998) Embryonic stem cell lines derived from human blastocysts. Science 282:1145–1147
2. Yu J, Vodyanik MA, Smuga-Otto K et al (2007) Induced pluripotent stem cell lines derived from human somatic cells. Science 318:1917–1920
3. Takahashi K, Tanabe K, Ohnuki M et al (2007) Induction of pluripotent stem cells from adult human fibroblasts by defined factors. Cell 131:861–872
4. Akopian V, Andrews PW, Beil S et al (2010) Comparison of defined culture systems for feeder cell free propagation of human embryonic

stem cells. In Vitro Cell Dev Biol Anim 46:247–258
5. Mallon B, Park KY, Chen KG et al (2006) Toward xeno-free culture of human embryonic stem cells. Int J Biochem Cell Biol 38:1063–1075
6. Xu C, Inokuma MS, Denham J et al (2001) Feeder-free growth of undifferentiated human embryonic stem cells. Nat Biotechnol 19:971–974
7. Amit M, Shariki C, Margulets V et al (2004) Feeder layer- and serum-free culture of human embryonic stem cells. Biol Reprod 70:837–845
8. Ludwig TE, Bergendahl V, Levenstein ME et al (2006) Feeder-independent culture of human embryonic stem cells. Nat Methods 3:637–646
9. Wagner KE, Vemuri MC (2010) Serum-free and feeder-free culture expansion of human embryonic stem cells. Methods Mol Biol 584:109–119
10. Vemuri MC, Schimmel T, Colls P et al (2007) Derivation of human embryonic stem cells in xeno-free condition. In: Vemuri MC (ed) Stem cell assays. Humana, Totowa, NJ, pp 1–10
11. Chen G, Gulbranson DR, Hou Z et al (2011) Chemically defined conditions for human iPSC derivation and culture. Nat Methods 8:424–429
12. Rajala K, Vaajasaari H, Suuronen R et al (2011) Effects of the physiochemical culture environment on the stemness and pluripotency of human embryonic stem cells. Stem Cell Stud 1:e3. doi:10.4081/scs.2011.e3
13. Watanabe K, Ueno M, Kamiya D et al (2007) A ROCK inhibitor permits survival of dissociated human embryonic stem cells. Nat Biotechnol 25:681–686
14. http://www.jove.com/video/1427/culture-and-maintenance-of-human-embryonic-stem-cells
15. Xu Y, Zhu X, Hahm HS et al (2010) Revealing a core signaling regulatory mechanism for pluripotent stem cell survival and self-renewal by small molecules. Proc Natl Acad Sci USA 107:8129–8134

Chapter 9

Reverse Engineering Life: Physical and Chemical Mimetics for Controlled Stem Cell Differentiation into Cardiomyocytes

Gary R. Skuse and Kathleen A. Lamkin-Kennard

Abstract

Our ability to manipulate stem cells in order to induce differentiation along a desired developmental pathway has improved immeasurably in recent years. That is in part because we have a better understanding of the intracellular and extracellular signals that regulate differentiation. However, there has also been a realization that stem cell differentiation is not regulated only by chemical signals but also by the physical milieu in which a particular stem cell exists. In this regard we are challenged to mimic both chemical and physical environments. Herein we describe a method to induce stem cell differentiation into cardiomyocytes using a combination of chemical and physical cues. This method can be applied to produce differentiated cells for research and potentially for cell-based therapy of cardiomyopathies.

Key words Stem cells, Embryonic stem cells, Hydrogels, Human induced pluripotent stem cells, Human embryonic stem cells, Polydimethylsiloxane molds, Bone morphogenic protein 4

1 Introduction

The ability to engineer functional cardiac tissue has great potential not only for the fields of tissue engineering but also for regenerative medicine. Cardiomyocytes do not regenerate after birth; thus the loss of cardiomyocytes can lead to significant cardiac dysfunction. Countless individuals suffer morbidity from cardiac cell death and nearly 200 people per 100,000 die from heart disease each year in the USA alone (CDC vital statistics). Currently the development of treatment modalities for cardiac disorders using cardiomyocytes is limited by the availability of donor cells. Thus, the controlled differentiation of stem cells into cardiomyocytes could be used as an alternative for studies for which cardiomyocytes are not readily available. Applications of differentiated cardiomyocytes in the laboratory include early drug screening and cytotoxicity

Joydeep Basu and John W. Ludlow (eds.), *Organ Regeneration: Methods and Protocols*, Methods in Molecular Biology, vol. 1001, DOI 10.1007/978-1-62703-363-3_9, © Springer Science+Business Media New York 2013

testing, studying the events of early cardiogenesis, developing models of human diseases and investigating new treatment options, cardiac cell transplantation therapy, and facilitation of heart repair.

Of course, any protocol designed to create biological tissues requires a full understanding of the natural processes that underlie cellular differentiation and tissue modeling. Despite the fact that our understanding is incomplete, many laboratories have successfully generated differentiated cells and reasonable approximations of biological tissues in the laboratory. Those efforts have employed naturally occurring molecules including morphogens such as bone morphogenic protein (BMP) and transforming growth factor β (TGFβ) (1), other growth factors, for example as produced by cocultured cells (2), and various small molecules such as cAMP (3).

In addition to chemical signals, many laboratories have employed physical cues such as calcium phosphate-based materials and fibronectin for bone morphogenesis (4), extracellular matrix (ECM) components (5), myocardial ECM (6), and fibrin (7) for cardiomyocyte differentiation. These approaches have been modified through the use of synthetic materials such as hydrogels (8) and polyethylene terephthalate (PET), a plastic commonly used in consumer goods (9) in order to model the three-dimensional biological structures of interest.

The following protocol can be used as a means of providing both physical and biochemical cues for differentiating stem cells into cardiomyocytes. The protocol couples standard procedures for differentiating ESCs with microfabrication techniques to create reproducible tissue geometries (10, 11, 12). The first portion of the procedure begins with the differentiation process of hESCs into cardiomyocytes. The second portion of the procedure employs cell-hydrogel micro-molding to create tissue networks with specifically designed architectures (11). The molded microarchitectures are used to guide tissue porosity and cell alignment and could be used to generate tissue samples with differentiated cardiomyocytes that closely model cardiac tissue histology in vivo (12).

2 Materials

All solutions should be prepared with ultrapure deionized distilled water and all cells are cultured under standard conditions of 37°C with an atmosphere of 5% CO_2 and 85% relative humidity.

2.1 Cell Culture

1. DMEM-F12 medium (Invitrogen, Grand Island, NY, USA) is stored at between 2 and 8°C.
2. DMEM high glucose medium with GlutaMAX (Invitrogen) is stored at between 2 and 8°C.

3. RPMI-PVA: RPMI 1640 (Invitrogen), 400 μM 1-thioglycerol (Sigma-Aldrich, St. Louis, MO, USA), 4 mg/ml poly vinyl alcohol (PVA, Sigma-Aldrich), 10 μg/ml human recombinant insulin (hr-Insulin, Sigma-Aldrich), 25 ng/ml bone morphogenic protein 4 (BMP4, Invitrogen), 5 ng/ml human FGF2 (R&D Systems, Minneapolis, MN, USA), 1:100 dilution of chemically defined lipids (Invitrogen), 1 μM Y-27632 (EMD Serono, Rockland, MA, USA). RPMI-PVA should be stored at 4°C and used within 3 months.
4. Fetal bovine serum (FBS, Invitrogen) is stored frozen at −20°C.
5. 100 mM (100×) L-glutamine (Invitrogen) is stored at between 2 and 8°C.
6. Trypsin–EDTA (Invitrogen) is stored frozen at between −5 and −20°C.
7. Knockout serum replacer (KnockOut SR, Invitrogen) is stored at −5 to −20°C and protected from light. Prior to use the frozen medium should be placed at 2 to 8°C (i.e., in a refrigerator) overnight. In the morning the partially frozen medium should be placed in a 37°C water bath, with occasional swirling, until completely thawed (see Note 1). Once thawed KnockOut SR can either be stored in a refrigerator for up to 4 weeks or it can be divided into working volumes and stored frozen at −5 to −20°C.
8. 100× non-essential amino acids (Invitrogen) are stored at between 2 and 8°C.
9. β-Mercaptoethanol (Sigma-Aldrich) is stored at room temperature. Frequent opening of the bottle may reduce the purity by up to 2% per year.
10. Human FGF2 (R&D Systems, Minneapolis, MN, USA) is stored frozen between −20 and −70°C.
11. Geltrex (Invitrogen) is provided frozen in 1 ml aliquots that should be stored at between −20 and −80°C.
12. Matrigel (BD Biosciences, Sparks, MD, USA) is stored at −20°C (see Note 2). Frozen Matrigel should be thawed overnight at 4°C on ice (see Note 3) and kept on ice prior to use. Chilled pipettes should be used to prevent premature gelling. Gelled Matrigel can be reliquified by placing it on ice at 4°C for 24–48 h.
13. Dulbecco's phosphate buffered saline (DPBS, Invitrogen) is stored at between 2 and 8°C.
14. Fibroblast growth factor-basic human (bFGF, Sigma-Aldrich) is provided as a powder. The reconstituted solution should be stored at −20°C.

15. Human embryonic stem cells (hESC, line SI-233, Stemride International, London, UK).
16. Human induced pluripotent stem cells (hiPSC, lines iPS (IMR90)-1 and iPS (IMR90)-4, WiCell Research Institute, Madison, WI, USA).
17. The mouse embryonic fibroblast (MEF) feeder layer is prepared by growing irradiated MEFs (Invitrogen) in standard 6 well culture plates (BD Falcon, San Jose, CA, USA) in a medium consisting of DMEM supplemented with 10% FBS, as per supplier's instructions.
18. T-225 tissue culture flasks (BD Falcon).
19. Cell scrapers (Thomas Scientific, Swedesboro, NJ, USA).
20. Aggregation medium: RPMI 1640 supplemented with 400 μM 1-thioglycerol and 20% FBS or human serum.
21. Cardiomyocyte differentiation medium: RPMI 1640 supplemented with 400 μM 1-thioglycerol, 10 μg per ml hr-insulin, 1× chemically defined lipids, and 20% FBS or human serum.
22. Undifferentiated human embryonic stem cell (ESC) culture medium (ESC medium): DMEM-F12 supplemented with 20% KnockOut serum replacer, 0.1 mM nonessential amino acids, 1 mM L-glutamine, 0.1 mM β-mercaptoethanol, and 4 ng/ml human FGF2.

2.2 Mold Fabrication

1. 95–98% (vol/vol) Sulfuric acid (ACS grade, Sigma-Aldrich).
2. 30% (vol/vol) hydrogen peroxide solution (ACS grade, VWR, Arlington Heights, IL, USA).
3. SU-8 100 photoresist (Microchem, Newton, MA). SU-8 should be stored upright in a tightly closed container at a temperature of 40–70°F (4–21°C). Store away from light, acids, heat, and any potential ignition sources. The shelf life of SU-8 is 13 months from the date of manufacture.
4. Polypropylene glycol monomethyl ether acetate (PGMEA; Sigma-Aldrich).
5. Isopropyl alcohol.
6. Silane(Tridecafluoro-1,1,2,2-tetrahydrooctyl)-1-trichlorosilane (United Chemical Technologies, Bristol, PA, USA). Store in a cool and dry place. Protect from moisture. Protect from heat, direct sunlight, and sources of ignition.
7. Sylgard® 184 Silicone Elastomer Kit (Dow Corning, Midland, MI, USA).
8. Xiameter PMX-200 Silicone Fluid 20 cSt (Dow Corning).
9. Silicon wafers (3″ diameter) (1182, Wafer World, West Palm Beach, FL, USA).

10. 60 mm nontissue culture treated Petri dishes (BD Falcon).
11. 10 mm nontissue culture treated Petri dishes (BD Falcon).
12. Wafer tweezers (0S4WF-XD, SPI Supplies, West Chester, PA, USA).
13. Transparencies for Inkjet printers.
14. Flat quartz plate glass (Fisher Scientific, Pittsburgh, PA, USA).
15. Karl Suss MA/BA6 Mask Aligner (Suss Microtech, Sunnyvale, CA).
16. 2-Methoxy-1-methylethylacetate (Eastman Chemical Company, Kingsport, TN, USA).
17. Trichloro(1,1,2,2-perfluorocytl) silane (Sigma-Aldrich).
18. Vacuum desiccator (Cole Parmer, Vernon Hills, IL, USA).
19. Plasma asher (Electron Microscopy Services, Hatfield, PA, USA).

2.3 Hydrogel Patterning and Cell Culture

1. Fibrinogen from bovine plasma (Sigma-Aldrich). Store at –20°C.
2. Thrombin from bovine plasma (Sigma-Aldrich). Store at –20°C.
3. Pluronic F-127 (Invitrogen). Store at room temperature, do not freeze. Stable for 6 months.
4. 70% (vol/vol) Ethanol.
5. Tissue culture water (Sigma-Aldrich).
6. Dulbecco Modified Eagle Medium (DMEM) Powder, Low Glucose, Pyruvate (Invitrogen). Store at 2–8°C.
7. 1× liquid DMEM (Invitrogen). Store at 2–8°C.
8. Penicillin–streptomycin, liquid (Invitrogen).
9. Fetal bovine serum (Invitrogen). May be stored at room temperature.
10. Heat-inactivated horse serum (Sigma-Aldrich). May be stored at room temperature.
11. Chicken embryo extract ultrafiltrate (C3999, US Biologicals, Swampscott, MA, USA). May be stored at room temperature. Stability is greater than 1 year. Storage at 4°C extends the stability for an additional 3–6 months. Some precipitate has been observed when frozen.
12. 6-Aminocaproic acid (Sigma-Aldrich).
13. 6- or 12-well tissue culture-treated plates (sterile) (BD Falcon).
14. Sulfuric acid (Sigma-Aldrich).
15. 30% Hydrogen peroxide (Sigma-Aldrich).
16. L-Ascorbic acid (Sigma-Aldrich).

17. Human Insulin (Sigma-Aldrich). Store at 2–8°C. Light sensitive.
18. HEPES Buffer Solution (Life Technologies, Carlsbad, CA, USA). May be stored at room temperature.
19. Piranha etching solution (concentrated sulfuric acid:hydrogen peroxide (30%) 3:1) is prepared in a teflon beaker by pouring 150 ml of concentrated sulfuric acid into the beaker followed by 50 ml of hydrogen peroxide (purchased not more than 30% concentration in water). The mixture will begin to boil and foam after a few seconds (see Note 4).
20. Headway spinner (model PWM32-PS-CB15, Headway Research, Garland, TX, USA).
21. Mix the PDMS base, PDMS curing agent, and Dow Corning 200 Fluid (10:1:1) (Dow Corning) to prepare 20 g of PDMS solution. Use a spoon to dispense the base and a pipette for the curing agent. Add the base first, then the curing agent (see Note 5).
22. 2-Methoxy-1-methylethylacetate (Eastman Chemical Company).
23. Trichloro (1,1,2,2-perfluorocytl) silane (Sigma-Aldrich).
24. Vacuum desiccator (Cole Parmer).
25. Plasma asher (Electron Microscopy Services).

3 Methods

3.1 Stem Cell Maintenance Cultures (Fig. 1)

1. MEFs are thawed quickly by partial immersion in a 37°C water bath until completely thawed.
2. The thawed cells are resuspended in a sufficient volume of DMEM + 10% FBS to achieve an initial plating density of 6×10^4 per cm^2 or approximately 5.6×10^5 cells per well of a 6-well tissue culture plate.
3. Once the MEFs are confluent, the growth medium is removed by aspiration and the cells are washed three times with DPBS.
4. Either hESC or hiPSC are plated at an initial density of 2×10^4 cells per cm^2 and maintained with ESC medium.
5. Change the growth medium daily by gently aspirating the old medium and replacing it with fresh medium.

3.2 Preparation of Geltrex-Coated Plates (Adapted from Manufacturer's Instructions, See Note 6)

1. Thaw Geltrex on ice in a refrigerator overnight (see Note 7).
2. Mix by slowly pipetting up and down without introducing air bubbles.
3. Dilute 1 ml of Geltrex 1:30 by adding 1–29 ml of chilled DMEM-F12 (at between 2 and 8°C).

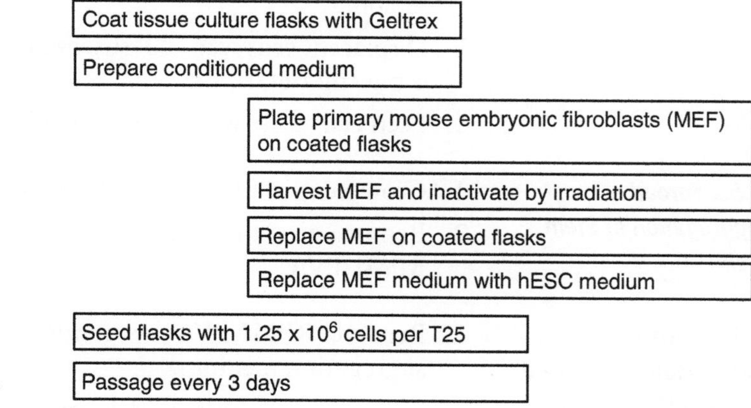

Fig. 1 Overall workflow

4. Cover the growth surface with a sufficient amount of Geltrex. Typically this is approximately 1.5 ml for a 35 mm plate and 3 ml for a 60 mm plate.
5. Place the coated dish at 37°C for at least 60 min to ensure complete gelling (see Note 8).
6. When needed, coated plates should be warmed to room temperature for 1 h before aspirating excess Geltrex.
7. Cells should be added to the coated plates immediately using culture medium that has been warmed to 37°C.

3.3 Preparation of MEF Conditioned Medium

1. Coat tissue culture flasks with Geltrex as described above for plates. A T-225 flask typically requires approximately 8 ml of Geltrex.
2. Plate MEFs at a density of 2.25×10^4 cells per cm^2 on Geltrex-coated flasks.
3. Cells are passaged by aspirating off the growth medium, washing cells three times with DPBS, and treating the adherent cells with trypsin–EDTA for 2–5 min (9 ml for T-225 flask).
4. Once a sufficient number of cells have been generated, they should be plated on Geltrex-coated T-225 at a density of 5.6 x 10^4 cells per cm^2 (see Note 9).
5. Cells should be passaged every 3 days.

3.4 Growth of Human Embryonic Stem Cells

1. Remove the MEF conditioned medium (MEF-CM) by aspiration and rinse cells with DPBS.
2. Remove DPBS and replace with 9 ml trypsin–EDTA.

3. When edges of the hESC colonies begin to cure, gently remove the trypsin–EDTA by aspiration.
4. Cells should be displaced from the substrate with gentle scraping.
5. Displaced cells can be suspended in MEF-CM containing 4 mg/ml bFGF.
6. Feed cells daily with growth medium.

3.5 Forced Aggregation of Stem Cells

1. Grow stem cells in MEF-CM with 4 mg/ml bFGF until needed.
2. Remove growth medium by gentle aspiration and replace with RPMI-PVA.

3.6 Cardiac Specification (After ~5 Days in RPMI-PVA)

1. Remove RPMI-PVA by gentle aspiration and replace with aggregation medium.
2. Feed cells on alternate days by gently removing the aggregation medium and replacing it with fresh medium.

3.7 Cardiomyocyte Development

1. Remove aggregation medium by gentle aspiration.
2. Replace medium with cardiomyocyte differentiation medium.
3. Cells should be fed on alternate days by gently aspirating the cardiomyocyte differentiation medium and replacing it with fresh cardiomyocyte differentiation medium.

3.8 Physical Signals to Enhance Cardiomyocyte Differentiation

3.8.1 Cleaning Silicon Wafers to Remove Organic Contaminants

1. Carefully place the beaker of piranha etch on a hot plate. Heating piranha increases the solution's reactivity and consequently the dangers associated with its use. Using Teflon tweezers, slowly add the wafers, one at a time (see Note 10).
2. After 15 min, carefully remove the samples using the tweezers and rinse the wafers by sequentially submerging the wafers in two, 1 l deionized water baths. After removing the wafers from the rinse, allow the wafers to air-dry.
3. Cool the piranha etching solution to <30°C and slowly add to a waste container in accordance with laboratory safety guidelines. Heat the hot plate to 200°C and place the etched wafers on the hot plate face up to dehydrate the wafers and to maximize the adhesion of the photoresist. Wafers should be dehydrated for 10 min.

3.8.2 Photoresist Coating (Fig. 2a)

1. Place the silicon wafer on a clean sheet of aluminum foil and pour 2–3 ml SU-8 100 photoresist reagent onto the wafer (see Note 11).
2. Spin-coat the SU-8 100 at room temperature. Ramp up to 500 rpm at 100 rpm/s and maintain for 10 s. Then ramp up to 1,000 rpm at 300 rpm/s, and maintain for 30 s. Finally ramp down to 0 rpm at 300 rpm/s. The procedure should result in a 250 μm thick photoresist layer (see Note 12).

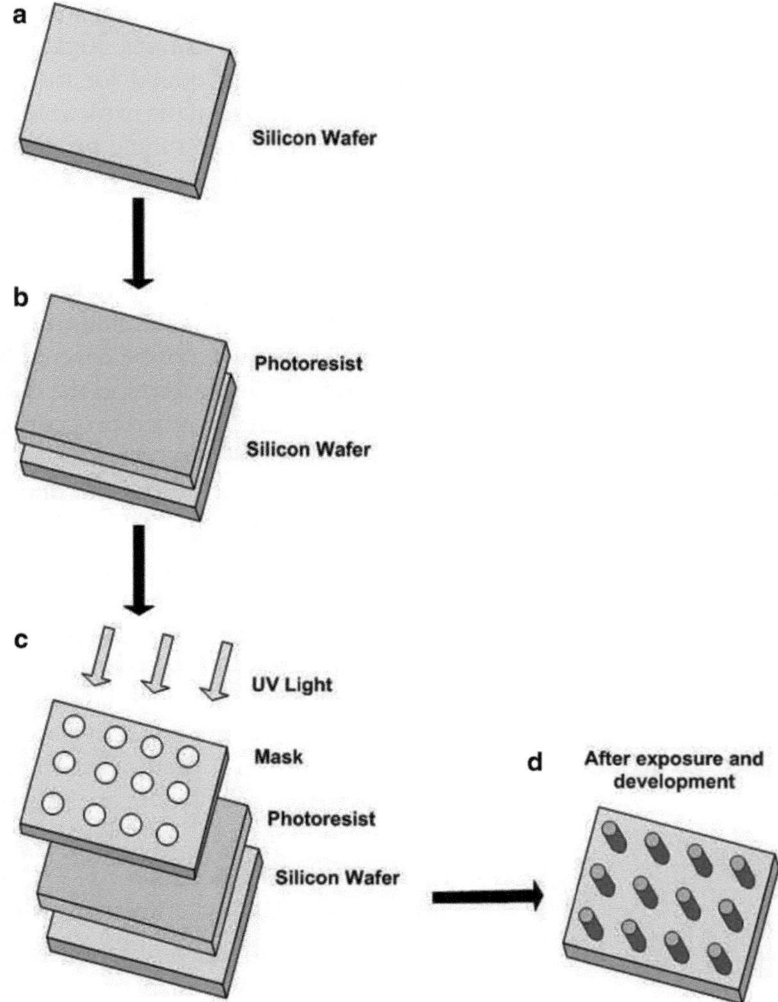

Fig. 2 Fabrication of microarchitecture on silicon wafer

3. Keep the SU-8-coated glass substrate on a perfectly leveled surface for 10–15 min to allow the resist to rest and acquire uniform thickness. Remove the wafer from the spinner using dedicated SU-8 tweezers.
4. Prebake the wafer on a hot plate at 65°C for 15 min to remove solvents, followed by a softbake for 2 h at 95°C (see Note 13). Turn off the hot plate and slowly cool the wafer and hot plate to room temperature. The photoresist layer should be hard at this point and should not be indented by tweezers.
5. For taller features (up to 2.5 mm), SU-8 can be layered sequentially, but on spin-coating the final layer onto the wafer, carry out a final, longer soft-bake at 95°C for at least 10 h (or overnight). Allow the coating to cool to room temperature before exposing the photoresist layer.

3.8.3 Photomask Creation and Alignment (Fig. 2b)

1. A transparency should be prepared ahead of time to yield the desired architectural features. Rigid or glass masks could also be used, but are not required for the resolution needed for this application. The size of the mask needed will depend on the mask aligner to be used and should be taken into consideration when designing the mask. Thicker masks are stronger and more rigid, but absorb proportionally more light and will also heat more.
2. Tape the photomask, using clear tape, to a flat piece of quartz plate glass, ensuring that the tape does not go over the edges of the glass. Use pieces of aluminum foil to protect all of the areas of the wafer that will not be covered by the photomasks. Place the photomask (and glass) and the coated silicon wafer into the mask aligner using standard procedures for the aligner.

3.8.4 Exposure, Baking, and Development of the Photoresist Layer (Fig. 2c, d)

1. In the mask aligner, expose the photoresist with 365 nm UV light at a power density of 12 mW/cm^2 (see Note 14).
2. Once the wafer has been exposed, cross-link the exposed photoresist by baking on a hot plate at 40°C for a minimum of 24 h. After baking, slowly cool the wafer by ramping down the temperature. Ramping down the temperature will minimize stress in the coating due to the thermal mismatch between the coating and substrate and can prevent cracking in the wafer. Soak the wafer in 2-methoxy-1-methylethylacetate (PGMEA developer overnight). Repeat the soak for 1 additional hour using fresh PGMEA.
3. Rinse in isopropyl alcohol for 1–2 h, changing for fresh alcohol after 10 min. Allow the wafer to air-dry (see Note 15).
4. Under a hood and using wafer tweezers, place the SU-8-coated wafer along with a silanizing agent (e.g. 500 μl trichloro (1,1,2,2-perfluocytl) silane) in a vacuum desiccator to passivate the wafer surface (see Note 16).

3.8.5 Preparing Negative Polydimethylsiloxane Template (Fig. 3)

1. Degas the polydimethylsiloxane (PDMS) solution thoroughly in a vacuum desiccator until no bubbles appear (~1 h) (see Note 17).
2. Place the SU-8 master in a new 100 mm Petri dish. Pour 35 ml PDMS solution over the previously prepared wafer. The wafer should be completely submerged. Degas thoroughly in a vacuum desiccator and remove any bubbles using a fine gage needle.
3. Cure the treated wafer at 80°C in an oven for at least 4 h and then cool to room temperature.
4. Remove the negative replica PDMS template from the wafer and cut the template to the desired size. Be sure to keep the mold clean during this process. Removal is accomplished most easily by breaking the edges of the Petri dish.

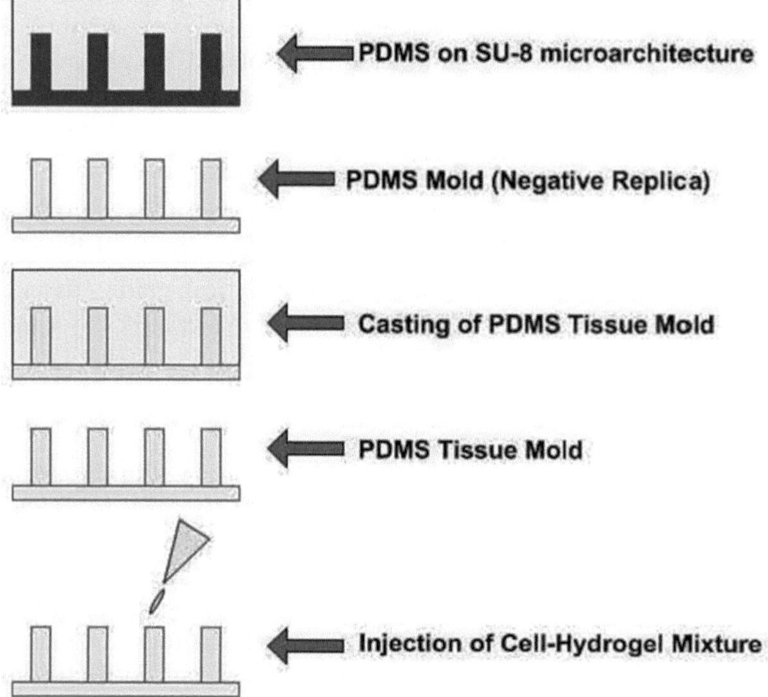

Fig. 3 Creation of PDMS mold for cell-hydrogel molding

5. Place the PDMS template on a clean glass slide and repeat the silanization procedure above.

3.8.6 Cast Tissue Molds

1. Prepare another 35 g of PDMS solution, as described above, by mixing the PDMS base, curing agent, and Dow Corning 200 Fluid.
2. Place the PDMS template in a new 60 mm Petri dish, submerge in approximately 15 ml of fresh PDMS solution, and degas thoroughly (see Note 18).
3. Cure the PDMS in an oven at 80°C for 4 h and allow to cool to room temperature before peeling the PDMS off the template. Store PDMS components under clean, dry conditions.

3.8.7 Prepare PDMS for Cell-Hydrogel Molding

1. Once the PDMS has been peeled from the template, place the PDMS molds in a plasma asher and treat at 100 W under O_2 for 1 min according to the instructions provided by the manufacturer. Immerse in deionized H_2O. Plasma processing at low power for a short time duration will make the PDMS molds hydrophilic. Over-ashing should be avoided.
2. Prepare a square frame of hook and loop tape (e.g., Velcro) and affix it to the PDMS mold assuring that the frame is touching the mold without gaps at all points.

3. Place the framed PDMS molds in a new 100 mm Petri dish filled with 70% (vol/vol) ethanol. Place in a biosafety hood under antimicrobial UV light overnight.
4. Rinse the PDMS molds with sterile H_2O and dry thoroughly using a nitrogen gun. The PDMS must be kept sterile at all times from this point forward.
5. Immerse the PDMS molds in a 0.2% (wt/vol) pluronic (serving as a surfactant) solution for 1 h.
6. Rinse with analytical grade (tissue culture grade) H_2O and store the molds in sterile H_2O for up to 24 h at room temperature before molding.

3.8.8 Molding the Cell-Hydrogel Mixture

Prepare cells for culture as described previously.

1. Begin by preparing a solution of fibrinogen (33 mg/ml) in HEPES buffer (20 mM in 0.9% saline). Slowly mix the fibrinogen into the buffered saline over a period of several hours at 37°C. Leave the solution overnight at 2–8°C and then rewarm to 37°C. Filter the solution through a series of sterile filters: 40 μm cell strainers, 0.45 μm bottle-top filters with glass pre-filters, and 0.2 μm bottle-top filters with glass pre-filters. Aliquot the solution into 1 and 3 ml volumes and store at -20°C. Solutions should be made prior to adding cells and stored at the described conditions.
2. Add 500 U of thrombin to 18 ml of 0.9% saline and 2 ml of sterile tissue culture-grade water to create a 25 U/ml solution. Filter through a 0.2 μm filter. Aliquot into 500 and 250 μl volumes and freeze at -80°C.
3. Dissolve 50 g of Pluronics F-127 in 700 ml of sterile culture-grade water (will yield 5% w/v Pluronics F-127 solution). Add additional water to bring the total volume up to 1 l.
4. Add 10% horse serum, 2% FBS, 1% penicillin–streptomycin, and 6 mg/ml aminocaproic acid into DMEM. Add 50 μg/ml of ascorbic acid, and 2 μg/ml of insulin in 25 μM HEPES before feeding.
5. Using a 0.2 μm filter, sterile filter the Pluronics F-127 solution. Place the PDMS molds into 5% Pluronics solution for 2–3 h in a biological hood to maintain sterility (see Note 19).
6. Remove the Pluronics solution and rinse and store the PDMS molds in sterile culture-grade water. Place sterile drapes on the surface of the hood to maintain sterility (see Note 20).
7. In a conical tube, add 112 μl of fibrinogen stock to 558 μl of 20 mM HEPES buffer in 0.9% saline solution (yields 3.3 mg/ml final fibrinogen concentration, 25 U/ml final thrombin concentration).

8. In a second conical tube, add 17 μl of thrombin stock and 1.3 μl of 2N Ca^{2+} solution to 135 μl of DMEM.
9. Resuspend the differentiated cardiomyocytes in 1× culture media to obtain six times the final concentration of cells desired in the construct.
10. Dry the PDMS molds with nitrogen and place them in a 6- or a 12-well tissue culture plate.
11. Using a pipette to mix, add 667 μl of the solution from step 7 into a sterile 50 ml centrifuge tube, followed by 167 μl of the solution from step 8 (4:1:1 ratio) (see Note 21). Mix together, being careful not to introduce bubbles.
12. Add several drops of sterile culture medium to each well, adjacent to the mold, in order to maintain the proper humidity to prevent drying.
13. Incubate the tissue culture plates at 37°C and 5% CO_2 for 45 min.
14. Add a sufficient amount of culture medium to each well to fully immerse the PDMS mold. Return to incubator.

4 Notes

1. Any flocculent material present during the thawing process should redissolve while swirling at 37°C. For maximum effectiveness the time allowed for thawing at 37°C should be minimized.
2. Multiple freeze/thaw cycles should be avoided to maintain the integrity of the reagent. Matrigel will gel rapidly at 22–35°C. The color of frozen Matrigel will vary from yellow to dark red due to interactions between CO_2 the bicarbonate buffer in the reagent and the phenol red pH indicator. Once thawed and equilibrated in a 5% CO_2 atmosphere the color will return to normal.
3. It is important to thaw Matrigel on ice because temperature variation in a refrigerator that results in temperatures above 4°C will cause the solution to gel.
4. Add the sulfuric acid slowly so that the solution does not boil over the edge of the container. Teflon tweezers can be used to stir the solution.
5. The raw silicone material is extremely sticky. Tape a piece of aluminum foil down inside the fume hood to act as a disposable work space. Thorough mixing is required for proper curing.
6. Geltrex should be thawed at 2–8°C overnight on ice, to buffer temperature variations, in a refrigerator. Geltrex gels in 5–10 min

at temperatures above 15°C. When working from volumes of 5 ml or more, it is not necessary to keep the basement membrane matrix on ice nor it is necessary to prechill pipette tips, plates, tubes, etc. that will contact the mixture. When using smaller volumes that may warm more quickly, it should be kept on ice to prevent unwanted gelling.

7. Matrigel is an alternative ECM substitute that behaves like a basement membrane does in vivo. It can be substituted for Geltrex with minimal or no effects on the cells cultured thereon.
8. Plates are stable for up to 2 weeks when stored at between 2 and 8°C and sealed with Parafilm. Dried plates should be discarded.
9. This requires approximately 12.5×10^6 cells per T-225 flask.
10. The solution will foam with the addition of each wafer. Wait for the foaming to subside before adding the next wafer. Up to four wafers can be submerged at a time. The temperature of all piranha solutions should be monitored at all times.
11. The photoresist reagent is highly viscous; thus, the reagent is poured directly onto the wafer to reduce waste. SU-8 is hard to remove and can easily contaminate glassware, hot plates, ovens, and tools.
12. For thicker SU-8 (>20 μm) or high-aspect-ratio feature (height size:feaure size > 2:1) processes, removal of the edge bead may provide better contact between the photomask and the photoresist layer. Spin rate (RPM) should be determined based on the SU-8 viscosity and the final thickness needed. Spin speed curves are available for each SU-8 resist formulation from the manufacturer.
13. Prebake and softbake times vary by formulation and desired photoresist thickness. Consult the manufacturers' datasheets to confirm baking times. Removing as much solvent as possible during the pre-exposure bake is important so that the amount of swelling in the SU-8 membranes is reduced later. The hot plate should be leveled to ensure that the layer of SU-8 has a uniform thickness. Any air bubbles can be removed during the bake at 95°C by gently tapping the air bubbles with a 24-gauge needle.
14. Be sure to confirm the intensity of the UV light prior to setting exposure times for the wafers. Exposure times are a function of the mold thickness, exposure energy, and mean exposure intensity. Using a 360 nm long-pass (LP) filter on the glass plate (or the chrome mask) may reduce the short wavelength effect and improve the feature profile. Add 20–40% more exposure time to compensate for the intensity loss due to the filter. Correct exposure dose is strongly dependent on substrate. Also, exposure dose will vary for different thicknesses and SU-8

formulations. Consult the manufacturer datasheets. If overheating occurs, the exposure may need to be carried out in short increments (e.g.. 1 min, followed by 2-min breaks) until the total exposure time is obtained.

15. If a white film is produced during the alcohol rinse, it is an indication of underdevelopment of the unexposed photoresist. Immerse or spray the substrate with additional SU-8 developer to complete the exposure process and then rinse again.
16. The silanizing agent can be placed in a Petri dish, in a vial, or on two microscope slides. Attach the desiccator to a vacuum hose on the hood vacuum port. Open the vacuum valve. Leave the wafer for at least 1 h, preferably overnight. The silanizing agent will evaporate and will prevent the PDMS from adhering to the master wafer. Avoid inhalation of silane vapors.
17. Make sure that the mixture does not foam out of the container. Vent the vacuum to pop any large bubbles that form at the surface.
18. Bubbles can alter the optical properties of the cured PDMS and should be removed before curing.
19. Soaking the molds in the solution will help prevent the gel from adhering to the PDMS molds.
20. Sterile gloves should be worn to construct to the tissue-gel molds. If necessary, the PDMS molds may be stored in sterile culture-grade water for up to 24 h at room temperature in a biological hood.
21. The fibrin gel will begin to polymerize quickly; thus the injection of the constructs should be done immediately. Injection into the molds in a liquid state is critical for forming a uniform construct. The volume of gel needed will be dependent on the mold size and number of molds. The rapid cross-linking of the fibrinogen will likely limit the maximum number of molds to 4. Calculate the volume of ingredients needed for the particular mold configuration prior to mixing to ensure adequate volumes of gel since the reaction occurs too rapidly to mix additional volumes of gel.

References

1. Ripamonti U (2010) Soluble and insoluble signals sculpt osteogenesis in angionenesis. W J Biol Chem 1:109–132
2. Sahoo S, The TKH, He P et al (2011) Interface tissue engineering: next phase in musculoskeletal tissue repair. Ann Acad Med Singapore 40:245–251
3. Boink GJ, Verkerk AO, van Amersfoorth SC et al (2008) Engineering physiological controlled pacemaker cells with lentiviral HCN4 gene transfer. J Gene Med 10:487–497
4. Chan WD, Perinpanayagam H, Goldberg HA et al (2009) Tissue engineering scaffolds for the regeneration of craniofacial bone. J Can Dent Assoc 75:373–377
5. Feng Q, Chai C, Jiang XS et al (2006) Expansion of engrafting human hematopoietic stem/progenitor cells in three-dimensional

scaffolds with surface-immobilized fibronectin. J Biomed Mater Res 78A:781–791
6. Singelyn JM, DeQuach JA, Seif-Naraghi SB et al (2009) Naturally derived myocardial matrix as an injectable scaffold for cardiac tissue engineering. Biomaterials 29:5409–5416
7. Jockenhoevel S, Zund G, Hoerstrup SP et al (2001) Fibrin gel—advantages of a new scaffold in cardiovascular tissue engineering. Eur J Cardiothorac Surg 19:424–430
8. Newman KD, McBurney MW (2004) Poly(d, l-lactic-co-glycolic acid) micropheres as biodegradable microcarriers for pluripotent stem cells. Biomaterials 25:5763–5771
9. Cetinkaya G, Turkoglu H, Arat S et al (2007) LIF-immobilized nonwoven polyester fabrics for cultivation of murine embryonic stem cells. J Biomed Mater Res 81:911–919
10. Burridge PW, Thompson S, Millrod MA et al (2011) A universal system for highly efficient cardiac differentiation of human induced pluripotent stem cells that eliminates interline variability. PLoS One 6:e18293. doi:10.1371/journal.pone.0018293
11. Bian W, Liau B, Badie N et al (2009) Mesoscopic hydrogel molding to control the 3D geometry of bioartificial muscle cells. Nat Protoc 4:1522–1534
12. Ye KY, Sullivan KE, Black LD (2011) Encapsulation of cardiomyocytes in a fibrin hydrogel for cardiac tissue engineering. J Vis Exp 55:e3251

Chapter 10

Phenotypic Analysis of Bioactive Cells for Application in Regenerative Medicine

Teresa Bora Burnette and Andrew T. Bruce

Abstract

The following chapter outlines methodologies to phenotypically characterize primary cells for the use in tissue-engineered and regenerative medicine applications. Methods covered include analyzing cells using immunocytochemistry, fluorescence-activated cell sorting, and confocal microscopy of adherent and suspended cells, as well as combinations of formulated cell–biomaterial constructs.

Key words Phenotype, Primary tissue culture, Immunocytochemistry, FACS, Confocal microscopy

1 Introduction

Analyzing cells using immuno-phenotyping (antibody staining) provides a highly sensitive method to characterize regenerative cell-based tissue-engineered products. Three main approaches to evaluate phenotype include immunocytochemistry, fluorescence-activated cell sorting (FACS), and confocal microscopy. Each approach provides valuable information depending on the process step and the formulation (cells only versus cell(s)–biomaterial construct). Using phenotypic analysis throughout the cell isolation and expansion procedure enables process checks and monitors cell stability. This chapter gives step-by-step methods that can help determine isolated cell phenotype from the initial processing through formulated prototype. Examples cited will reference Tengion's Neo-Bladder Augment™ (NBA), Neo-Urinary Conduit™ (NUC), and Neo-Kidney Augment™ (NKA) products.

Joydeep Basu and John W. Ludlow (eds.), *Organ Regeneration: Methods and Protocols*, Methods in Molecular Biology, vol. 1001, DOI 10.1007/978-1-62703-363-3_10, © Springer Science+Business Media New York 2013

2 Materials

2.1 Reagents

1. D-PBS without magnesium chloride and calcium chloride.
2. Goat Serum (G9023 Sigma), Horse Serum (H0146 Sigma).
3. Triton-X 100.
4. 40% (w/v) paraformaldehyde.
5. Hoechst nuclear dye.
6. Biologically inert antibodies: Mouse IgG, Rabbit IgG, Goat IgG. Store for 2–3 weeks at 4°C. For longer term storage, keep at –20°C.
7. Cell type-specific primary antibodies (Table 1).
8. Isotype-matched secondary antibodies (Table 1). Anti-mouse/-rabbit/-goat conjugated to either Alexa Fluor 488/ Alexa Fluor 594/or Alexa Fluor 647 (Invitrogen). Store at 4°C or according to manufacturer's recommendations.

2.2 Solutions

1. Fixation solution for adherent cells: Dilute stock 4% (w/v) paraformaldehyde in D-PBS. Dilute 40% (w/v) stock solution tenfold with D-PBS, solution can be stored at room temperature for 1 month.
2. Fixation solution for suspended cells: 2% (w/v) paraformaldehyde in D-PBS. Dilute 40% (w/v) stock solution tenfold with D-PBS.
3. Surface blocking solution: 2% FBS, 0.09% sodium azide in D-PBS.
4. Surface staining solution: 0.2% BSA, 0.09% sodium azide in D-PBS.
5. Permeabilization and blocking solution for intracellular staining: 0.2% Triton X-100, 10% goat or horse serum in D-PBS.
6. Alcohol-based permeabilization solution: 90% ice-cold methanol. Store at –20°C.
7. Intracellular staining buffer: 0.2% Triton X-100, 2% goat/ horse serum in D-PBS.
8. Vectashield Mounting Medium with DAPI (H-1200 Vector Labs).

2.3 Equipment

1. Millicell EZ 8-chambered slide (PEZGS0816 Millipore).
2. Fluorescent microscope (Leica DMIL5000B or equivalent) or confocal microscope (BD Pathway 855 or equivalent).
3. 96-well black/clear tissue culture-treated imaging plate, flat bottom.
4. Centrifuge capable of 500 × g centrifugal force.

Table 1
Example antibodies used in FACS analysis and immunocytochemistry (NBA™, NUC™, and NKA™)

Tissue	Step	Target	Antibody	Manufacturer
Kidney	Primary	Epithelial[a]	CK8/18/19 (mouse IgG_1)	Abcam (ab41825)
		Proximal tubular[b]	Aquaporin 1 (mouse IgG_{2b})	Abcam (ab9566)
		Distal tubular[b]	Ecahedrin (mouse IgG_{2a})	BD (610182)
		Proximal tubular[b]	Lotus tetragonolobus agglutinin (LTA)	Vector (B1325)
		Collecting duct[b]	Dolichos biflorus agglutinin	Vector (B1035)
		Endothelial[b]	PE-CAM (mouse IgG1)	BD (555025)
Bladder and AdSMC		Epithelial[a]	CK(AE1/AE3) (mouse IgG1)	DAKO (M35150)
		Smooth muscle[a]	Smooth muscle actin (mouse IgG2a)	DAKO (M0851)
		Smooth muscle[a]	Caldesmon (mouse IgG1)	DAKO (M3557)
		Smooth muscle[a]	Calponin (mouse IgG1)	DAKO (M3556)
	Isotype control		Mouse IgG1	DAKO (X0931)
			Mouse IgG2a	DAKO (X0932)
			Mouse IgG2b	Dako (X0944)
			Rabbit IgG	Invitrogen (02-1202)
	Secondary		Goat anti-mouse IgG1 (A488/A647)	Invitrogen (A211121/ A21240)
			Goat anti-mouse IgG2a (A488/A647)	Invitrogen (A211131/ A21241)
			Goat anti-mouse IgG2b (A488/A647)	Invitrogen (A211141/ A21242)
			Streptavidin (A488/A647)	Invitrogen (S32354/ S32357)

[a]Intracellular expression
[b]Surface expression

5. 15 ml conical polypropylene centrifuge tubes.
6. 1.5 ml polypropylene microcentrifuge tubes.
7. 12 × 75 mm round-bottom polystyrene centrifuge tubes.
8. Flow cytometer (BD FACS Aria or equivalent).
9. FlowJo (Treestar) flow cytometry analysis software.
10. Cytocentrifuge (7620 Cytopro Wescor) capable of 300 × *g* for spinning cells onto polylysine-coated microscope slides (SS-218 dual-chamber slides and SS-213 dual-sample chambers Cytopro Wescor).
11. Liquid Blocker PAP pen (EMS).
12. Cover glass 24 × 55 mm.

3 Methods

3.1 Initial Identification of Target Cell Types from Primary Tissue Source

The analysis of cells post-isolation from mammalian primary tissues using enzymatic digestion or tissue disassociation techniques allows for the identification of starting material for certain regenerative medicine/tissue-engineered products. Since single-cell suspensions are difficult to achieve with enzymatic digestion (for example, kidney may generate kidney tubules and glomeruli), flow cytometric analysis may not be the right choice for initial screening. The following steps can be applied to any tissue where the target cells need to be quantitated and identified at the initial phase of the process even if there are aggregates and clusters post-isolation.

Immunocytochemistry can be a useful tool for evaluating and analyzing cytocentrifuged cell aggregates and clusters from the initial cell isolation from multiple species. Although direct staining is usually faster and less complicated, utilizing an indirect single-color method for immunocytochemistry that targets either surface or intracellular antigens removes limitations of species-specific antibody availability.

1. Obtain cell suspension.
2. Fix cells in a final concentration of 2% paraformaldehyde in D-PBS for 20 min at room temperature.
3. Wash cells twice by centrifugation at $500 \times g$ and then resuspend in D-PBS (1 ml per 1×10^6 cells).
4. Determine cell concentration with a hemocytometer (see Note 1).
5. Adjust concentration to 1×10^5 cells/ml. For intracellular targets proceed to **step 6**; for surface targets proceed to **step 8**.
6. For intracellular antigens, wash once more and resuspend 1×10^6 cells in 0.5 ml blocking buffer.
7. Incubate for 30 min at room temperature to permeabilize cell membrane and block nonspecific binding.
8. For surface antigens, wash once and resuspend in surface blocking solution for 5–15 min at room temperature.
9. After blocking, add 200 μl of cell suspension into the single Cytopro chamber with glass slide containing a fast white Cytopad.
10. Centrifuge for 4 min at $300 \times g$.
11. After centrifugation, remove slide and mark the perimeter of the cell spot with a liquid blocker PAP pen.
12. Rehydrate cells on the slide by adding 200 μl of D-PBS for 5 min at room temperature.

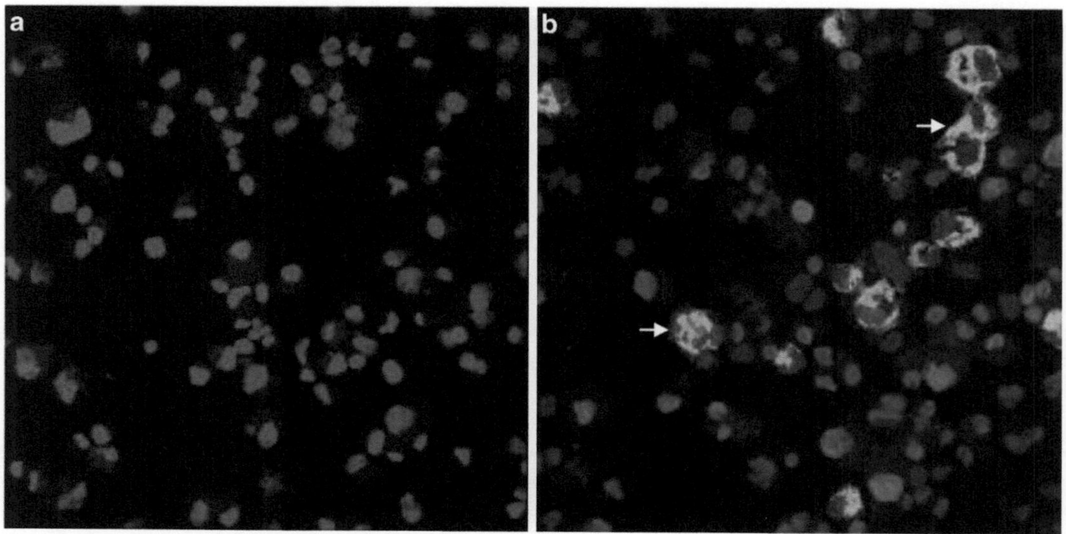

Fig. 1 Fluorescent image of cytocentrifuged kidney cells at 200× magnification. (**a**) Negative control with nuclear dye. (**b**) DBA-positive cells (*arrows*) with nuclear dye

13. Remove D-PBS by aspiration, and add desired primary antibody, diluted in surface or intracellular staining buffer (0.2% Trion X-100 with 2% serum of the species from which the secondary antibodies were generated from) as well as mouse isotype control and stain for 90 min to overnight at 4°C.
14. After staining with primary antibody, carefully wash cells twice using aspiration with D-PBS.
15. Following washing, add isotype-matched secondary antibody for 30 min at room temperature in the dark.
16. Wash cells twice with D-PBS and then add a drop of mounting medium to each spot with a coverslip.
17. Analyze slides using fluorescent microscope using image analysis software (Fig. 1) (see Note 2).

Isolated aggregates and clusters such as glomeruli and intact tubules from post-enzymatically digested kidney can be identified using immunofluorescence and confocal microscopy (1). Phenotypic analysis of these structures can confirm the specificity of antibodies that may be used later in the expansion process. As cells expand, antibody specificity may change due to protein regulation and functional differences caused by culture conditions (2). Having a compartmental specific antibody or genetic tracer that confirms the identity of cells prior to culture is helpful in monitoring any changes that may be associated with the expansion process (3, 4).

1. Obtain a sample of live cells from isolated tissue suspended in D-PBS.
2. For single-color cell surface phenotypic markers, use 0.5×10^6 live cells suspended in 0.5 ml of D-PBS in a 1.5 ml microcentrifuge tube. Divide the number of tubes per antibodies and include isotype-matched controls for each antibody (see Note 3).
3. Block cells for nonspecific binding with 1 ml surface blocking solution for 10 min at 4°C.
4. After blocking, wash by centrifugation at $500 \times g$ for 3 min.
5. Resuspend cells with 0.5 ml surface staining buffer containing primary antibody at a concentration of (1 μg/ml/1×10^6 cells) (see Note 4).
6. Incubate for 30 min at 4°C.
7. After 30 min, wash cells twice with D-PBS by centrifugation at $500 \times g$ for 3 min.
8. Add secondary antibody diluted in surface staining solution at a concentration of 1 μg/ml/1×10^6 cells and incubate for 30 min at 4°C.
9. Following secondary incubation, wash cells twice with D-PBS by centrifugation at $500 \times g$ for 3 min.
10. Fix cells for 10 min with fixation solution containing Hoechst nuclear dye at a concentration of 1 μg/ml. Transfer 100 μl from each stained sample to a corresponding well of the 96-well imaging plate and select the correct dye and light path prior to imaging.
11. Use image analysis software and the confocal microscope to view stained structures in 3D (Fig. 2) (see Note 5).

3.2 Phenotypic Analysis of Expanded Cells Using Immunofluorescent Microscopy

Expansion of cells from initial cell isolates or primary tissue may be required for immunofluorescence and FACS analysis. Phenotypic analysis using immunofluorescent microscopy can be performed using techniques on adherent cells in many formats. Test plates can range from 6- to 96-wells and should be chosen based on the adherence and growth characteristics that best mimic the expansion culture. The format can also be dictated by the number of cells available and antibodies to be targeted including technical replicates. This can be minimized by multicolor staining referred to later in the chapter (see Note 6).

3.2.1 Characterization of Surface/Membrane Antigens

1. Obtain adhered cultured cells in a plated format.
2. Wash cultured cells by removing medium and adding appropriate volume of D-PBS to each well and or dish: 3 ml/well for 6-well dish, 2 ml/well for 12-well dish, and 250 μl/well of chamber slide.

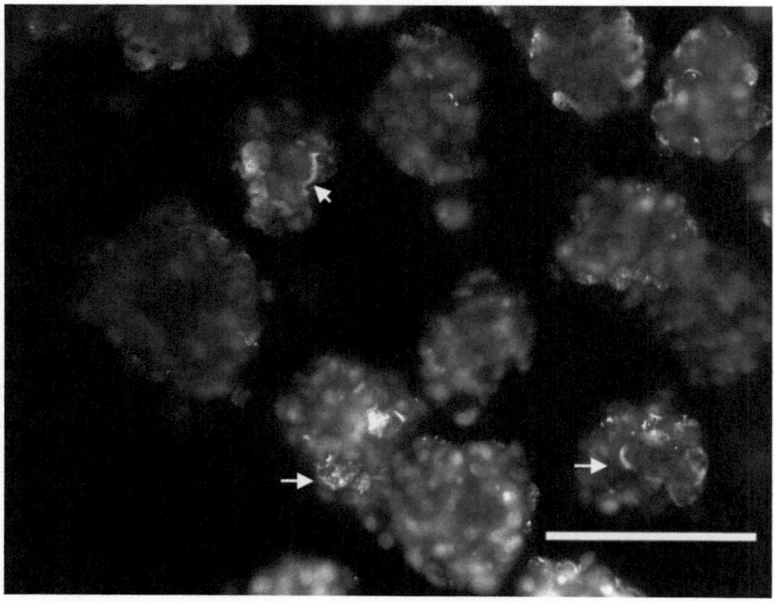

Fig. 2 Confocal collapsed Z-stack image at 100× magnification of immunostained kidney tubular clusters. Bright areas within clusters are positive for anti-mouse pan cadherin antibody (*arrows*). Nuclear staining are dim spots within clusters. Scale bar equals 100 μm

3. After washing with D-PBS, add 0.5 ml/well of 2% paraformaldehyde to fix cells. Incubate at 4°C for 20 min (see Note 7).
4. Remove fixative and gently wash cells twice with D-PBS.
5. Block any nonspecific binding sites with surface blocking solution for 30 min at room temperature.
6. After 30 min, remove blocking solution.
7. Add primary antibody of interest as well as isotype-matched control at a previously titered concentration in a volume of 0.5 ml surface staining solution per well.
8. Incubate overnight at 4°C.
9. Wash cells twice with D-PBS.
10. After washing, add secondary antibody at a previously titered concentration (starting at 1 μg/ml) in a volume of 0.5 ml surface staining solution per well for 30 min at room temperature protected from light.
11. Wash cells twice with D-PBS.
12. Add 0.5 ml/well of Hoechst nuclear dye 1 μg/ml for 10 min at room temperature protected from light.
13. Wash cells twice with D-PBS.
14. Analyze using inverted fluorescent microscope with image analysis software.

3.2.2 Characterization of Intracellular/Cytoplasmic Antigens

1. Obtain adherent cultured cells in a plated format (6-, 12-, 24-, or 48-well tissue culture dish).
2. Wash cultured cells by removing medium and adding appropriate volume of D-PBS to each well and or dish (3 ml/well for 6-well dish, 2 ml/well for 12-well dish, 250 μl/well of chamber slide).
3. After washing with D-PBS, add 2% paraformaldehyde to wells to fix cells. Incubate at 4°C for 20 min.
4. Remove fixative and wash cells twice with D-PBS.
5. Permeabilize and block cells by adding permeabilization and blocking solution (0.2% Triton X-100 with 2% serum of the species from which the secondary antibodies were generated from) for 30 min at room temperature.
6. Once cells are blocked, add desired primary antibody diluted in intracellular staining buffer, as well as isotype-matched control, and stain overnight at 4°C.
7. Wash cells twice with D-PBS.
8. Following washing, add secondary antibody (diluted in intracellular staining buffer) for 30 min at room temperature protected from light.
9. Wash cells twice with D-PBS.
10. Add 0.5 ml/well of Hoechst nuclear dye (1 μg/ml) for 10 min at room temperature protected from light.
11. Wash cells twice with D-PBS.
12. Analyze using inverted fluorescent microscope with image analysis software (Fig. 3).

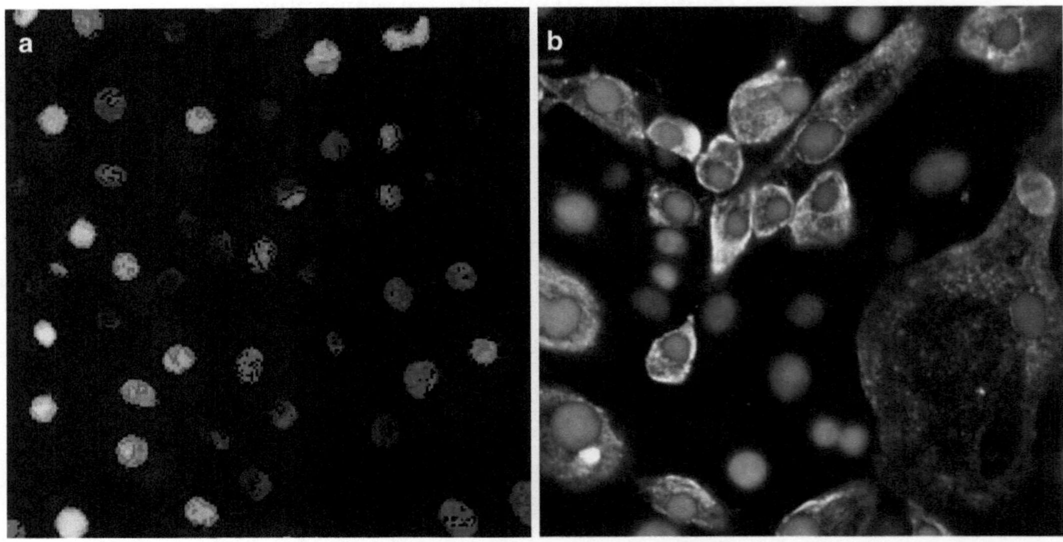

Fig. 3 Immunofluorescence image of expanded cultured kidney cells 200× magnification; (**a**) negative control with nuclear dye. (**b**) Cell membrane positive for a proximal tubule marker aquaporin 1

3.3 FACS Characterization of Primary Cultured Cells Using Single-Color Staining

FACS analysis is a very effective method for generating quantitative results from a given population of expanded, subcultured cells. Direct versus indirect staining depends on the availability and reactivity of the specific antibodies for the species of interest (see Note 8). The following section details an indirect method.

3.3.1 Characterization of Surface/Membrane Antigens

1. Obtain sample of live cells from isolated tissue washed in D-PBS.
2. For surface targets, use live cells in suspension at a starting concentration of 1×0^6 cells/ml in a 12×75 mm tube. Include isotype-matched controls for each antibody (see Note **9**).
3. Block cells for nonspecific binding with 1 ml surface blocking solution for 10 min at 4°C.
4. After blocking any nonspecific binding, wash by centrifugation at $500 \times g$ for 3 min.
5. Resuspend cells with 1 ml surface staining buffer containing primary antibody of interest at a concentration of 1 μg/ml/1×10^6 cells (see Note 4).
6. Incubate for 30 min at 4°C.
7. After 30 min, wash cells twice with D-PBS by centrifugation at $500 \times g$ for 3 min.
8. Add secondary antibody diluted in surface staining solution at a concentration of 1 μg/ml/1×10^6 cells and incubate for 30 min at 4°C.
9. Following secondary incubation, wash cells twice with D-PBS by centrifugation ($500 \times g$ for 3 min) and then fix cells with 2% paraformaldehyde in D-PBS prior to analysis on the flow cytometer.

3.3.2 Characterization of Intracellular/Cytoplasmic Antigens

1. Obtain sample of live cells from isolated tissue washed in D-PBS.
2. Determine the number of tubes to be used, including antibody targets and isotype-matched controls, as well as an unlabeled sample to measure intrinsic autofluorescence using a starting cell concentration of 1×10^6/ml/tube.
3. For intracellular targets, cells must be fixed and cell membranes permeabilized. Fix cells at a concentration of 1×10^6/ml in 2% paraformaldehyde in D-PBS for 20 min at room temperature.
4. After fixation, wash cells thoroughly in D-PBS by centrifugation at $500 \times g$ for 3 min at room temperature. Prior to last wash, distribute 1×10^6 cells into 1.5 ml microcentrifuge tubes for staining.
5. Resuspend cells in a concentration of 1×10^6 cells/0.2 ml of permeabilization and blocking solution (10% serum of the species from which the secondary antibodies were generated from in 0.2% Triton X-100 in D-PBS) and incubate for 30 min at room temperature.

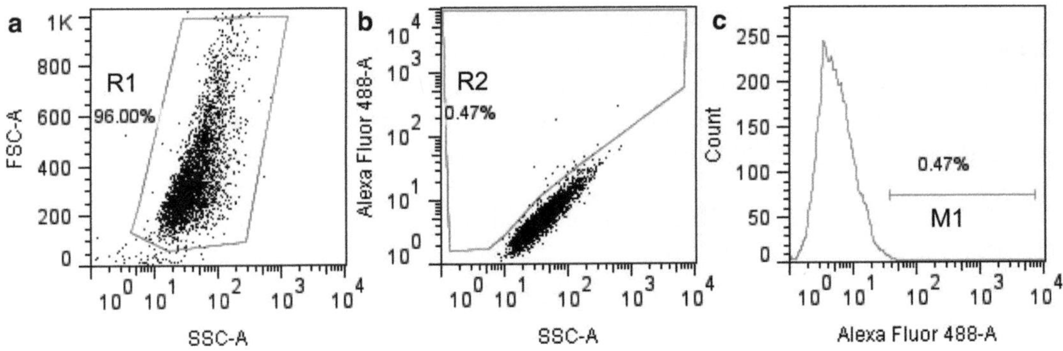

Fig. 4 Recommended starting dot plots for single-color FACS analysis. (**a**) Selected cell population of interest (R1). (**b**) The R1 gate from the first plot displayed in a second dot plot where a second region (R2) is drawn using the isotype-matched control as a negative population. (**c**) Setting the positive marker (M1) for fluorescence using a histogram

6. After permeabilization, add 0.8 ml of 0.2% Triton X-100 in D-PBS containing primary antibody and isotyped-matched controls to each tube. Final concentration of antibodies should be determined beforehand using optimal titration assays. A general starting point is 1 μg/1 × 10^6 cells/ml of intracellular staining solution.
7. Following blocking, wash by centrifugation at 500 × *g* for 3 min.
8. Incubate for 90 min to overnight at 4°C.
9. After 90 min, wash cells twice with D-PBS by centrifugation at 500 × *g* for 3 min.
10. Add secondary antibody diluted in intracellular staining buffer at a concentration of 1 μg/ml/1 × 10^6 cells and incubate for 30 min at 4°C.
11. Following secondary incubation, wash cells twice with D-PBS by centrifugation at 500 × *g* for 3 min and then resuspend cells in 0.5 ml of D-PBS and filter if necessary through a 40 μm filter (BD Falcon) to remove clumps prior to analysis on the flow cytometer.

3.3.3 Flow Cytometer Plot Setup for Basic Single-Color Analysis (see Note 10)

1. Make three starting plots: two dot plots and one histogram. The first dot plot measures size and granularity of the entire population; set up as SSC area (log) on *X*-axis and FSC area (linear) on the *Y*-axis. Make a capture gate for the region of interest (Fig. 4a) (see Note 11).
2. The second dot plot should be set up to capture the region of interest. Set the parameter for the *X*-axis to be SSC-area (log) vs. fluorophore-area (log) on *Y*-axis.
3. The third plot can be a histogram showing fluorophore such as A488-Area (log) vs. counts (Fig. 4c).

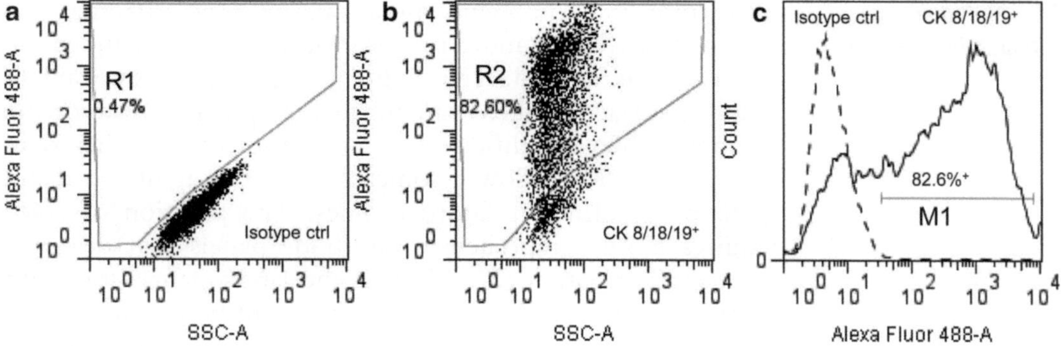

Fig. 5 FACS analysis of cultured kidney cells stained with cytokeratin (CK) 8/18/19: (**a**) Isotype-matched control with gated population (R1) of positive stained cells, (**b**) percentage positive CK8/18/19 (R2) by dot plot, (**c**) overlay histogram of the same data showing percentage positive CK8/18/19 (M1) compared to control

4. Place unlabeled or isotyped-matched control tube in loader and start acquisition. Adjust the PMT voltages for the forward side scatter and side scatter detectors to capture the entire cell population in the plot window. Since the initial cell culture population at early passage might have a mixture of cell types with varied sizes and granularities, start off by placing the entire population within the window.
5. Adjust the capture gate (R1) to exclude debris (Fig. 4a).
6. Select the gated population (R1) from the first plot to show in plot two and adjust the fluorophore PMT voltage (such as AlexaFluor 488-Area) by placing the population between the first and second decade of the *Y*-axis scale.
7. Make a second polygon-shaped gate in plot 2 (R2) (Fig. 4b) that captures the positively stained cells. The gate from the unlabeled and isotype-matched control should be similar if the antibodies have been titered properly (5) and can be adjusted to include 0.5% or less of the control population in order to capture any of the "low bright" positively stained cells of interest. The histogram marker (M1) can also reflect the percentage of positive cells that shifts to the right of the control peak (6) (Fig. 4c).
8. Sample acquisition can begin once the regions of interest have been gated. The total number of cells to be acquired and recorded can range between 5 and 50,000 events and can depend on the frequency of the rarest cell population of interest.
9. Percentage positivity using single-color fluorophores can be determined by subtracting the percentage of cells from the isotyped-matched controls from the positive stained cells in the capture gate. Percentage positive equals 82.15% in the example of Fig. 5 (see Note 12).

3.4 Multicolor Immunophenotyping

Multicolor immunophenotyping can be used effectively on passaged cells when more information from the expanded cultured cells is needed. The ability to simultaneously stain for multiple surface antigens can enhance cell characterization and distinguish between different cell populations during the expansion process. Most flow cytometers have the ability to distinguish many different fluorochromes. The addition of colors requires more comprehensive antibody panels that deal with compensation issues (7). Most of the larger panels for multicolor staining use directly conjugated antibodies for surface antigens on lymphocytes (8). This section describes a basic method for FACS that can distinguish two-color immunophenotyping targeting two surface antigens.

3.4.1 Multicolor Indirect Immunophenotyping of Surface Antigens Utilizing Different Fluorescent Spectra on a Flow Cytometer

1. Obtain sample of live cells from isolated tissue (i.e., kidney) washed in D-PBS.
2. Determine the number of tubes to be used including antibody targets and isotype-matched controls. For multicolor experiments greater than three colors, the use of "fluorescence minus one" (FMO) is recommended (see Note 13).
3. Set up 1.5 ml microcentrifuge tubes with a starting concentration of 1×10^6 cells per tube.
4. Block cells for nonspecific binding with 1 ml surface blocking solution for 10 min at 4°C.
5. After blocking, wash by centrifugation at $500 \times g$ for 3 min.
6. Resuspend cells with 1 ml surface staining buffer containing unconjugated primary monoclonal antibodies of interest with different isotypes at a concentration of 1 μg/ml/1×10^6 cells, for example, mouse IgG1 and mouse IgG_2b.
7. Incubate for at least 30 min at 4°C.
8. After primary incubation, wash cells twice with D-PBS by centrifugation at $500 \times g$ for 3 min.
9. Add secondary antibodies such as Alexa Fluor 488 or Alexa Fluor 647 that can be excited by either the blue or red laser to minimize bleed over from one channel to another and to minimize compensation issues. Dilute antibodies in surface staining buffer to a concentration of 1 μg/ml/1×10^6 cells and incubate for 30 min at 4°C.
10. After secondary incubation, wash cells twice with D-PBS by centrifugation at $500 \times g$ for 3 min and then fix cells in 1% paraformaldehyde in D-PBS for 10 min.
11. Filter cells if necessary through a 40 μm filter to remove clumps prior to analysis on the flow cytometer.

3.4.2 Flow Cytometer Plot Setup and Basic Two-Color Analysis of Culture Cells

1. Make two starting plots: The first dot plot measures size and granularity of the entire population; set up as SSC area (log) on *X*-axis and FSC area (linear) on the *Y*-axis. Make a capture gate for the population of interest (R1).
2. The second dot plot should be set up to show both fluorophores, one on each axis. Placing a four-quadrant region within the plot will enable the distinction of single- as well as dual-positive cells. Set the parameter for the *X*-axis to be A488-area (log) vs. A647-area (log) on *Y*-axis (Fig. 6).
3. Place the control tube for each fluorophore in the flow cytometer cell loader and start acquisition. Adjust the PMT voltages for the forward side scatter and side scatter detectors to capture the entire cell population in the plot window. Since the initial cell culture population at early passage might have a mixture of cell types with varied sizes and granularities, start off by placing the entire population within the window.
4. Adjust the capture gate (R1) to include nucleated cells, excluding debris (Fig. 6a).
5. Select the gated population from plot one to show in plot two. Adjust both of the fluorophores' PMT voltages, placing the population between the first and second decade of the *X*- and *Y*-axis scale and within the quadrant 1 region (lower left).
6. Adjust PMT voltages from controls to ensure that the positive controls fall within the upper region on the scale (Fig. 6b).
7. Sample acquisition can begin once the regions of interest have been gated. The total number of cells to be acquired and recorded can range between 5 and 50,000 events and can depend on the frequency of the rarest cell population of interest.

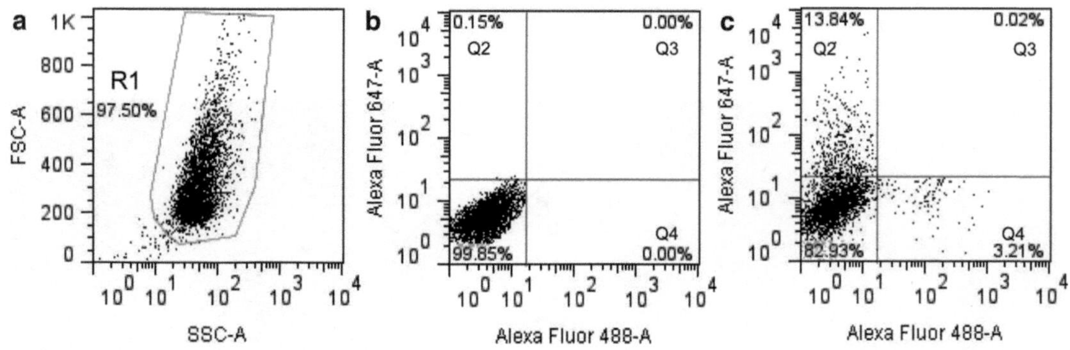

Fig. 6 Multicolored analysis of cultured human kidney cells stained with an endothelial marker CD31-A488 and a proximal tubular marker LTA- A647; (**a**) selected population. (**b**) Quadrant adjustment based on placing the isotype-matched control in lower left quadrant 1. (**c**) Bivariate 2-color analysis showing 13.84% positive for LTA and 3.21% for CD31

8. Percentage positivity using multicolor fluorophores can be determined by the percentage of cells located in each quadrant. Cells located in Q2 (upper left) for this example would be positive for LTA-A647, Q3 (upper right) would be positive for both CD31 A488 and LTA-A647, and Q4 (lower right) would be positive for CD31-A488 only (Fig. 6c).

3.5 Three-Dimensional Phenotypic Analysis of Cell/Biomaterial Combinations Using Confocal Microscopy

The ability to analyze cells seeded onto or embedded within biomaterials can be helpful in characterizing combination products for the use in regenerative medicine (9). The following section describes immunophenotyping methodology using a cell–biomaterial combination analyzed by spinning disk confocal microscopy. Examples of cell-biomaterial combinations include adopose-derived smooth muscle cells (Ad-SMC) seeded onto PGA/PLGA scaffolds, kidney tubular epithelial cells seeded on Collagen Type 1 beads and embedded in gelatin-based hydrogels (Figure 7).

3.5.1 Immunophenotyping Cell/Biomaterial Combinations

Different biomaterials may require separate procedures for targeting membrane or cytoplasmic antigens. Considering the biomaterials' natural autofluorescence will help in deciding which fluorophores to use to maximize the signal-to-noise ratios. Other considerations may include fixation with paraformaldehyde which may increase cross-linking, depth of seeded cells within the biomaterial, and cell morphology. This section provides a method for single-color immunophenotyping of cells on paraformaldehyde-fixable biomaterials using an indirect method for a cytoplasmic antigen target. Multiple colors can be added using the same principles if autofluorescence, isotype, and compensation can be accounted for.

1. Obtain sample of a cell-seeded biomaterial (see Note 14).
2. Fix cells seeded onto biomaterial with 2% paraformaldehyde in D-PBS for 20 min at room temperature.

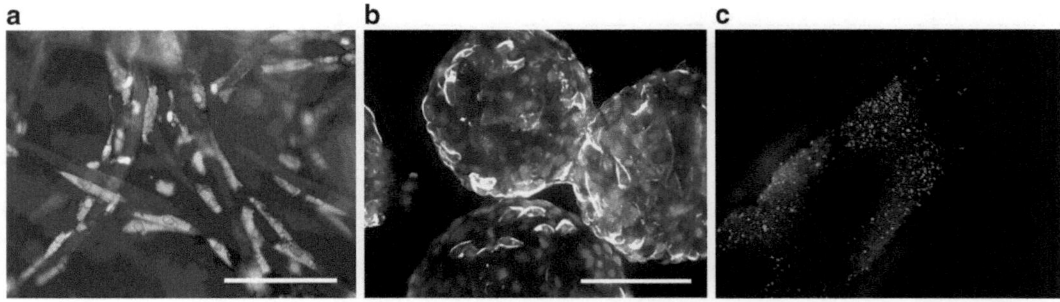

Fig. 7 Immunophenotyping cell/biomaterial combinations: (**a**) Smooth muscle actin-positive cells stained on PGA/PLGA scaffold, 200× collapsed z-stack image with 100 μm scale bar. (**b**) Cultured kidney epithelial cells stained with cytokeratin 8/18/19 on type 1 collagen beads, 200× collapsed Z-stack image with 100 μm scale bar. (**c**) Cytokeratin 8/18/19-positive epithelial cells embedded in gelatin-based hydrogel, 100× collapsed Z-stack 3 × 3 montage image

3. After fixation, wash twice with D-PBS and for cell membrane targets, block nonspecific binding using surface blocking solution for 30 min at room temperature.
4. Add primary antibody in 0.5 ml surface staining buffer at a starting concentration of 3 μg/ml.
5. For intracellular targets, the cells need to be permeabilized. Add 1 ml of permeabilization and blocking solution (10% serum of the species from which the secondary antibodies were generated from) and incubate for 30 min at room temperature.
6. After permeabilization, aspirate permeabilization and blocking solution, and add 0.5 ml of intracellular staining solution containing primary antibody and isotype-matched controls to each piece of cell–biomaterial combination. Final concentration of antibodies should be determined beforehand using optimal titration assays. A general starting point is 3 μg/ml of intracellular staining solution.
7. Incubate for 90 min to overnight at 4°C.
8. Following primary antibody incubation, wash cell–biomaterial combination twice with D-PBS by aspiration.
9. Add secondary antibody diluted in intracellular staining buffer at a concentration of 1 μg/ml and incubate for 30 min at 4°C.
10. After secondary incubation, wash cells twice with D-PBS by aspiration, then add 0.5 ml of D-PBS containing 1 μg/ml of Hoechst nuclear dye, and incubate for 10 min.
11. Following incubation, wash twice using aspiration and then transfer cell–biomaterial combination to an appropriate imaging plate to match the size of biomaterial (see Note 15).
12. Image cell-seeded biomaterial using confocal microscope with image analysis software such as Attovision™ (BD) for the Pathway 855 microscope (BD). Depending on the material, manually setting the depth at which the signal is brightest in the Z-plane is recommended; this will vary with material composition. Generally, for the Pathway 855, 70 microns is the maximum depth for PGA/PLGA materials. 3D analysis can be done using post-acquisition collapsed Z-Stack processing.

4 Notes

1. Determining cell count using a standard hemocytometer on a cluster of cells can be difficult and not always as accurate as counting a single-cell suspension. For this reason, an approximation is reasonable when performing a cytocentrifugation to ensure that there is separation of clusters and aggregates on the slides prior to staining.

2. Image analysis can be performed on a variety of fluorescent microscopes. Both inverted and upright microscopes are suitable for viewing immunostained slides. The most common filters used are the ones that will excite fluorescein or rhodamine as well as DAPI. Quantification can be completed by using image analysis software that recognizes individual cells and uses segmentation masks to differentiate a positive signal over background.
3. Prior to confocal microscopy, the immunostaining procedure can be completed first in 1.5 ml microcentrifuge tubes. After the last wash, the cells can be concentrated and resuspended in 100 μl and transferred into the 96-well thin plastic imaging plate.
4. Antibody titration may be needed to maximize signal-to-noise ratios. A general rule of thumb is to use a known positive control cell and use a matrix to test a twofold dilution series for both primary and secondary antibodies. If there is not a positive control cell available, a general starting concentration of 1–3 μg/ml may suffice. Determining the best signal-to-noise ratios may increase sensitivity and decrease false positives.
5. Image acquisition using a confocal microscope needs to be performed by someone trained in imaging to ensure that the most optimal image parameters are set correctly. Parameters such as image processing, flat field correction, background subtraction, and segmentation are some of the things to consider. For the Pathway 855™, AttoVision™ is the preferred software for image acquisition.
6. Deciding on which plating format to use to maximize the number of antibodies to screen (including controls) can depend on target cell number frequency. For rare cells with low frequencies, a larger plating format may be required to capture enough cells for statistical significance such as a 6-well dish or a 100 mm plate. For larger cell target frequencies greater than 10%, a smaller plating format could be used such as a 12- or a 24-well plate. This will enable the researcher to use a minimum volume of antibody solution to stain the cells. For multiplexing and high-throughput studies, it is possible with some cell types to use a 96- or 384-well format.
7. Fixation solutions can vary depending on the localization of epitopes. For most applications, a low percentage of 1–4% paraformaldehyde will suffice and maintain expression consistency. Zinc salt-based fixation for analysis of intracellular and surface epitopes has been recently described (10) to allow for long-term storage without changing labeling. Detergent-based versus ethanol-based permeabilization depends on whether the surface epitopes need to be stained prior to fixation. Using paraformaldehyde or zinc-based fixative that preserves surface epitopes for initial staining followed by a detergent-based permeabilization solution allows for multiple analyses of surface and intracellular antigen targets.

8. Direct versus indirect staining mainly depends on the availability of species-specific conjugated antibodies. Direct staining removes steps due to the primary antibody being conjugated to a fluorophore. Whenever possible, try to select conjugated antibodies. In many instances, when either multiplexing or dealing with species other than human, unconjugated primary antibodies may be the only choice. Recognition of both primary and secondary isotype has to be considered and more planning may be required to ensure specificity or nonspecific binding of target antigen. Sources for more information on cytometry can be found on the Internet at Purdue University Cytometry Laboratory Web site (www.cyto.purdue.edu) as well as the International Society for Analytical Cytometry (www.isac-net.org).
9. Staining for surface epitopes on live cells can be accomplished using the 12×75 mm polystyrene round bottom tubes (BD Falcon). For intracellular targets post-fixation and permeabilization, there can be a significant loss of cells when using the 12×75 mm tubes. The 1.5 ml microcentrifuge tubes work well when centrifuging at 300–500×*g* for 3 min; the losses are minimal. Upon the last wash in the process, the cells can then be transferred to the 12×75 mm tubes for acquisition.
10. Prior to acquiring cells on the flow cytometer, follow manufacturer's recommendations on instrument start-up procedures and instrument QC.
11. When analyzing whole-organ primary cultures, cell size (FSC) and granularity (SSC) are important parameters in determining the heterogeneity of the population. Placing either of these parameters on the *X*- or *Y*-axis is a personal preference. Starting with broad markers that characterize epithelial, mesenchymal, and endothelial cells is a good way to divide the population into categories. Once those percentages are determined, more selective antibodies can be used to further specify the population.
12. The release criteria for cell-based regenerative medicine/tissue-engineered products may vary from product to product and is represented by the phenotypic characterization of cells that best defines the population of interest prior to being released for product use. There can be more than one marker to base the criteria on and usually these markers have to meet a certain percentage of positive cells in order for the product to meet acceptance for release.
13. FMO control refers to controls that measure the fluorescence in multicolor experiments of all reagents, except the one of interest. They are necessary in determining which cells do not express the given target antigen. FMOs are usually run on peripheral blood mononuclear cell (PBMC) analysis using directly conjugated antibodies. For experiments that require a two-color analysis, using isotype-matched controls and a

separate wavelength where fluorescence spillover from one channel into the other channel does not occur is acceptable.

14. Cell–biomaterial combinations can be stained in many formats depending on the size of the sample. A larger scaffold may fit into a 6-well dish, whereas a smaller cell-seeded bead sample can fit into one well of a 96-well imaging plate. Picking a format that optimizes the staining surface and limits the reagents may be the best choice.
15. Choosing imaging plates for confocal microscopy can depend on a few factors such as size of sample, cost, and background fluorescence. Imaging plates are available in different materials such as glass (<200 μm), thick plastic (>500 μm), and thin plastic (<250 μm). Each has its pros and cons but a general rule is to choose one that is flat and allows the use of high-resolution objectives.

Acknowledgments

The authors acknowledge Tengion Inc., where all work was carried out. The analysis of the cell–biomaterial combinations would not have been possible without help of the Biomaterials Development group at Tengion, who provided testing material.

References

1. Dean P (1998) Confocal microscopy: principles and practices. Curr Protocols Cytometry 2:8.1–2.8.12
2. Dugina V, Alexandrova A, Chaponnier C, Vasiliev J, Gabbiani G (1998) Rat fibroblasts cultured from various organs exhibit differences in alpha-smooth muscle actin expression, cytoskeletal pattern, and adhesive structure organization. Exp Cell Res 238(2):481–90
3. Jennette C, Olson JL, Schwartz MM, Silva FG (ed) (2006) Heptinstall's pathology of the kidney, vol 1. Lippincott Williams & Wilkins
4. Humphreys BD, Valerius MT, Kobayashi A, Mugford JW, Soeung S, Duffield JS, McMahon AP, Bonventre JV (2008) Intrinsic epithelial cells repair the kidney after injury. Cell Stem Cell 2(3):284–291
5. Stewart C, Stewart S (2001) Cell Preparation for the Identification of Leukocytes. Methods in Cell Biology (ed: Darzynkiewicz Z,Crissman H, Robinson J) 63; Cytometry, 3rd ed. Part A. 244–249
6. Overton WR (1989) Modified histogram subtraction technique for analysis of flow cytometry data. Cytometry 9:619–626, Erratum in Cytometry 10:492–494
7. Roederer M (2001) Spectral compensation for flow cytometry: visualization artifacts, limitations, and caveats. Cytometry 45:194–205
8. Chattopadhyay PK, Perfetto SP, Roederer M (2004) The colorful future of cell analysis by flow cytometry. Discov Med 4(23):255–262
9. Basu J, Genheimer CW, Rivera EA, Payne R, Mihalko K, Guthrie K, Bruce AT, Robbins N, McCoy D, Sangha N, Ilagan R, Knight T, Spencer T, Wagner BJ, Jayo MJ, Jain D, Ludlow JW, Halberstadt C (2011) Functional evaluation of primary renal cell/biomaterial Neo-Kidney augment prototypes for renal tissue engineering. Cell Transplant 20:1771–1790
10. Christensen R, Owens D, Thomsen A, Pedersen S, Jensen U (2011) Zinc fixation for flow cytometry analysis of intracellular and surface epitopes, DNA content, and cell proliferation. Curr Protocols Cytometry 7:40.1–7.40.9

Chapter 11

Preparation and Evaluation of Natural Scaffold Materials for Kidney Regenerative Applications

Monica A. Serban, Toyin A. Knight, and Richard G. Payne

Abstract

Tissue engineering involves the concerted action of biomaterials, cells, and growth factors. Kidney regeneration relies on the same combination of ingredients. Here, we describe an example of gelatin-based biomaterial preparation and its evaluation in the context of kidney biocompatibility and integration. This biomaterial manufacturing technique is simple, cost-effective, highly reproducible and the in vivo evaluation procedure highly informative on the biocompatibility and regenerative potential of the tested construct.

Key words Gelatin, Microspheres, Injectable, Renal

1 Introduction

In the past few decades, tissue engineering emerged as a new field with highly promising therapeutic potential (1). Concepts that appeared unattainable not long ago, such as organ regeneration and de novo organogenesis, have now been successfully translated into concrete clinical cases and tissue engineering and regeneration products are slowly progressing toward commercialization (1). One of the initial accomplishments in the field was bladder engineering, where a biodegradable synthetic scaffold and autologous cells were combined successfully to create a brand-new, functional organ (2). The same design approach widely applies to tubular of hollow organs: a biocompatible, persistent scaffold; cells capable of repopulating the scaffold surface and interacting with their endogenous counterparts; and a proper biological milieu that exposes the "organ to be" to physiologically similar conditions (temperature, pressure, pH, nutrient gradients, etc.) (3). While the number of success stories with hollow organ regeneration increases, the design and successful achievement of functional organs with more complicated architectures (i.e., heart, kidney, liver, etc.) still remains a

Joydeep Basu and John W. Ludlow (eds.), *Organ Regeneration: Methods and Protocols*, Methods in Molecular Biology, vol. 1001, DOI 10.1007/978-1-62703-363-3_11,

challenge. One crucial parameter of every tissue engineering and regenerative process is the choice of the proper scaffold or biomaterial. Synthetic, semisynthetic, and natural scaffolds are widely available and each class features a comparable array of advantages and disadvantages (4). Synthetics are typically more cost-effective, involve highly controllable manufacturing processes, have longer in vivo persistence which is required for certain applications but often are associated with toxic effects and bio-rejection (5). In contrast, natural scaffolds are characterized by great biological compatibility and controllable and predictable biodegradation rates, but present risks of disease transmission, lot-to-lot variability, and high manufacturing costs. Semisynthetics represent a hybrid of the aforementioned categories, and typically leverage the positive features of their natural and synthetic counterparts (i.e., cross-linked gelatin-based biomaterials—maintain the biocompatibility and biological effects associated with collagens (6–10) and in addition they have improved lot-to-lot consistency, longer in vivo residence, and tailorable stiffness) (4).

For our kidney regeneration-targeted applications, the design of the system was focused on the development and consistent production of biomaterials that (a) would deliver attached cells to the desired sites and create space for regeneration; (b) would be able to persist at the site long enough to allow cells to establish, function, remodel their microenvironment, and secrete their own extracellular matrix (ECM); and (c) would promote integration of the transplanted cells with the surrounding tissue. In other words, process reproducibility, biocompatibility, and biodegradability were key demands. A number of scaffolds were considered: gelatin, hyaluronic acid, alginate, polyglycolic acid (PGA), and poly-lactic-*co*-glycolic acid (PLGA) (11). For this specific application, gelatin-based scaffolds featured most positive parameters, but the long bioresorption rate of commercially available variants appeared to hinder tissue integration and regeneration.

Here, we detail a simple, straightforward method to obtain gelatin-based microspheres with controllable proteolytic susceptibility. By using a carbodiimide widely employed in the manufacturing of collagen-based FDA-approved devices and a physiological chemical cross-linking process, one can manufacture gelatin beads with biodegradation rates spanning across a significant range. In addition, the resulting constructs (cellularized scaffolds) are injectable and easy to evaluate both in vitro and in vivo for their cytocompatibility and biocompatibility.

2 Materials

All solutions should be prepared with deionized water (resistivity of 18 MΩ cm at 25°C). Follow all handling, storage, and disposal recommendation provided by the manufacturer for all chemicals.

2.1 Biomaterial Preparation

1. 10% (w/v) gelatin solution: 10 g of gelatin (low endotoxin, Gelita Inc., Sioux City, IA, USA) into deionized water to a final volume of 100 ml.
2. Heat/stir plate with temperature control.
3. Stir bar.
4. Support stand with rod.
5. Clamp and clamp holder.
6. Deep metal tray.
7. Thin layer chromatography reagent sprayer.
8. Liquid nitrogen.
9. Aluminum foil to line the metal tray.

2.1.1 Microsphere Cross-Linking

1. Cross-linking buffer: 0.1 M MES, 0.9% NaCl pH 4.7. 2-(morpholino) ethanesulfonic acid (MES) buffered saline (Fisher Scientific, Pittsburg, PA, USA).
2. Cross-linking reagent stock solution: 1 M *N*-ethyl-*N*′-(3-dimethylaminopropyl) carbodiimide hydrochloride (EDC, Sigma-Aldrich, St. Louis, MO, USA) in cross-linking buffer. If desired, for differential cross-linking a concentration range of cross-linking reagent solutions (i.e., 10–100 mM) can be obtained from the 1 M EDC stock solution via serial dilutions.
3. Conical tubes (50 ml).
4. Tabletop vortex.

2.1.2 Microsphere Purification

1. Cross-linking buffer.
2. Deionized water.
3. Vacuum flask.
4. Buchner funnel.
5. Filter paper.

2.2 Biomaterial Evaluation

2.2.1 Cross-Linking Efficiency Determination

1. Lyophilized cross-linked gelatin microspheres.
2. Phosphate-buffered saline (1× PBS).
3. Collagenase/dispase digestion mix: 30 U/ml collagenase IV (Worthington Biochemical Corp., Lakewood, NJ, USA), 4 U/ml dispase I (STEMCELL Technologies, Vancouver, BC), 0.5 mM $CaCl_2$ in PBS.
4. Coomassie Plus (Bradford) protein assay solution (Fisher Scientific, Pittsburg, PA, USA).
5. Picrylsulfonic acid (TNBS) 5% w/v in H_2O (Sigma-Aldrich).
6. 1 M NaOH solution
7. pH paper strips (0–14 pH range).
8. Eppendorf tubes (1.5 ml).
9. Multi-well spectroscopy plate (96-well).

2.2.2 Microsphere Sizing

1. Nylon meshes of desired pore size (Small Parts, Inc., Atlanta, GA, USA).
2. 70% v/v ethanol solution.
3. Buchner funnel.
4. Vacuum flask.

2.2.3 Biocompatibility

1. Mammalian primary kidney cells (12, 13).
2. Basal medium: 1:1 Dulbecco's Modified Eagle Medium-High Glucose (DMEM-HG; Invitrogen, Carlsbad, CA, USA): Keratinocyte-Serum Free Medium (K-SFM; Invitrogen).
3. Kidney growth medium: Basal medium containing 5% v/v fetal bovine serum (FBS), 5.0 μg/l epidermal growth factor (EGF), 50 mg/l bovine pituitary extract (BPE), 1× insulin-transferrin-sodium selenite medium supplement (ITS; Invitrogen), 1× antibiotic–antimycotic (Invitrogen).
4. LIVE/DEAD mammalian cell viability assay kit (Invitrogen).
5. Injection needle or catheter and 1 ml syringe.

3 Methods

3.1 Gelatin Microspheres Preparation

3.1.1 Bead Production

1. Prepare a concentrated gelatin solution (i.e., 10% w/v) in water and allow the protein to dissolve at 40°C under stirring for 1 h.
2. Set up the necessary equipment for bead spraying (support stand with rod, clamp holder, clamp) (Fig. 1).
3. Rinse tubing and TLC reagent sprayer with hot water for 5–10 min to warm them up. Alternatively, equilibrate in a 50°C oven.
4. Fix the gelatin solution flask onto the stand with a clamp, connect the TLC reagent sprayer to the flask on one end and a forced air source on the other end.
5. Position the TLC reagent sprayer tip perpendicularly to the tray surface at a distance of 25–30 cm (see Note 1).
6. Pour liquid nitrogen into a metal tray lined with aluminum foil to an approximate liquid depth of 1 in.
7. Air spray gelatin solution into liquid nitrogen until the desired amount of gelatin solution is sprayed (see Note 2).
8. Place the tray in a fume hood to evaporate the liquid nitrogen.
9. Collect the beads and lyophilize.

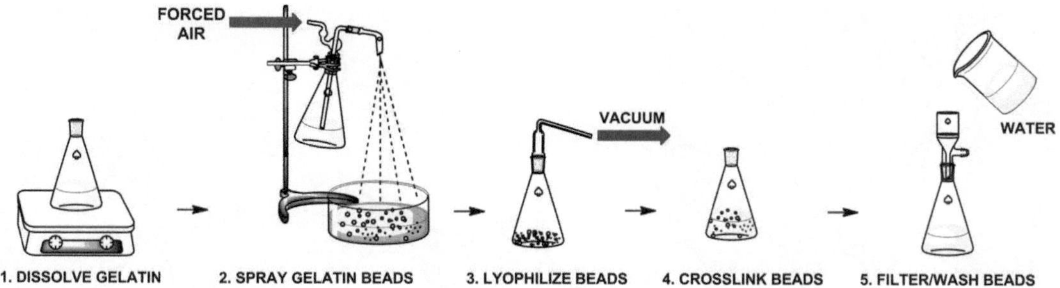

Fig. 1 Schematic illustration of the gelatin microsphere production and cross-linking process

3.1.2 Bead Cross-Linking

Soak 1 g of lyophilized beads in (20 – X) ml prechilled cross-linking buffer (where X = cross-linking reagent stock solution volume that needs to be added to the beads in order to obtain the final desired cross-linking reagent solution concentration) for at least 1 h at 4°C (see Notes 3 and 4).

1. Briefly vortex samples.
2. Incubate for 24 h at 4°C under static conditions.
3. Add the X ml cross-linking reagent stock solution to the beads, so that the total solution volume is 20 ml.
4. Briefly vortex samples to ensure homogeneity.
5. Incubate for 24 h at 4°C under static conditions.
6. Filter beads and wash extensively with deionized water.
7. Freeze and lyophilize for long-term storage or size separation (Fig. 1).
8. Alternatively, purified beads can be kept in PBS or ethanol solution at 4°C and used within a few days.

3.1.3 Optional: Bead Sizing (see Notes 5 and 6)

1. Transfer beads into 70% ethanol solution.
2. Prepare a filtering flask with a Buchner funnel with the larger pore size mesh (see Note 7).
3. Filter beads through.
4. Remove mesh, resuspend captured beads in ethanol solution, and repeat filtering step two more times (see Note 8).
5. The beads that filtered through the mesh are now all ≤250 μm.
6. Remove the large pore size mesh from the funnel. Thoroughly rinse funnel with hot water, connect to a clean filtering flask, and insert the small pore size mesh (i.e., 100 μm).
7. Filter the ≤250 μm beads through the mesh three times as described above.
8. Collect the beads that get retained on the mesh: these beads will be 100–250 μm in diameter (see Note 9).

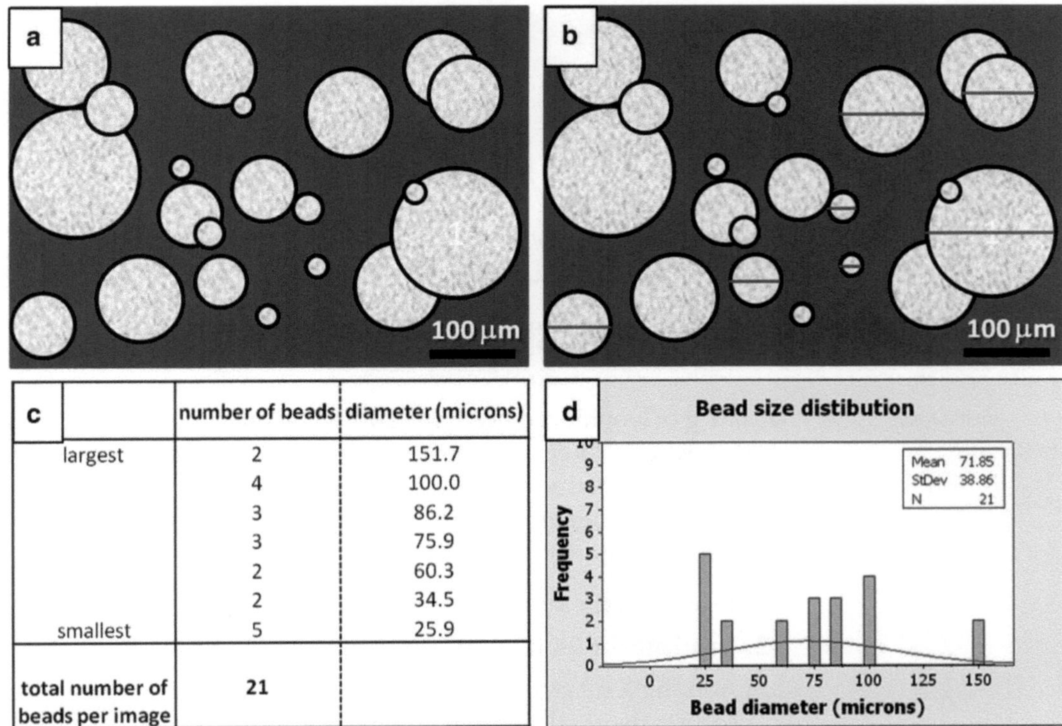

Fig. 2 Methodology for bead size distribution determination. (**a**) Schematic of SEM image of beads; (**b**) illustration of bead diameter measuring procedure; (**c**) example of bead size and diameter compilation; (**d**) determination of bead size distribution

3.2 Gelatin Microsphere Characterization

3.2.1 Morphology and Size Distribution (see Note 10)

1. Cover a scanning electron microscopy (SEM) stub with carbon tape.
2. Apply lyophilized beads onto carbon-taped SEM stubs.
3. Sputter-coat with gold (see Note 11).
4. Image.
5. Verify sizes by measuring the bead diameters from at least ten different images using a software of choice (i.e., ImageJ).
6. Compile bead sizes and numbers, then determine the bead size distribution by using a software of choice (i.e., MiniTab) (Fig. 2).

3.2.2 In Vitro Enzymatic Degradation Assay (see Notes 12 and 13)

7. Weigh out dry cross-linked beads.
8. Suspend in PBS, pH 7.4 to a concentration of ~20 mg/ml.
9. Add 50 μl of collagenase/dispase digestion mix to 0.5 ml bead suspension.
10. Vortex samples.
11. Incubate for 1 h at 37°C on a rocker (use $n=3$ for each cross-linked concentration used) to yield partially digested samples.

12. Subsequently, collect 20 μl supernatant from the partially digested samples and evaluate for soluble protein content via Bradford assay (see Note 14).
13. Further incubate the remaining digestion mix overnight as described above to yield totally digested samples.
14. Determine total protein content by Bradford assay (see Note 15).
15. Determine the percent degradation per hour for each sample by using the formula: ((amount of protein detected in the supernatants of partially digested sample)/(amount of protein in the totally digested samples)) × 100.

3.2.3 Quantification of Primary Amines (see Note 16): Option 1 (see Note 17)

1. Add 5 μl of 1 M NaOH to each vial of fully digested cross-linked beads solution (see Subheading 3.1, step 2) to increase the pH.
2. Verify by pH paper strip that the final pH value is ~8.5.
3. Add TNBS to each sample to a final concentration of 0.25% w/v.
4. Incubate vials at 37°C for 2 h on a rocker (see Note 18).
5. Determine A_{335} values of the resulting colored reaction product with a plate reader.
6. Normalize values per milligram protein in each sample as determined previously (see step 8 of Subheading 3.2).

3.2.4 Quantification of Primary Amines: Option 2 (see Note 19)

1. Weigh out uncross-linked (see Subheading 3.1, step 1) and cross-linked beads (see Subheading 3.1, step 2).
2. Suspend in PBS, pH 7.4 to a concentration of ~20 mg/ml.
3. Add 50 μl of collagenase/dispase digestion mix to 0.5 ml bead suspension.
4. Vortex samples.
5. Incubate for overnight at 37°C on a rocker.
6. Add 5 μl of 1 M NaOH to each vial of beads solution to increase the pH.
7. Verify by pH paper strip that the final pH value is ~8.5.
8. Add TNBS to each sample to a final concentration of 0.25% w/v.
9. Incubate vials at 37°C for 2 h on a rocker.
10. Determine A_{335} values of the resulting colored reaction product with a plate reader.
11. Determine percent of primary amines in the cross-linked samples according to the formula: (A_{335} for cross-linked sample/A_{335} reference sample) × 100.

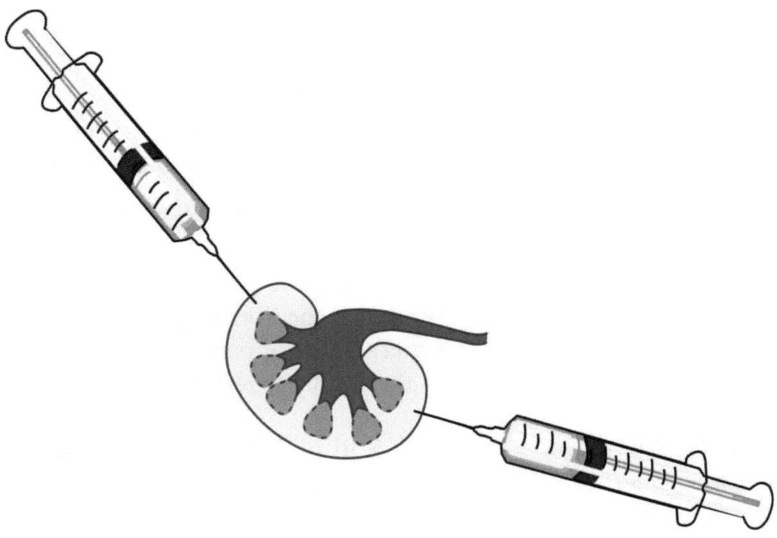

Fig. 3 Illustration of the construct microinjection technique

3.3 Gelatin Microsphere Cytocompatibility Evaluation (see Note 20)

1. Seed kidney cells onto cross-linked beads at a density of 7.15×10^6 cells/100 μl of beads (packed volume).
2. Incubate overnight at 37°C/5% CO_2 with 10 ml basal medium under dynamic conditions (see Note 21).
3. Assess cell viability 24 h later by using a LIVE/DEAD® viability/cytotoxicity kit (see Note 22).
4. Cell laden beads can be easily visualized with bright field and fluorescent microscopes.

3.4 Gelatin Microsphere Biocompatibility

3.4.1 Construct Preparation

1. Sterilize sized beads in 70% ethanol solution for 24 h.
2. Wash three times with sterile PBS.
3. Wash beads one time with sterile basal medium.
4. Seed renal cells at a density of 71.5×10^6 cells/ml packed beads under dynamic conditions (syringe placed on rotating device at 1 rpm, spinner flasks, etc.) in kidney growth medium.

3.4.2 Construct Microinjection (Fig. 3)

1. Expose kidney through midline incision.
2. Inject constructs from each pole into the cortex.
3. Suture incision and allow animal to recover.
4. Sacrifice at required time points and collect kidney for histological evaluation (i.e., hematoxylin–eosin (H&E), trichrome staining, and macrophage phenotyping (M1/M2)) (see Note 23).

4 Notes

1. By decreasing or increasing the distance between the TLC sprayer and the metal tray, the size distribution of the sprayed beads can be varied. Also, various types of TLC reagent sprayers are available, either standalone or bottle-type (Fig. 1)—either one is suitable for this application as long as the nozzle can be positioned perpendicularly to the tray and there is an uninterrupted flow of gelatin solution during spraying.
2. If the sprayer clogs with solidified gelatin, disconnect from air source, rinse rapidly with hot water, verify flow with hot water, reconnect, and continue procedure.
3. This temperature will prevent beads from melting and sticking together.
4. When adding the buffer to the dry beads account for the cross-linking agent solution volume that needs to be added so that the final cross-linking volume is 20 ml (i.e., if the cross-linking stock solution is 1 M EDC and the desired cross-linking concentration is 100 mM, then presoak beads in 18 ml prechilled buffer and subsequently add in 2 ml of the stock cross-linking solution. The final solution volume will be 20 ml of 100 mM EDC per gram of beads).
5. While optional, this bead sizing step narrows the diameter range of the beads and is recommended to facilitate injection and analysis of the in vivo data.
6. Gelatin beads tend to clump together but having them in 70% ethanol solution will diminish clumping.
7. Identify the desired bead size range for a specific application and select nylon meshes accordingly, i.e., 250 μm—for the upper limit and 100 μm for the lower limit. Keep in mind that sizing will decrease the final bead yield (i.e., if 1 g of beads is used for sizing, after this process the bead mass will be less than 1 g, in certain cases significantly less than 1 g depending on the initial size distribution of the beads).
8. The larger size beads that are retained after the filtration steps can be discarded or collected for different applications, as desired.
9. The beads that passed though the mesh pores (≤100 μm in diameter) can be discarded or collected for different applications as desired.
10. The size (before or after sizing) and morphology of the gelatin beads can be visualized in detail by scanning electron microscope (SEM).

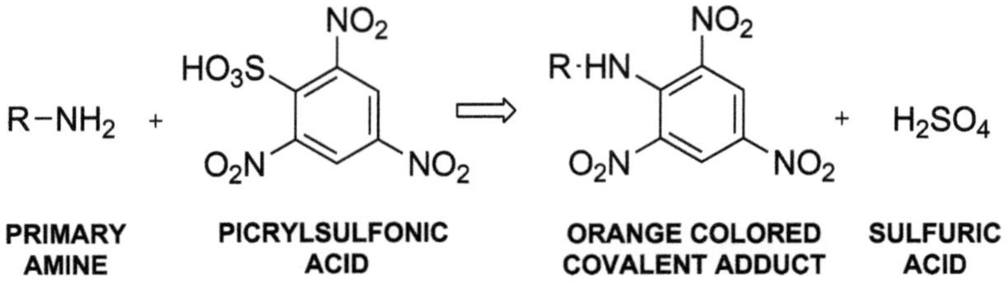

$R{-}NH_2$ + HO_3S–$C_6H_2(NO_2)_3$ ⟹ $R{\cdot}HN$–$C_6H_2(NO_2)_3$ + H_2SO_4

PRIMARY AMINE | PICRYLSULFONIC ACID | ORANGE COLORED COVALENT ADDUCT | SULFURIC ACID

Fig. 4 Reaction scheme illustrating the formation of a covalent adduct between available primary amine functionalities and TNBS

11. Other sputter-coating metals, compatible with SEM, can be used.
12. This assay is highly informative, especially if the production of differentially cross-linked beads was sought (i.e., a range of EDC solution concentrations were used for cross-linking).
13. This assay has a dual function: (a) confirms that cross-linking was successful and (b) differentiates between dissimilar degrees of cross-linking.
14. For gelatin-based biomaterials the Bradford assay needs to be modified, specifically the ratio of the Bradford reagent to protein solution needs to be 1:9 v/v in order to increase sensitivity.
15. The amounts of protein in the samples can be calculated from a gelatin standard curve obtained by plotting the A_{595} values for solutions made from known amounts of gelatin that were fully digested then assayed with Bradford reagent.
16. The EDC cross-linking reaction involves the formation of a covalent bond between adjacent inter- and intramolecular primary amine and carboxyl functionalities. Accordingly, increasing the extent of cross-linking will result in a decrease of free primary amines on the beads. Therefore by determining the number of free primary amines present on the beads, the extent of cross-linking can be determined (Fig. 4).
17. This amine quantification method should be used when analyzing differentially cross-linked beads.
18. Solutions will turn yellow with color intensities inversely proportional to the degree of cross-linking.
19. This amine quantification method should be used to generally detect cross-linking and the percentage of free primary amines in the cross-linked sample compared to the reference (uncross-linked) sample.
20. Because of the three-dimensional nature of the microspheres cell seeding onto the beads should be carried out under dynamic conditions.

21. For incubation under dynamic conditions a syringe placed on rotating device at 1 rpm or a spinner flask can be used.
22. LIVE/DEAD mammalian cell viability kits comprise of a dual calcein AM/ethidium homodimer mix that stains live cells green (calcein AM gets cleaved by the esterases of live cells to yield fluorescent calcein) and dead cells red (ethidium homodimer diffuses through the membranes of dead cells and upon binding to nucleic acids fluoresces red).
23. The histological analysis of the injection sites will provide information on the biocompatibility of the construct and its resorption/tissue integration rate. The H&E staining makes the gelatin beads easily detectable by staining them purple. Additional macrophage phenotyping helps estimate the inflammatory (M1 phenotype) or remodeling/regeneration (M2) phenotypes that are induced by the construct.

References

1. Place ES, Evans ND, Stevens MM (2009) Complexity in biomaterials for tissue engineering. Nat Mater 8:457–470
2. Atala A, Bauer SB, Soker S et al (2006) Tissue-engineered autologous bladders for patients needing cystoplasty. Lancet 367:1241–1246
3. Basu J, Ludlow JW (2010) Platform technologies for tubular organ regeneration. Trends Biotechnol 28:526–533
4. Serban MA, Prestwich GD (2008) Modular extracellular matrices: solutions for the puzzle. Methods 45:93–98
5. Lutolf MP, Hubbell JA (2005) Synthetic biomaterials as instructive extracellular microenvironments for morphogenesis in tissue engineering. Nat Biotechnol 23:47–55
6. Cenni E, Ciapetti G, Stea S et al (2000) Biocompatibility and performance *in vitro* of a hemostatic gelatin sponge. J Biomater Sci Polym Ed 11:685–699
7. Lee CH, Singla A, Lee Y (2001) Biomedical applications of collagen. Int J Pharm 221: 1–22
8. Waksman BH, Mason HL (1949) The antigenicity of collagen. J Immunol 63:427–433
9. Engvall E, Ruoslahti E (1977) Binding of soluble form of fibroblast surface protein, fibronectin, to collagen. Int J Cancer 20:1–5
10. Kim NR, Lee DH, Chung PH et al (2009) Distinct differentiation properties of human dental pulp cells on collagen, gelatin, and chitosan scaffolds. Oral Surg Oral Med Oral Pathol Oral Radiol Endod 108:e94–e100
11. Basu J, Genheimer CW, Rivera EA et al (2011) Functional evaluation of primary renal cell/biomaterial Neo-Kidney Augment prototypes for renal tissue engineering. Cell Transplant. doi:10.3727/096368911X566172
12. Kelley R, Werdin ES, Bruce AT et al (2010) Tubular cell-enriched subpopulation of primary renal cells improves survival and augments kidney function in rodent model of chronic kidney disease. Am J Physiol Renal Physiol 299:F1026–F1039
13. Presnell SC, Bruce AT, Wallace SM et al (2011) Isolation, characterization and expansion methods for defined primary renal cell populations from rodent, canine and human normal and diseased kidneys. Tissue Eng Part C Methods 17:261–273

Chapter 12

Rapid Production of Autologous Fibrin Hydrogels for Cellular Encapsulation in Organ Regeneration

Adelola O. Oseni, Peter E. Butler, and Alexander M. Seifalian

Abstract

Autologous hydrogel manufacture is an exciting technique within the field of regenerative medicine. Fibrin is a protein with many biocompatible and regenerative features. The ability to generate fibrin scaffolds with the necessary matrix topography for cell integration, from a patient's autologous tissue, could improve the translation of many tissue engineering efforts from bench to clinical application. Here we describe the rapid extraction and production of fibrin hydrogels for development of organs, using a simple low-cost series of centrifugations and ethanol precipitation, which produces hydrogels of a more predictable amount and morphology.

Key words Fibrin, Fibrinogen, Encapsulation, Hydrogels

1 Introduction

Fibrinogen is a soluble protein component of blood and is fundamental to the process of wound healing and repair. Whole blood consists of 0.2% fibrinogen by volume, and it is present as a product of blood clotting cascades. When injury occurs, a biological cascade of responses is activated at the wound bed. The first is the influx of inflammatory cells that release cytokine molecules. These cytokines encourage the removal of dead tissue, and the deposition of macromolecules, one of which is fibrin (1, 2). Fibrinogen itself is a dimeric structure consisting of three pairs of polypeptide chains which are joined by disulphide bonds (2, 3). In the presence of thrombin these disulphide bonds are broken to form monomeric units that eventually undergo covalent cross-linking to form insoluble fibrin.

Fibrin is commonly used as a hemostatic agent in coagulopathic and heparinized patients to reduce peri- and postoperative bleeding (4, 5). It was officially approved for clinical use by the FDA in 1998 and as a result, manufactured fibrin gel systems have

Joydeep Basu and John W. Ludlow (eds.), *Organ Regeneration: Methods and Protocols*, Methods in Molecular Biology, vol. 1001, DOI 10.1007/978-1-62703-363-3_12, © Springer Science+Business Media New York 2013

been commercially available for over two decades (6, 7). Their application has however been hindered by the potential for viral contamination and foreign body responses, especially if sourced from animals.

To combat the risk of foreign body reactions, methods for preparing fibrin hydrogels from autologous blood have been described in the literature (8, 9). Fibrinogen isolation from autologous blood has been achieved using four different methodologies, namely, (1) cryoprecipitation, the "gold standard" method established for many years. In such instances, the blood sample is centrifuged to remove the cruor, and the remaining plasma fraction is taken through a series of freezing and thawing cycles designed to precipitate the fibrin protein (temperatures range between −20 and −80°C). Further centrifugation produces a fibrin pellet. However there are numerous variables in the freezing/thawing times, temperatures, and number of cycles. This has caused vast discrepancies in the amounts of fibrinogen that can be isolated and studies have shown a tendency for increased difficulty of gelation after entering extreme freeze–thaw cycles (10, 11). (2) Ammonium sulfate precipitation associated with precipitation of small amounts of fibrinogen. The solubility of a protein is highly dependent on the ionic strength of the solution into which it is to be dissolved. Ammonium sulfate is used to increase the ionic strength of the plasma solution thereby decrease the solubility of the fibrin protein, which therefore precipitates out. This is a delicate process which is dependent on transitory ionic concentrations and therefore requires vast volumes of blood to get sufficient quantities of fibrinogen (12). (3) Poly(ethylene glycol) precipitation uses PEG molecules to increase the osmotic pressure within the plasma solution, reducing the solubility of the fibrin protein and causing it to precipitate out of solution. This method is complex and there is little and inconsistent literature to use (13). (4) Ethanol precipitation is a technique largely based on the established precipitation of nucleic material, DNA, RNA, etc. Ethanol disrupts the polar attraction of water to solutes. This causes the negative ends of the proteins to be attracted to positively charged ions in solution. Stable ionic bonds are formed, causing the protein, in this case fibrin to precipitate out as a solid. The application of this method to fibrin isolation is relatively new; however it is a low-cost method for extraction (14). The lack of comparative data in the literature, overexaggeration of potential yields, and poor descriptions of methodology lead many groups to struggle for many years with inconsistency in the gels that they produce especially when being used in vivo.

Our lab specializes in the translation and application of bench top practices into the clinical setting. This has been shown in recent years the development of lachrymal duct construct, vascular bypass grafts, and trachea translation using nanocomposite materials and stem cells (15–19). Currently we use fibrin hydrogels to encapsulate bone marrow-derived mesenchymal stem cells/chondrocytes for producing cartilaginous constructs for facial reconstructing. For fibrin hydrogels to have the greatest clinical application, a simple, consistent, and low-cost method needs to be used. By including a series of centrifugations combined with carefully timed ethanol precipitation, we produce fibrin hydrogels with a more consistent morphology that can be scaled up or down depending on the tissue target. Most importantly this method produces the autologous hydrogel in 24 h and is readily translated into the clinical setting.

2 Materials

1. BD Vacutainers® (BD Dickinson, Franklin Lakes, NJ, USA).
2. Sodium citrate anticoagulant blood tubes (363083, 10.5 M, 3.2%, Beckton Dickinson, Franklin Lakes, NJ, USA).
3. 100% ethanol.
4. Thrombin solution: Add thrombin (Tisseel® Fibrin sealant, Baxter, Deerfield, IL, USA) to 10 mM CaCl2 to reach a final concentration of 1,200 U/ml. Gentle swirl the mixture to aid the dissolving. Use immediately. Thrombin is added at a ratio of 1:3 fibrinogen. The average harvest from a single participant would need no more than 1–2 ml of thrombin to be prepared.
5. Aprotinin solution (A6279, Sigma Aldrich, St. Louis, MO, USA).
6. Un-supplemented cell culture medium: For BMSCs a-MEM or DMEM. For chondrocytic cells DMEM HamF12 is preferred, whereas other cells types such as epithelial cells tend to prefer a simple medium of RPMI.
7. Supplemented cell culture medium: 10% (v/v) fetal bovine serum (12763025, Invitrogen, Life technologies, Grand Island, NY, USA), 1% (v/v) penicillin-streptomycin (P4333, Sigma Aldrich, St. Louis, MO, USA), 1% (v/v) amphotericin B (A2942, Sigma Aldrich, St. Louis, MO, USA, Sterile 10 mM).
8. Sterile10 mM CaCl2 solution.
9. Centrifuge.
10. 50 ml centrifuge tubes.
11. Disposable 3 ml sterile pipettes.
12. Standard phosphate-buffered saline.
13. Standard sterile tissue culture plates (6 well).

3 Methods

All procedures are to be performed at room temperature (21 ± 1°C) unless otherwise specified. The use of aseptic technique is of paramount importance for the production of pathogen-free fibrinogen hydrogels and should be employed at all times especially during manufacture for clinical use. The volumes used in this method are based on a 2–3 ml harvest of fibrin glue from 25 ml of whole blood (Table 1). The volumes and concentrations used are scalable for larger quantities of blood.

3.1 Whole Blood Plasma Separation

1. Collect whole blood by venipuncture samples using vacutainers and sodium citrate tubes. The first blood tube can be discarded as this is considered an unsterile blood sample. Invert the tubes a few times and store on a rack. These can be refrigerated at 4°C until ready to use (see Note 1).
2. Under sterile conditions (sterile flow hood) remove the cap of the blood tubes and pour the contents into a 50 ml centrifuge tube (Fig. 1a).
3. Centrifuge at 200 × g for 20 min at 4°C to obtain platelet-rich plasma fraction.

Table 1
Based on 2–4 ml of fibrinogen harvested from 25 ml of the whole blood of eight human subjects (unpublished data)

Subject no.	Wet weight of fibrin pellet (mg)	Concentration (mg/ml)	Gelation time (min)	Degradation time (days)
1	570	25	<3	>7
2	456	33	<3	>7
3	653	46	<3	>7
4	554	28	<3	>7
5	521	34	<3	>7
6	756	75	<3	>7
7	481	22	<3	>7
8	334	24	<3	>7
Average	540.63			

Fibrin pellets were isolated from eight human subjects, using the techniques described in this chapter. The weight of each pellet was calculated and there was significant variation in the weight of samples from different participants as would be expected. ELISA (enzyme-linked immunosorbent assay) was also used to determine fibrinogen concentrations. All samples took less than 3 min to gel in the thrombin solution producing the hydrogel, and had a degradation rate of less than 7 days

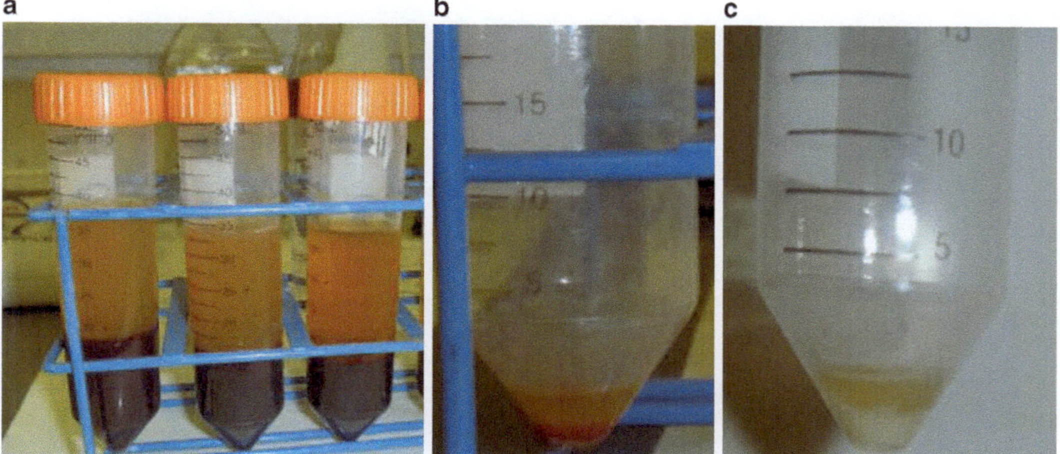

Fig. 1 (**a**) Whole blood plasma separation, (**b**, **c**) fibrinogen concentration step one and two

4. Carefully remove tubes from the centrifuge onto a rack without disturbing the cruor (the cellular fraction of the blood) and plasma fractions (c. 55% of total blood volume).
5. Observe the separation in straw-colored plasma fractions and cruor.
6. Weigh a 50 ml tube and note the reading for use at a later stage. Aspirate the straw-colored serum and transfer to another 50 ml centrifuge tube.
7. Centrifuge again at 400 ×*g* for 20 min to obtain a platelet poor plasma fraction.
8. Discard the supernatant (clear liquid) using a pipette leaving behind the straw-colored plasma (fibrinogen-rich fraction). Do not invert the tube (see Note 2).

3.2 Fibrinogen Concentration

1. Using a pipette add cold ethanol (0°C) to a final concentration 10% v/v in the plasma solution (c. 1–3 ml).
2. Pipette the mixture up and down to ensure the solutions are well mixed.
3. Incubate for 45 min at 4°C. The ethanol causes a chemical precipitation of the fibrinogen out of the plasma solution, so take care to not disturb the tube needlessly. Shaking can redissolve the precipitated fibrinogen (see Note 3).
4. Centrifuge the solution at 800 ×*g* for 15 min at 4°C to pellet the precipitated fibrinogen (Fig. 1b).
5. Carefully aspirate the supernatant using a pipette, taking care not to disturb the pellet. This pellet can be stored at 4°C for up to 24 h in phosphate-buffered saline until ready for use, at which point the centrifugation should be repeated to re-pellet all the fibrinogen (see Note 4).

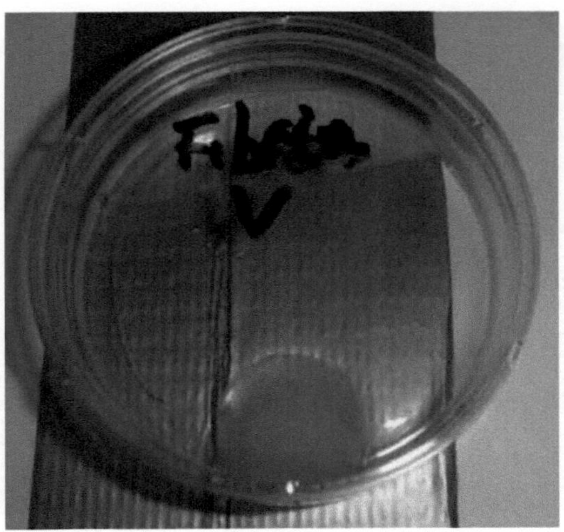

Fig. 2 Human fibrin hydrogel made from a 30 ml sample of blood

3.3 Cell Encapsulation and Gelation

1. Weigh the 50 ml tube and deduct the previous measurement taken to find out the weight of the fibrinogen pellet (see Note 4).
2. In a sterile flow hood, vigorously resuspend pellet in un-supplemented cell culture medium (warmed to 37°C, specific to the cells to be cultured). Depending on the amount of fibrinogen the final concentration should be circa 20–40 mg/ml. Avoid passing excess air through the solution, rendering it more effervescence than solution (see Note 5).
3. Cap the tube and warm the fibrinogen solution to in a sterile incubator set at 37°C for <10 min.
4. In a sterile flow hood, add concentrated cell suspension to the fibrinogen solution, taking care not to dilute the fibrinogen concentration excessively. (For 1 ml of fibrin gel we have successfully encapsulated up to 5×10^5 BMSCs, but this does depend on cell type, with respect to size and preferred seeding density.) Resuspend the cell pellet in the fibrinogen solution and minimize the passage of air into the solution (see Note 6).
5. Add aprotinin solution (c. 3,000 KIU/ml or 2–3 TIU/ml). The addition of aprotinin solution can turn the sample to a milky colored cloudy suspension when fully dissolved (see Note 7).
6. Coat sterile culture plates/tubes with thrombin solution (1,200 U/ml in 10 mM $CaCl_2$) at a ratio of 1 thrombin:3 fibrinogen/cell fraction. Add the fibrinogen/cell solution in a drop wise manner. Swirl/mix twice clockwise and anticlockwise.
7. Incubate for 3 min in a sterile cell culture incubator at 37°C and 5% CO_2. Once the gels have formed (Fig. 2), aspirate any remaining supernatant under sterile conditions and add 1/2 ml of supplemented cell culture medium to cover the gels and re-incubate until ready to use (assay for viability or ECM protein production or tissue engineering) (see Note 8).

4 Notes

1. Using clinical grade sodium citrate anticoagulant tubes is easier and removes the need for preparing syringe tubes with heparin. The authors find that greater consistency can be maintained between samples when using anticoagulant blood tubes as compared with syringes. However when harvesting larger volumes of blood >50 ml the disparity is less so. The syringes can be coated with sodium citrate (7–10% v/v, 10.5 M in distilled water) or heparin at 1,000 U/ml.
2. During the first centrifugation, the cruor of the blood sample will be red colored and settled toward the bottom of the centrifuge tube. The serum (platelet-rich plasma) will be at the top and straw colored. The second centrifugation will produce a platelet poor plasma fraction at the bottom of the tube, which is to be used to isolate fibrinogen.
3. Maintain at cool temperature, and avoid excessive handling which can reduce the sterility of the product and the effectiveness of precipitation.
4. If one is not ready to use the fibrinogen immediately it can be kept for up to 24 h at 4°C. For long-term storage, freezing at −80°C maintains the protein at its optimum.
5. Weigh the fibrinogen pellet in the tube to maintain sterility and reduce the risk of infection. Avoid removing the fibrinogen from the tube as this can result in substantial loss in the amount of fibrinogen available for use. The 50 ml tube must be weighed each time as there can be significant variations even if they are from the same supplier.
6. The authors prefer concentrations of 1×10^5 to 1×10^6 cells/ml for optimum cell growth (bone marrow-derived mesenchymal stem cells, chondrocytes, and fibroblasts); however this value should be optimized for each cell type as they are likely to differ. Cells are pelleted and re-suspended in <2 ml of supplemented media.
7. Aprotinin is an inhibitor of protease activity and slows down the rate of degradation of the fibrinogen, allowing for the temporary extracellular matrix to be maintained for longer. Traditionally it has been sources from bovine lung tissue and has been associated with anaphylaxis in a small subset of the population, where it was used during cardiac surgery. Therefore, the authors recommend animal-free aprotinin, especially if for clinical use, such as the recombinant aprotinin available from *Nicotiana* sp.
8. Encapsulated cells can be used to coat polymer matrices, or can be used to engineer tissues independently without an additional scaffold.

References

1. Hedelin H, Lundholm K, Teger-Nilsson AC et al (1983) Influence of local fibrin deposition on granulation tissue formation. A biochemical study in the rat. Eur Surg Res 15:312–316
2. Mosesson MW (1990) Fibrin polymerization and its regulatory role in hemostasis. J Lab Clin Med 116:8–17
3. Mosesson MW, Kaminski M (1990) The acceleratory effect of thrombin on fibrin clot assembly. Blood Coagul Fibrin 1:475–478
4. Spotnitz WD (2001) Commercial fibrin sealants in surgical care. Am J Surg 182:8S–14S
5. Spotnitz WD (1995) Fibrin sealant in the United States: clinical use at the University of Virginia. Thromb Haemost 74:482–485
6. Spotnitz WD (1997) New developments in the use of fibrin sealant: a surgeon's perspective. J Long Term Eff Med Implants 7:243–253
7. Spotnitz WD (2010) Fibrin sealant: past, present, and future: a brief review. World J Surg 34:632–634
8. Alston SM, Solen KA, Sukavaneshvar S et al (2008) In vivo efficacy of a new autologous fibrin sealant. J Surg Res 146:143–148
9. Alston SM, Solen KA, Broderick AH et al (2007) New method to prepare autologous fibrin glue on demand. Transl Res 149:187–195
10. Dresdale A, Rose EA, Jeevanandam V et al (1985) Preparation of fibrin glue from single-donor fresh-frozen plasma. Surgery 97:750–755
11. Spotnitz WD, Mintz PD, Avery N, Bithell TC, Kaul S, Nolan SP (1987) Fibrin glue from stored human plasma. An inexpensive and efficient method for local blood bank preparation. Am Surg 53:460–462
12. Park MS, Cha CI (1993) Biochemical aspects of autologous fibrin glue derived from ammonium sulfate precipitation. Laryngoscope 103:193–196
13. Weisman RA, Torsiglieri AJ, Schreiber AD et al (1987) Biochemical characterization of autologous fibrinogen adhesive. Laryngoscope 97:1186–1190
14. Kjaergard HK, Weis-Fogh US, Thiis JJ (1993) Preparation of autologous fibrin glue from pericardial blood. Ann Thorac Surg 55: 543–544
15. Chaloupka K, Motwani M, Seifalian AM (2011) Development of a new lacrimal drainage conduit using POSS nanocomposite. Biotechnol Appl Biochem 58:363–370
16. Ahmed M, Ghanbari H, Cousins BG et al (2011) Small calibre polyhedral oligomeric silsesquioxane nanocomposite cardiovascular grafts: influence of porosity on the structure, haemocompatibility and mechanical properties. Acta Biomater 7:3857–3867
17. Ghanbari H, de Mel MA, Seifalian AM (2011) Cardiovascular application of polyhedral oligomeric silsesquioxane nanomaterials: a glimpse into prospective horizons. Int J Nanomed 6:775–786
18. Baguneid M, de Mel MA, Yildirimer L et al (2011) In vivo study of a model tissue-engineered small-diameter vascular bypass graft. Biotechnol Appl Biochem 58:14–24
19. Jungebluth P, Alici E, Baiguera S et al (2011) Tracheobronchial transplantation with a stem-cell-seeded bioartificial nanocomposite: a proof-of-concept study. Lancet 378:1997–2004

Chapter 13

Electrospinning Tubular Scaffolds with Tissue-Like Mechanical Properties and Biomimetic Surface Features

Scott Rapoport

Abstract

Electrospinning is a useful technique for generating scaffolds composed of nano- and/or microfibers. Through an innovative modification of the cylindrical target mandrel it is possible to create macrostructural or microstructural invaginations. The former provides complex bulk mechanical behavior reminiscent of native soft tissue; the latter leads to a surface topography reminiscent of intestinal villi.

Key words Electrospinning, Biomimetic, Scaffold, J-shaped, Expanding mandrel, Tubes

1 Introduction

Some of the first patents describing electrospinning originate from the early twentieth century (1, 2). For example, the Cooley patent describes using "electricity at high tension" to cause "filaments to start out, which quickly set or harden because of the dispersion of their volatile liquid component." To wit, electrospinning is the use of a high-voltage electric field to liberate small-diameter fibers of dry polymer from a formerly homogeneous mixture of polymer and volatile solvent. The key features of the process are an electric field that places a positive charge on a polymer solution and a negatively charged grounded target, which leads to the formation of a stable electrospinning jet of polymer in solution exiting from a needle orifice.

Despite its long history, electrospinning (a term credited to Reneker (3) in 1995) has only recently experienced resurgence in the past 30 years due to applications largely centered in the field of biotechnology. Specifically, there has been great interest in this technique's ability to generate with defined fiber characteristics tubular scaffolds, patches, and other structures, possessing high degrees of biological and physiological relevance (4).

Joydeep Basu and John W. Ludlow (eds.), *Organ Regeneration: Methods and Protocols*, Methods in Molecular Biology, vol. 1001, DOI 10.1007/978-1-62703-363-3_13,

Recently, some of the more novel techniques with wide-ranging applicability have involved generating complex fibers through coaxial spinning (5), the controlled alignment of fibers during deposition (6), and mixtures that incorporate living cells and polymer (7).

The technique described in this chapter was created in the spirit of developing novel electrospinning techniques in order to meet the burgeoning complexity of scaffold requirements attributed to the desire to create scaffolds that are more tissue-like, instructive to cells that interact with them, and in possession of mechanical properties reminiscent of the native tissues that they are designed to augment (8).

In example, some research groups are working with multilaminate (9) and composite tubular scaffolds (10) whose purpose is to mimic soft tissue complex mechanics found in certain native tissues such as blood vessels. In these cases the researchers are utilizing novel techniques of scaffold assembly to form mechanically responsive materials with the hope of recapitulating the sequential activation of both collagen and elastin laminas seen expressed in the aforementioned tissue.

Similarly, modifications to the electrospinning setup are possible in such a way as to create complex laminated tubular scaffolds whose constituent laminas are activated sequentially, rather than in parallel during distention of the scaffold. In this fashion, a way to approach native tissue mechanical behavior which is based on complex interactions of constituent materials and laminas has been developed.

Here I show how an innovative modification of the target mandrel—the ability to expand—can be employed in a two-material system to create both topological features of biological relevance, as well as the aforementioned sequential material activation.

1.1 What This Technique Accomplishes

The technique employed here can serve two functions: (1) it can generate small scale invaginations that mimic the size and gross appearance of intestinal villi (Fig. 1a–d), and (2) it can create macroscale invaginations (Fig. 2e, f) that can be employed to impart complex mechanical behavior in response to applied strain. Schematically, the latter case is presented specifically in Fig. 2 although the technique is largely identical for generation of any scale of invaginations.

2 Materials

Although one polymer system is utilized here, a wide range of polymers, both synthetic and naturally occurring, are amenable to the technique described in this chapter. Electrospinning solvents

Fig. 1 Representative prototypes of tubular scaffolds with strain-induced invaginations. Panels (**a–d**) represent an early prototype where created invaginations were much finer (scale bar in **a/b** is 700 μm (abluminal surface to the *right*, luminal surface to the *left*); scale bars in **c/d** are ~6 mm). Panels (**e** and **f**) represent the current prototypes in which the surface features are much more pronounced (scale in **e** is 8 cm, **f** is 6 mm)

tend to be carcinogens and should be handled by trained personnel within approved fume hood devices. Bulk polymers have differing storage requirements, but biodegradable polymers tend to experience moisture-catalyzed degradation and should be handled accordingly. Prepared polymer-solvent solutions maintained at room temperature should be utilized as soon as possible to avoid polymer degradation as evidenced by a change in color of the original solution.

2.1 Selection of Primary Component (Elastic)

The first electrospun coating must consist of an elastic material. Depending on the application, a wide variety of materials (see examples in Table 1) are available from synthetics to even some naturally occurring materials—although the naturally occurring materials do not tend to reconstitute with sufficiently robust elastic

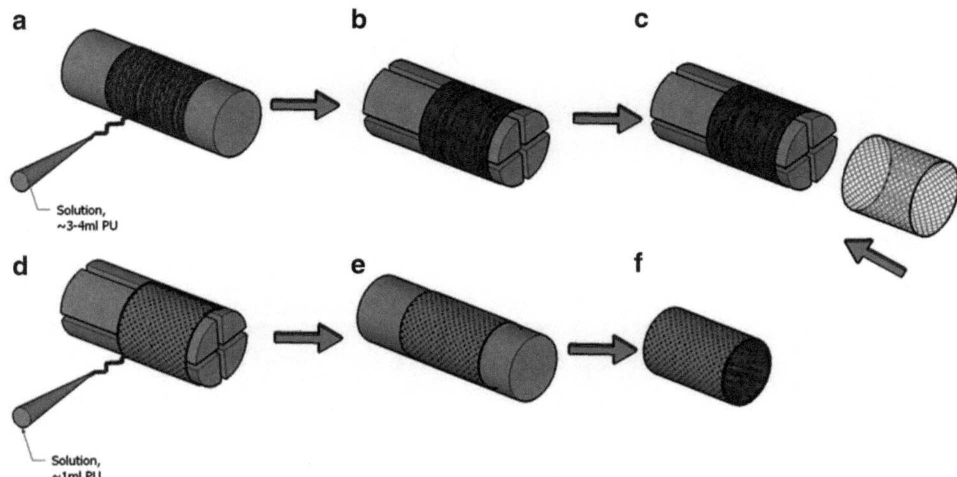

Fig. 2 Formation of the tubular structures seen in Fig. 1e, f proceeded as follows: (**a**) An elastic material is electrospun onto a rotating mandrel at a diameter of Do. (**b**) The mandrel's diameter is increased to ~120% of its original value, Df. (**c**) Polymeric mesh tubing of internal diameter, Df, is now placed around the mandrel and the elastic material. (**d**) To bind the mesh with the underlying tube, a thin coat of elastic material is applied via electrospinning. (**e**) The final step involves returning the mandrel to initial diameter, Do, and removing the scaffold. In this fashion, a prestressed laminate is created. (**f**) Invaginations (kinking) form in the circumferential direction caused by a forced diameter reduction in the larger diameter mesh tube by entraining it with the smaller-diameter elastic inner tube

Table 1
Some sample elastic materials

Naturally occurring	Synthetic
Elastin	Polyurethane (PU)
Resilin	Poly(lactide-co-epsilon-caprolactone) (PLCL)
Abductin	Polydioxanone (PDO)
Some silk varieties	Polyester urethane urea (PEUU)

properties. It's important to select an elastic material that has a good restorative/recoil characteristic once electrospun. Some trial and error may be required to identify the desired material and properties.

For the purposes of this exercise, medical grade polyurethane (PU), Tecothane TT1074A, was obtained from Lubrizol, Inc. (Cleveland, OH).

2.2 Creation of Electrospinning Working Solution of Elastic Component

Working solutions for electrospinning consist of a desired polymer at a percent concentration in a volatile solvent or in a mixture of volatile solvents (see examples of common solvents in Table 2). Like material selection, there are numerous solvents available with

Table 2
Some solvents commonly utilized in electrospinning

Solvent	Abbreviation
1,1,1,3,3,3-Hexafluoro-2-propanol	HFIP
Methylene chloride	MC
Tetrahydrofuran	THF
Toluene	T
Formic acid	FA
Chloroform	CHLF

various polarities, vapor pressures, and other characteristics. Some experimentation may be required to find a solvent that works best for the desired application.

For the purposes of this exercise, HFIP (Sigma-Aldrich, St. Louis, MO) was utilized. Begin by mixing a 10% concentration of PU (which comes in pellet form) in 10 ml HFIP. For utility in mixing, a suitable receptacle is a 20 ml glass scintillation vial with lid (Sigma-Aldrich, St. Louis, MO). Seal the lid and affix the vial to a nutating platform to facilitate rapid dissolution of the PU. It may take up to an hour to obtain a homogenous solution.

2.3 Selection of Secondary Component

Depending on the intended outcome, the second material utilized can be tensile or elastic in nature, something that is either added by electrospinning or through another technique. For example, in one embodiment of the technique a tensile mesh was utilized as a secondary material and then attached to the initial material with a sandwich of the primary electrospun material. For the purposes of this exercise, PU was also utilized as the secondary component, mixed as described in the preceding section.

2.4 Electrospinning Target

The key feature of an electrospinning target is that it should be grounded in order to complete the circuit involving the positively charged polymer solution. Targets can be a variety of surfaces including cylindrical forms and flat plates. The technique detailed in this chapter involves a custom-made cylindrical mandrel device (Custom Design & Fabrication, Richmond, VA) with independently controlled motors for rotational speed and translation speed with respect to the axis of the impinging jet of electrospun fibers (Fig. 3a). The cylindrical mandrel was further modified to be expandable (Fig. 3d) by Omni Tool, Inc. (Winston-Salem, NC).

2.5 High-Voltage Power Source

A high-voltage (DC) low-amperage power supply is required (Spellman, Hauppauge, NY). Upper-range voltage need be at most

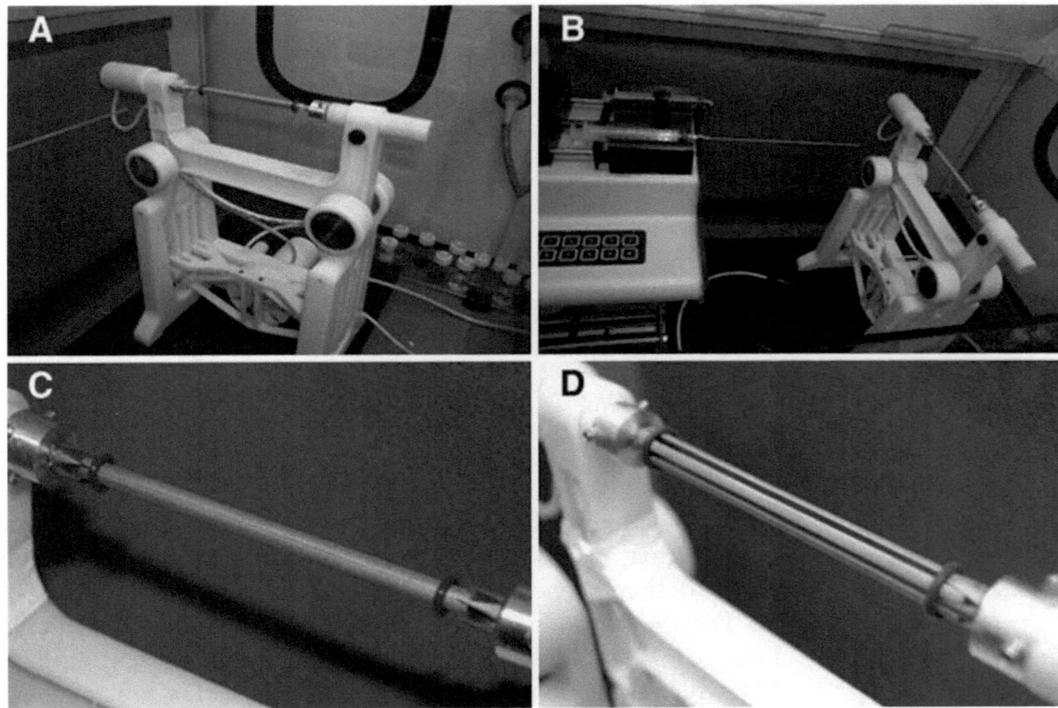

Fig. 3 Electrospinning setup and expanding mandrel device. (**a**) Target holding device that is capable of rotation and translation (mandrel is attached) while maintaining an electrically grounded connection. (**b**) Electrospinning setup consisting of a syringe pump and grounded target. (**c**) Close-up of expanding mandrel in collapsed state (OD ~6 mm). (**d**) Close-up of mandrel in expanded state (OD ~13 mm). The wedge-shaped end pieces are capable of driving the expanding mandrel's segments apart thus increasing the diameter by ~120%

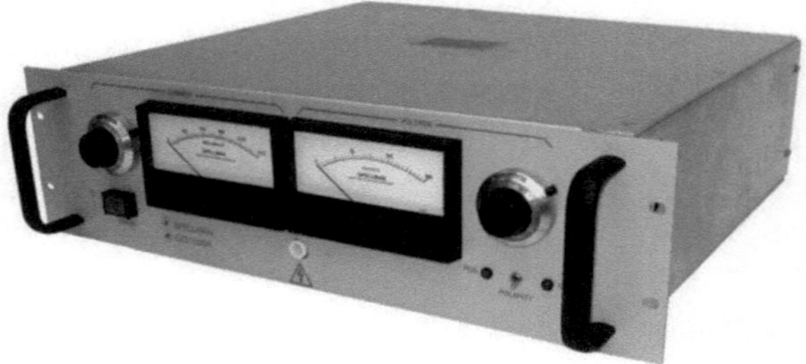

Fig. 4 Spellman's CZE100R rack mountable power supply

30 keV. Units that possess a degree of resolution through the utilization of a dial-based system tend to be highly operator-friendly (Fig. 4).

2.6 Syringe Pump/ Stand

A simplistic system only requires the delivery of one syringe's contents (i.e., not mixes of solutions), so a single syringe pump would suffice (Cole-Parmer, Vernon Hills, IL; Harvard Apparatus, Holliston, MA; and others). Highly viscous solutions might require

Fig. 5 Basic 6″ × 6″ telescoping lab stand

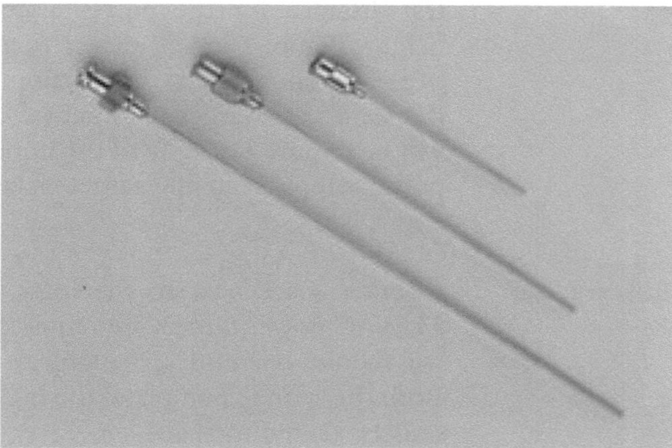

Fig. 6 Standard blunt end, hypodermic needles are ideal for electrospinning

a high-pressure syringe pump. A telescoping stand is recommended to support the syringe pump (Fig. 5) and allow vertical adjustment of the electrospinning jet. VWR (Radnor, PA) is a typical supplier.

2.7 Delivery Needle/Syringe Selection

Most articles involving electrospinning in the literature utilize 18G hypodermic needles. Different sizes may be utilized, but smaller bores may lead to difficulties with hyper-viscous solutions. Larger bore needles may experience excessive flow that will likely overwhelm the draw of the electric field. Hypodermic needles should be metal and blunt end. The former guarantees that a proper circuit is formed; the latter avoids asymmetries applied to the exiting solution that could disrupt your electrospinning jet (Cadence, Inc., Staunton, VA) (Fig. 6).

For exploratory work, standard tuberculin 10 ml syringes are suitable (e.g., BD Biosciences, Sparks, MD). Transparency of the syringe is useful for allowing visualization of your solution volume during electrospinning. For experiments with more stringent processing requirements, a glass syringe (no polymeric parts) is recommended (KD Scientific, Holliston, MA) as solvent-induced leaching from the syringe into your solution is a danger.

3 Methods

Those experienced with electrospinning will recognize that numerous variables affect the quality of your outcome: humidity, viscosity, electric field strength, pump speed, hypodermic needle size, and others. As such, the operator following these directions must be prepared to make adjustments to the system in order to achieve favorable results.

3.1 Preparation: Setup

A good setup requires some iteration to determine the best neutral distance for the pump-target distance and pump-ground height. Figure 7 is a crude cartoon showing placement of several of the components: syringe pump, syringe and needle, telescoping stand, and target mandrel (also, see Fig. 3b). Counter space will be required outside of the hood to support a power strip, the power supply, and the target mandrel's rotational/translational controls.

3.1.1 Equipment Positioning

Counter space next to the fume hood and sufficient access to 115VAC power outlets are required. The grounding of the target will require one outlet; the target's motors require another outlet. Both the syringe pump and the high-voltage power supply require outlets. Outlet total is 4.

The syringe pump should be raised slightly higher than the target mandrel to account for the downward trajectory of fibers travelling through the field under the influence of gravity. This can be adjusted by turning the knob on the telescoping stand. An increase or decrease in the distance from the syringe to the mandrel will require "tuning" of the syringe pump height.

Place the mandrel 11 cm from the tip of the hypodermic needle. This can be accomplished by sliding the telescoping base.

3.2 Electrospinning Inner Elastic Tube

1. Load 10 ml of solution mentioned in Subheading 2.2 into a 10 ml syringe.
2. Attach hypodermic needle.

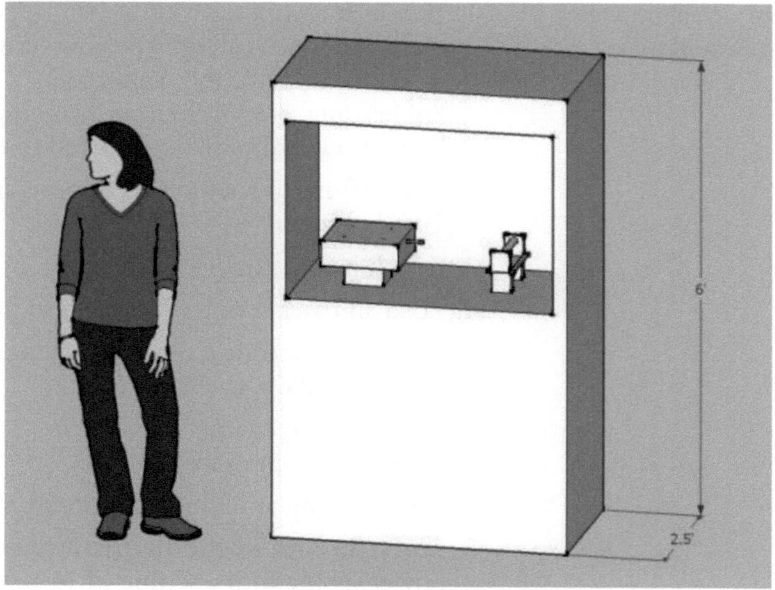

Fig. 7 Positioning of syringe pump (L), telescoping stand (that supports syringe pump), and target mandrel (R)

3. Load and lock syringe + needle into pump fixture.
4. Connect electrode from high-voltage power supply to hypodermic needle via alligator clip.
5. Set and engage target mandrel at a rotation of 2,000 rpm and a translation speed of 100 Hz.
6. Set the pump at 15 ml/h, and start infusion.
7. Leave the area of the electric field and press the turn on the power supply with a setting of 14 keV.
8. See Subheading 4 on typical issues that arise and make corrections accordingly ensuring a stable electrospinning jet on the outflow of the hypodermic needle.
9. Electrospin 3 ml of polymer solution before turning off power supply, pump, and mandrel device (in order as listed).
10. Allow some drying time (1–2 h) for any potential solvent residuals.

3.3 Preparation for the Addition of Outer/Second Layer

1. If you have the expanding mandrel device, turn the tightener to the mandrel fastener clockwise until the blades of the mandrel spread to 13 mm (~120% increase from original 6 mm diameter seen in Fig. 3c).
2. If you do not have the expanding mandrel, you can obtain the same effect by carefully removing the electrospun tube and stretching it onto a larger diameter mandrel.

3.4 Electrospinning of Outer/Second Layer

1. If you are working with a material in the form of a precast mesh tube, you can now position it onto the expanded elastic tube described in Subheading 3.3 and proceed to Subheading 3.5. If you are planning on adding the second layer via electrospinning, proceed to number 2.
2. Load 10 ml of solution mentioned in Subheading 2.2 into a 10 ml syringe.
3. Attach hypodermic needle.
4. Load and lock syringe + needle into pump fixture.
5. Connect electrode from high-voltage power supply to hypodermic needle via alligator clip.
6. Set and engage target mandrel at a rotation of 2,000 rpm and a translation speed of 100 Hz.
7. Set the pump at 15 ml/h, and start infusion.
8. Leave the area of the electric field and press the turn on the power supply with a setting of 14 keV.
9. See Subheading 4 on typical issues that arise and make corrections accordingly ensuring a stable electrospinning jet on the outflow of the hypodermic needle.
10. Electrospin 2 ml of polymer solution before turning off power supply, pump, and mandrel device (in order as listed).
11. Allow some drying time (1–2 h) for any potential solvent residuals.
12. Proceed to Subheading 3.5, step 8.

3.5 Final Processing of Scaffold

1. If arriving from Subheading 3.4, step 1, the mesh tube will need to be bound to the underlying tube with a thin layer of electrospun fiber deposited semi-wet.
2. Move the pump forward so that the needle-to-mandrel distance is now 6 cm.
3. Set the pump at 15 ml/h, and start infusion.
4. Leave the area of the electric field and press the turn on the power supply with a setting of 14 keV.
5. See Subheading 4 on typical issues that arise and make corrections accordingly ensuring a stable electrospinning jet on the outflow of the hypodermic needle.
6. Electrospin 2 ml of polymer solution before turning off power supply, pump, and mandrel device (in order as listed).
7. Allow some drying time (1–2 h) for any potential solvent residuals.
8. Loosen the tightener to the mandrel fastener. This will have the effect of allowing the expanded blades to collapse. The formation of invaginations should be immediately apparent.
9. Carefully remove the scaffold from the mandrel.

4 Notes

This section deals with many of the common issues that can affect the quality of electrospun fibers. Many issues are easily corrected, and as skill of the operator increases, they become more easily recognizable.

1. *Voltage Flux*: At times, a periodic flux in the voltage field will manifest as a repeating cycle consisting of a building in strength of the electric field followed by a sudden grounding. This will be evident in the voltage gage of the power supply which will begin to oscillate from low to high values. This indicates that your voltage is set too high and that dielectric breakdown is occurring somewhere in the system resulting in sparking. *Solution*: Turn the voltage knob down and then slowly bring it back up to a value less than the previous value. It should now stabilize.
2. *Syringe Fouling*: It is common to witness buildup of material at the tip of the hypodermic needle. This buildup can interfere with proper electrospinning jet formation and should be removed. *Solution*: The best way to remove this is to utilize a nonconductive material such as wood as a wiper. Kimwipes or the like can be taped to one end of the stick in order to give a clean wipe of the hypodermic needle. This stick should be kept in the fume hood as it will eventually contain solvent polymer-impregnated solvent. A rectangular material with a thin width is ideal. One skilled in the art will be able to disengage the electric field and swiftly wipe the hypodermic without causing a major interruption to the current scaffold construction session.
3. *Pump Power Loss*: At times, the syringe pump may suddenly stop working (i.e., the screen may freeze or go blank when power is still obviously on). In this case an electrical ground has traveled back through the pump incapacitating its circuitry. *Solution*: merely turn the pump off for a few seconds and then turn it back on. Normal pump function should return. In case this technique does not work, check the pump's fuse located near the power cord insertion plug. Replace as necessary. If the issue continues frequently, check the isolation of the pump from the electric field circuit.
4. *Sputtering/Dripping*: Sputtering of the solution into the electric field does not generate fibers, but rather blobs of material that often fall short of the target. This is caused by the following:
   a. Electric field is too low; *Solution*: Turn up to higher voltage.
   b. Nonhomogeneous mixture; *Solution*: ensure that no particulates are present in solution or prefilter solution.
   c. Low viscosity; *Solution*: try increasing the concentration in 2% increments until resolved.

5. *Spiderman*: This is a facetious name to describe an interesting situation in which conditions are hyper-optimal. The operator may notice that the airspace is filled with a high density of fibers. The effect is sudden and potentially dangerous because it is now possible that these threads can interrupt the electric field and ground the voltage to a nearby object (including the operator). The exact cause is believed to be related to a high extrusion speed coupled with a high voltage and optimal viscosity. *Solution*: Immediately cut the power and turn off the syringe pump. Clean the area of all fibers, and start again at lower voltage, or change one or more of the other parameters.
6. *Pump Grounding*: This situation differs from "pump power loss." In this situation, the fibers do not proceed to the desired target; rather, they blow back onto the syringe pump. This can also happen with other objects in and around the electrospinning workspace. *Solution*: One useful solution is to utilize longer hypodermics—6" for example. In this fashion the actual pump is further back from the target and less likely to compete as a ground source with the mandrel. Another option is to attempt to electrically isolate the pump and/or other undesired grounds by placing a layer of dielectric (e.g., plastic) onto the surface where the fibers are alighting.
7. *Fused Fiber Mat*: A fused fiber mat is the evidence that fibers are landing at the target still wet and consequently merging with other fibers that have previously landed. This occurs because the fibers have not fully dried before landing at the target.

   This can often be caused by too much relative humidity, or too small of a gap between the needle and the target. *Solution*: If the same setup has worked in the past, it's possible that humidity may be an issue. It might be useful to track temperature/humidity every time electrospinning is conducted. Persistent problems may require humidity control. A more simple solution is available if the difference in fiber quality can be traced to a change in the needle-to-target distance. Try increasing this distance (and the electric field as well) until fibers are dry upon encountering the target.
8. *Safety Concerns*: Two major safety concerns exist: electric shock and chemical exposure.

   The operator will be working near a high-voltage field. High voltage alone is not necessarily lethal in the absence of preexisting medical conditions or some sort of shock susceptibility (consider deaths by Taser).The amperage is of concern. Amperage should be set as low as possible in order to generate the desired field strength. A lethal dose of amperage can be lower than 100 mA. Any breaks in the electrode's insulation can lead to sparking from the dielectric breakdown of air at high enough voltages. The operator must be cautious around all electrical components. A containment box can be a good safety measure because a switch can be installed that cuts power to the power source when the lid is opened for user access. Lastly, a good investment would be a rubber floor mat to ensure that minimal grounding of the operator occurs.

Fumes from volatile solvents are also of concern. Some of the solvents currently utilized in electrospinning are as follows: ether, hexafluoroisopropanol, methylene chloride, and others, all possess varying degrees of deleterious effects to human biology. Some could cause loss of consciousness when inhaled; others can burn tissue, or create long-term carcinogenic dangers. In light of this, the user should never be in a position that would lead to exposure within the fume hood. Mixed solutions should be kept inside the fume hood at all times. Freshly electrospun materials should remain in the fume hood to ensure residual solvent has evaporated. Additionally, electrospun materials can also be placed in low-pressure chambers (e.g., desiccation chambers) to further encourage complete evaporation of solvent.

References

1. Cooley FJ (1902) US patent no. 692,631
2. Morton WJ (1902) US patent no. 705,691
3. Doshi J, Reneker DH (1995) Electrospinning process and applications of electrospun fibers. J Electrostat 35:151–160
4. Bhardwaj N, Kundu SC (2010) Electrospinning: a fascinating fiber fabrication technique. Biotechnol Adv 28:325–347
5. Jiang H, Hu Y, Li Y, Zhao P, Zhu K, Chen W (2005) A facile technique to prepare biodegradable coaxial electrospun nanofibers for controlled release of bioactive agents. J Control Release 108(2–3):237–243
6. Aviss KJ, Gough JE, Downes S (2010) Aligned electrospun polymer fibres for skeletal muscle regeneration. Eur Cell Mater 19:193–204
7. Townsend-Nicholson A, Jayasinghe SN (2006) Cell electrospinning: a unique biotechnique for encapsulating living organisms for generating active biological microthreads/scaffolds. Biomacromolecules 7(12):3364–3369
8. Rapoport HS, Fish J, Basu J, Campbell J, Genheimer C, Payne R, Jain D (2012) construction of a tubular scaffold that mimics J-shaped stress/strain mechanics using an innovative electrospinning technique. Tissue Eng Part C Methods. Epub ahead of print PMID: 22250785
9. Soletti L, Hong Y, Guan K, Stankus JJ, El-Kurdi MS, Wagner WR, Vorp DA (2010) A bilayered elastomeric scaffold for tissue engineering of small diameter vascular grafts. Acta Biomater 6(1):110–122
10. Zhu C, Fan D, Duan Z, Xue W, Shang Lm Chen F, Luo Y (2009) Initial investigation of novel human-like collagen/chitosan scaffold for vascular tissue engineering. J Biomed Res A 89(3):829–840

Chapter 14

Design, Fabrication, and Preparation of Synthetic Scaffolds for Urologic Tissue Engineering

Richard G. Payne and Toyin A. Knight

Abstract

This chapter describes the fabrication of a polyglycolic acid (PGA)-based scaffold used to tissue engineer a Neo-Urinary Conduit™.

Key words Nonwoven mesh, Polyglycolic acid, Urinary diversion

1 Introduction

1.1 Scaffolds

The role of implanted biomaterials has evolved from merely providing a passive structural framework within the body (as for example with the use of gold or porcelain to repair dental cavities) towards more directly facilitating repair or regeneration through the development of biologically active scaffolds. This objective can be accomplished by inclusion of defined cell populations to form regenerative cell/scaffold complexes capable of facilitating the deposition of extracellular matrix (ECM). Such bipartite combination constructs may additionally mediate formation of a regenerative milieu to catalyze induction of neo-tissues or neo-organs. Ultimately, scaffolds may replace cells altogether within regenerative constructs by incorporating elements of ECM, signaling factors, and cytokines capable of manipulating the behavior of host-derived cells in the absence of any ectopic cellular component.

As currently understood, the roles of scaffolds within tissue engineering/regenerative medicine (TE/RM) products today may include:

- Providing space for tissue repair and regeneration
- Providing a foundation for the expansion and delivery of therapeutic exogenous cell populations

Joydeep Basu and John W. Ludlow (eds.), *Organ Regeneration: Methods and Protocols*, Methods in Molecular Biology, vol. 1001, DOI 10.1007/978-1-62703-363-3_14, © Springer Science+Business Media New York 2013

- Serving as a framework for the deposition of ECM and paracrine signaling factors
- Providing a foundation for the regeneration of neo-tissue and neo-organs in a manner appropriate to the local microenvironment

In this manner, scaffolds act to recapitulate aspects of the endogenous ECM. In contrast to the inert tooth fillings mentioned previously, in the field of TE/RM, scaffolds are designed to be fully biodegradable upon implantation within the body with rates of degradation predetermined by the needs of the particular application under consideration.

A common classification of scaffold biomaterials is based on source, which may be of natural or synthetic origin. Examples of naturally occurring scaffold biomaterials include gelatin, fibrin, hyaluronic acid (HA), chitosan, silk, collagen, and alginate. Such naturally derived biomaterials are typically well tolerated upon introduction within the body and may possess many of the physiomechanical properties of native ECM. However, from a process development and manufacturing perspective, naturally derived biomaterials are potentially problematic, presenting difficulties in sourcing, quality control, reliability, and reproducibility from lot to lot. In contrast, synthetic scaffold biomaterials such as poly(glycolic acid) (PGA), poly(lactic-*co*-glycolic acid) (PLGA), and poly(l-lactic acid) (PLLA) offer better reliability, reproducibility, and tunability of physical properties, greatly facilitating process development and manufacturing.

1.2 Urologic Organ Regeneration

The bladder is a foundation platform for TE/RM methodologies related to tubular organs (1). At its most fundamental level, the bladder is responsible for mediating storage and subsequent efflux of urine within a dynamically expandable and contractable container. Though relatively simple in terms of overall histology and structural organization, diseases affecting the bladder have the potential to significantly affect individual quality of life, resulting in continual incontinence or inability to effectively void urine as needed. Several congenital anomalies may result in abnormal bladder development requiring surgical intervention, including posterior urethral valves, bilateral ectopic ureters, bladder extrophy, cloacal extrophy, and spina bifida (myelomeningocele). The resultant clinical outcomes include incontinence and increased risk of renal failure from high pressures in the genitourinary system. The current standard of therapy for pediatric patients is bladder augmentation through enterocystoplasty, a procedure which involves the surgical removal of a section of large bowel that is then connected to the existing bladder to increase compliance, decrease pressure, and improve overall urine capacity. These surgeries are relatively complex and costly. Even in patients with good technical outcomes, the procedure is associated with numerous immediate risks and potential chronic complications. A similar surgical procedure is performed in adults requiring bladder replacement, typically secondary to the onset of bladder-related malignancies. To this end, cancer of the bladder may be manifested as a broad spectrum of disease presenting across distinct bladder compartments.

In some cases of severe bladder cancer or other pelvic or abdominal cancers, removal of the entire bladder is indicated. In these circumstances, current standard of care also involves reconstruction of a bladderlike replacement using bowel tissue. Application of bowel tissue for reconstruction of urinary neo-organs is clearly problematic for a number of fundamental reasons, including the fact that bowel is a principally absorptive organ, whereas bladder is designed to store and excrete urine. Exposure of bowel tissue to urine or bladder tissue to bowel-derived microorganisms has the potential to trigger multiple secondary complications. These may include bowel complications, absorption issues, infection, stone formation, mucus secretion, and (ironically) induction of cancer. In addition to cancers and developmental abnormalities, patients may present with neurogenic bladder or dysfunctional bladder due to some form of neurologic disease or condition. Treatment may often require an augmentation of the bladder in order to relieve high pressures and incontinence. Current therapies for neurogenic bladder include medical management through a combination of medication and clean intermittent catheterization and, in advanced cases, surgery. Surgical procedures, such as bladder augmentation, are often considered when other medical and less-invasive treatments fail to adequately lower bladder pressure or reduce the frequency of incontinent episodes.

Ultimately, it is self-evident that the ideal unit of anastomosis for urinary-like tissue is other urinary-like tissue. However, the lack of such material has generally precluded the widespread leveraging of this option. There is therefore clearly a compelling medical need for an improved approach that would eliminate or at least substantially reduce the complications potentially associated with the current standard of care. To this end, identification of bladderlike materials that may be applied towards bladder reconstruction in place of bowel tissue has been attempted. Pilot experiments in 1917 to augment bladder in dogs leveraged fascia (2), since then numerous scaffold candidates have been evaluated including skin, bladder submucosa, small intestine submucosa, omentum, dura, peritoneum gelatin, collagen, polyvinyl sponge, Teflon, Vicryl™, and silicone (3). Failure to achieve successful outcomes with such scaffolds may be attributed to physical or mechanical failure, lack of biocompatibility, and the induction of fibrosis and scarring leading to contraction of the implant and reduction in effective volume over time. Compared to such scaffold candidates, use of the patients' bowel tissue, harvested intraoperatively, has resulted in the most satisfactory outcomes. This established standard of care requires the resection and manipulation of GI tissue, consequently exposing the principally absorptive GI tissue to urine. This considerably increases the potential for infection and additional acute and chronic complications. There is therefore a clear and present need for additional, novel technology platforms.

TE/RM approaches offer an alternative, potentially superior methodology to the use of bowel tissue for urinary diversion or replacement. In this methodology, the patient's own cells would be sourced from a bladder biopsy or an alternate source (4) and applied to an appropriate degradable scaffold to create a neo-organ or organ-like construct that, upon implantation within the patient and anastomosis to native components of the urinary system, would lead to regeneration of functional, urinary-like neo-tissue. Such a cell/biomaterial construct would catalyze the regeneration of urinary-like neo-tissue recapitulating native, laminarly organized bladder wall histo-architecture composed of a luminal urothelial layer and multiple smooth muscle layers, appropriately vascularized and innervated (1, 5). Regeneration of urinary-like neo-tissue would be accompanied by progressive degradation of the biomaterial, such that a seamless transition is achieved between the degrading biomaterial and the regenerating urinary-like neo-tissue.

From a scaffold development perspective, the application of synthetic biopolymers such as PGA for the seeding of urothelial and bladder-derived smooth muscle cells has permitted development of modified polymers with continuously tunable physical and mechanical characteristics. To this end, the rate of polymer hydrolysis may be manipulated by altering the nature and sequence of individual monomer units. In addition, coating by other polymers such as PLGA may be applied to further fine-tune the physical properties of the biomaterial scaffold. Finally, the open, fibrous networked structure of the biomaterial (see Fig. 1) facilitates angiogenesis and neovascularization of the developing neo-organ. Taken together, this binary cell/synthetic biopolymer construct represented the foundational technology platform needed for initiating large animal clinical trials and proof-of-concept trials in man.

In the first such study, seven pediatric patients presenting with myelomeningocele (a form of spina bifida) were recruited to receive the first ever human neo-organ implants. Both urothelial and smooth muscle cells were isolated and expanded from autologously sourced bladder biopsies. For each patient, both cell types were seeded onto scaffolds, which were implanted, attaching the neo-bladder to the patient's existing neurogenic bladder. Engineered neo-bladders were found to functionally rescue urologic dynamics and were associated with trilaminar bladder wall architectures upon histological examination of bladder biopsies recovered at 31 months post-implantation. A number of different scaffold

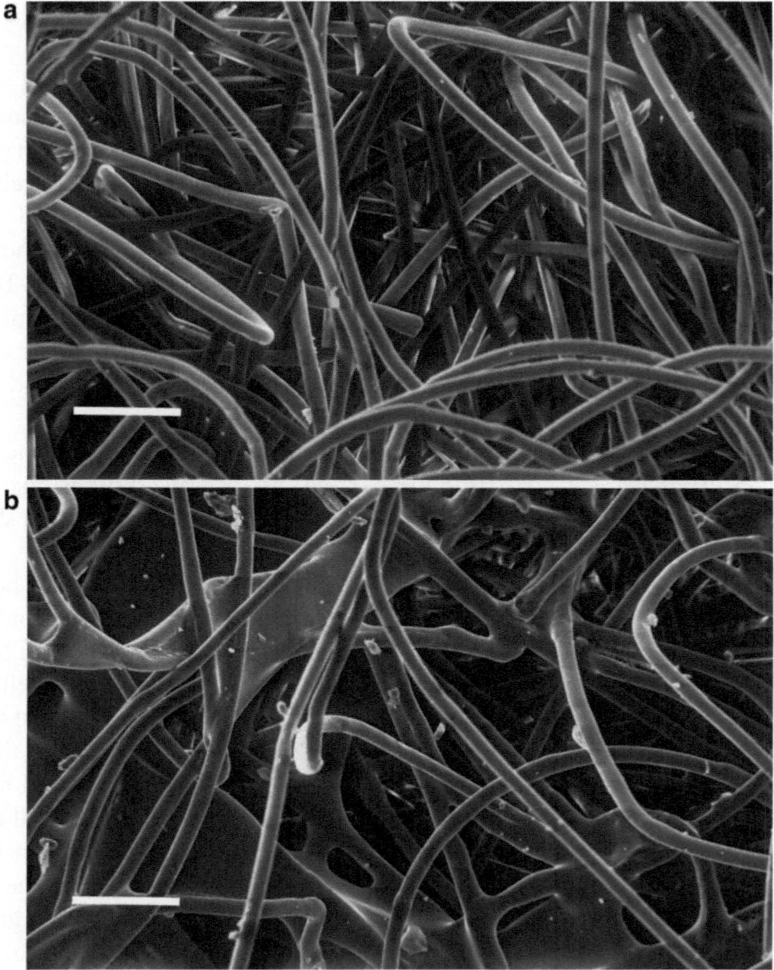

Fig. 1 Scanning electron micrographs of the uncoated PGA mesh (*left panel*), and PGA mesh coated with PLGA (*right panel*). The entangled, nonwoven nature of the fibers is apparent and the PLGA coating is visible mainly at the fiber junctions. Scale bar is 0.1 mm

iterations were evaluated, with changes being made over the course of the study to accommodate new data being made available from this and other related studies. Ultimately, an omentum wrapped, collagen/PGA scaffold was found to present best overall regenerative outcomes (6).

During efforts to transition this technology through the clinical trial process, alternative strategies for mediating efflux of urine from the body were considered. As we have discussed, current surgical treatment options for many urinary disorders caused by congenital conditions, injury, or cancer involve the application of gastrointestinal tissue (GI) to create an orthotopic neo-bladder or urinary diversion. Ideally, urinary diversion would be possible without the use of autologously sourced GI tissue; for this purpose, a TE/RM approach might provide one possible solution.

The possibility of creating a TE/RM product facilitating urinary diversion without utilizing native GI tissue was the rationale underlying the development of the Neo-Urinary Conduit (NUC™). In its simplest configuration, the NUC is a cell/scaffold construct that, upon implantation within the body and attachment to native ureters, allows efflux of urine from the kidneys directly to the external surface of the body. As with the Neo-Bladder Replacement and related TE/RM products, the NUC construct serves as a template to catalyze the regeneration of native-like urinary tissue concomitant to degradation of the biomaterial scaffold following implantation.

1.3 Scaffold Design Considerations

When designing a scaffold such as the one used in the NUC, attributes such as overall dimensions, mechanical properties, microstructure, and degradation profile should be considered, and each is discussed below.

The dimensions of the scaffold must be appropriate for the anatomy associated with the target application—the scaffold needs to fit in the area. Bear in mind that the overall dimensions may change as tissue is formed and remodeled. Some cell types, such as fibroblasts, can contract a scaffold, while mechanical stimulation during regeneration can lead to tissues that are larger than the initial scaffold (7). Dimensions of the tubular scaffold used for the NUC are 200 mm long, 20 mm diameter, with a 3 mm wall thickness.

There are many properties that can be controlled during the manufacture of the PGA felt used in the fabrication of the NUC. These include initial mechanical strength, overall thickness and bulk density, porosity, and surface area. These properties are all interrelated and although their interaction is not always straightforward, felt manufacturers are adept in meeting design specifications. Mechanical properties may be important when the scaffold is expected to have a load bearing application. Felt thickness will impact the ultimate thickness of the regenerated tissue and, along with porosity, influence diffusion to and from cells seeded throughout the scaffold. This diffusion is critical to the viability of the cells at different positions within the scaffold, both before and after implantation, as it allows for transport of nutrients to and waste products from the cells. The bulk density, expressed in units of mg/cc, measures the "loft" of the material, which is how tightly packed the fibers are. For a given fiber diameter, a lower bulk density corresponds to more space between the fibers. In addition to impacting the pore structure of the material, the bulk density will affect the quantity of acidic degradation products that is produced from the scaffold within a certain tissue volume. Fiber surface area will determine the number of cells that scaffold can be expected to support after seeding.

The degradation profile of the scaffold is another consideration. When discussing resorbable polymers, there are three separate measures of degradation—loss of strength, loss of molecular weight, and loss of mass. Unlike a foam or woven structure, where the mechanism of tensile failure is fracture of the material, the PGA felt (a nonwoven web) can also fail by individual fibers sliding past one another. As such, the PGA felt loses strength and mechanical integrity more quickly than mass or molecular weight. In other applications, changing base polymers to those with longer mass and molecular weight degradation times would typically correspond to lengthening the strength degradation profile. In this case, however, changes which increase the interfiber entanglement may have more of an impact on increasing the strength degradation time.

We have chosen to coat the PGA felt tube with PLGA. The purpose of the coating is twofold: it increases the stiffness of the felt and it extends the degradation time. The reason for this is that the coating method employed deposits the PLGA primarily at the fiber junctions, as seen in Fig. 1. This reduces the movement of the fibers in relation to one another (increasing stiffness and delaying strength loss due to fiber movement). Increasing the stiffness of the felt is important to ensure that the tubular conduit remains patent during initial tissue infiltration.

The remainder of this chapter describes the fabrication, sterilization, storage, and two methods of preparation of a PGA-based composite NUC scaffold for use in urinary tissue engineering. However, these methods are applicable to a variety of scaffold designs and organ types.

2 Materials

2.1 Scaffold Fabrication

1. Polyglycolic acid (PGA) BIOFELT® nonwoven tube (Biomedical Structures, Warwick, RI, USA) (see Note 1).
2. PLGA solution: 50:50 Poly(DL-lactide-*co*-glycolide) (PLGA) (Lactel Absorbable Polymers, Birmingham, AL, USA, catalog #B6010-2) (see Note 2) and Methylene Chloride (Spectrum Chemical, New Brunswick, NJ, USA).
3. Resorbable suture, such as Vicryl™ 4-0 undyed, braided (Ethicon Brand, McKesson J415H).
4. Scissors.

2.2 Prewetting (Preparation for Cell Seeding)

1. Ethanol (Aaper Alcohol & Chemical, Detroit, MI, USA, Ethyl Alcohol, 200 proof).
2. Phosphate-buffered saline (PBS) 1×.
3. Vacuum desiccator.
4. Cell culture medium—compatible with cells of interest.

3 Methods

3.1 Scaffold Fabrication

1. Remove BIOFELT tube from packaging (see Note 3). Cut and sew any seams using the resorbable suture (see Notes 4 and 5). In this application, one end is partially closed, to be fully closed by the surgeon after ureteral attachment (see Fig. 2). The other end of the tube is left open for urine drainage. These techniques are applicable to any seam construction, including formation of other shapes from flat sheets of felt.
2. Attach pieces of suture at least 10 cm long at four points on the tube, two on each end. These sutures will be used to secure the scaffold during coating, and may be used to suspend the scaffold during culture (see Fig. 2).
3. Coat PGA scaffold with PLGA solution.

3.2 Scaffold Preparation and Sterilization

1. Remove the scaffold from the vacuum. If necessary, reform scaffold to the desired shape. For the tubes, this involves exposing the scaffold to a heat lamp or placing it in an oven set to 42°C (slightly above the PLGA glass transition temperature).

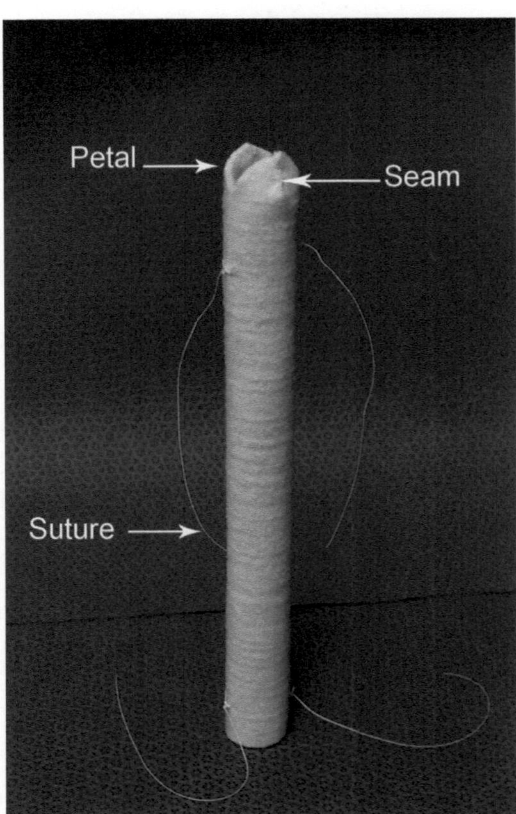

Fig. 2 Neo-Urinary Conduit, highlighting petals, seams, and attached sutures. At this point, only two of the 4 petal seams have been joined

Once softened, a rod that matches the tube inner diameter is inserted into the lumen. The rod is removed after the scaffold has cooled.

2. Confirm that the seams and four sutures are intact and firmly in place.
3. Place scaffold into an appropriate container and sterilize via ethylene oxide (EtO) exposure (see Note 6).
4. Following sterilization, store scaffold in a low-humidity (≤5% RH) environment.

3.3 Prewetting (Preparation for Cell Seeding)

3.3.1 Method 1: Vacuum Prewetting

1. In a biosafety cabinet (laminar flow hood) add sterile PBS to a vessel containing the sterile scaffold, immersing it. Scaffold will float, so ensure that it is secured and completely submerged. Seal the vessel such that gas will be able to move in and out of the vessel through a sterile filter.
2. Set vessel into a vacuum chamber, such as a vacuum desiccator.
3. Slowly apply vacuum to around 28.5 in Hg vacuum. The level of vacuum should be set so as to pull the air from the pores, but not cause the fluid to boil.
4. After 5 min at maximum vacuum, slowly release the vacuum. PBS will fill the pores and scaffold will sink.
5. Remove PBS and proceed to cell seeding by filling with desired cell culture medium for at least 5 min.

3.3.2 Method 2: Ethanol Prewetting

1. Fill vessel containing sterile scaffold with 60% ethanol (EtOH) in PBS for 10 min. Remove the 60% solution and replace with PBS alone for 5 min.
2. Proceed to cell seeding by filling with desired cell culture medium for at least 5 min.

4 Notes

1. Since PGA is susceptible to hydrolytic degradation, the felt and scaffold should always be stored in a low-humidity environment, such as a dry gas desiccator (RH <5%). Time out of a controlled environment should be minimized and tracked.
2. As PLGA is also hydrolytically labile, store PLGA in the freezer sealed in a poly bag with desiccant. The combination of low humidity and low temperature minimizes degradation. At the time of use, remove the bag from the freezer and allow it to come to room temperature prior to opening. This prevents condensation from forming on PLGA pellets.
3. The PGA fibers hold a static charge, and readily pick up dust, fibers, and other debris, which can become entangled in the

nonwoven mesh. Inspect the scaffold upon removal from the packaging. Extra care should be taken to handle the felt in a clean environment. This involves cleaning or covering the work surface and wearing gloves, lint-free sleeves, hair covering, etc.

4. In our experience, 3–7 stitches per cm produces an adequate seam. We use a continuous blanket stitch, knotted every cm or so. The suture ends should be captured in the seam and trimmed to minimize material protruding from the scaffold. We have found when constructing other scaffolds made from flat sheets of felt with multiple seams, the resulting final sizes could vary substantially depending on the stitch depth (bite) and the amount of tension in the seam. The choice of needle (straight, curved, etc.) is up to the operator, but in our hands the curved needle allows for a flatter seam. As a general rule, the seam should contain the minimum amount of felt while securely joining the pieces together.
5. Use of undyed or dyed suture makes no functional difference. Undyed suture is similar in color to the PGA material, and its use may result in a more visually pleasing outcome. Dyed suture will contrast with the felt. It is easier to see during the sewing process, and may be used to draw attention to the seam placement (e.g., to indicate where a surgeon should not cut during implantation).
6. All sterilization methods will have an effect on the scaffold, usually resulting in some amount of degradation, and literature is available discussing the relative merits of various sterilization methods (8, 9). This class of resorbable biomaterials is usually sterilized by exposure to ethylene oxide (10).

Acknowledgement

The authors would like to acknowledge Joydeep Basu for his thoughtful discussion and contributions to the introduction.

References

1. Basu J, Ludlow JW (2010) Platform technologies for tubular organ regeneration. Trends Biotechnol 28(10):526–533
2. Neuhof H (1917) Fascial transplantation into visceral defects: an experimental and clinical study. Surg Gynecol Obst 25:383
3. Shokeir AA (2002) Bladder regeneration: between the idea and reality. BJU Int 89:186–193
4. Basu J, Genheimer C, Guthrie KI, Sangha N et al (2011) Expansion of the human adipose derived stromal vascular cell fraction yields a population of smooth muscle cells with markedly distinct phenotypic and functional properties relative to mesenchymal stem cells. Tissue Eng Part C 17:843–860
5. Basu J, Jayo MJ, Ilagan RM, Guthrie KI et al (2012) Regeneration of native-like neo-urinary tissue from nonbladder cell sources. Tissue Eng Part A 18(9–10):1025–1034
6. Atala et al (2006) Tissue engineered autologous bladders for patients needing cystoplasty. Lancet 306:1241–1246

7. Seltzer E, Tillinger M, Jayo MJ, Bertram TA (2009) Role of biomechanical stimulation (cycling) in neobladder regeneration—translational basis for clinical outcomes. JUrol 181(4):282
8. Middleton JC, Tipton AJ (1998) Synthetic biodegradable polymers as medical devices. Med Plastics Biomater 5(2):30–39
9. Nuutinen JP, Clerc C, Virta T, Törmälä P (2002) Effect of gamma, ethylene oxide, electron beam, and plasma sterilization on the behaviour of SR-PLLA fibres in vitro. J Biomater Sci Polym Ed 13(12):1325–1336
10. Middleton JC, Tipton AJ (2000) Synthetic biodegradable polymers as orthopedic devices. Biomaterials 21(23):2335–2346

Chapter 15

Characterization of a PGA-Based Scaffold for Use in a Tissue-Engineered Neo-Urinary Conduit

Toyin A. Knight and Richard G. Payne

Abstract

A tissue-engineered product needs to be properly characterized in order to be used in vivo. Many methods can be used to characterize a scaffold during creation of a tissue-engineered product. This chapter looks at the mechanical (tensile testing) and biological characterization (cell viability and proliferation) of a polyglycolic acid-based scaffold used to tissue engineer a Neo-Urinary Conduit™. Such methods are more broadly applicable to characterization of other neo-organ product candidates.

Key words Degradation, Tensile strength, Cell viability, Cell proliferation, Mass loss

1 Introduction

There are a wide range of roles scaffolds/biomaterials play in tissue regeneration. Such roles include acting as a three-dimensional (3D) template for the regenerating tissue by being able to deliver/recruit the cells needed for regeneration. In this role, the scaffold supports the biological and mechanical needs of the regenerating tissue. In vitro characterization of the scaffold is useful in gaining an understanding of what may happen in vivo during contact with cells or tissue, thus assisting in improved scaffold design. There are many test methods used to characterize the properties of biomaterials, such as chemical tests (methods—molecular weight, inherent viscosity) and physical tests (method—porosimetry) (1). Depending on the final use of the biomaterial, multiple characterization methods may be used to understand the nature of the biomaterial as well as its interaction in vivo (2).

This chapter focuses on two tests used in the characterization of a polyglycolic acid (PGA)-based scaffold for use in tissue engineering a Neo-Urinary Conduit. The Neo-Urinary Conduit is a cell-seeded biodegradable PGA-based tubular scaffold used for de novo formation of urinary-like neo-tissue in vivo to help

Joydeep Basu and John W. Ludlow (eds.), *Organ Regeneration: Methods and Protocols*, Methods in Molecular Biology, vol. 1001, DOI 10.1007/978-1-62703-363-3_15, © Springer Science+Business Media New York 2013

alleviate complications associated with urinary diversion or bladder reconstruction/replacement surgeries (3).

For the two characterization tests of the PGA-based scaffold, one test examines the mechanical properties of the biomaterial through tensile testing and mass loss, while the other test looks at the biological properties of the biomaterial via cell–material interaction specifically through cell viability and cell proliferation. Tensile strength measured through tensile testing is the stress calculated from the maximum load a material experiences before undergoing failure (4). This tensile test usually records the stress versus strain and it is used to describe the mechanical property of the material while in the environment for which it will be used. This chapter looks at the in vitro tensile strength of the PGA-based scaffold over time while it undergoes degradation during sterilization as well as mimicking incubation in vivo. Generation of a degradation profile will provide better insight as to how the mechanical properties of the PGA-based scaffold change over time. Cell viability and proliferation will provide insight as to the biocompatibility of the PGA-based scaffold and if the cells are biologically active on the material. For cell viability, a LIVE/DEAD viability/cytotoxicity assay kit, which provides a two-color fluorescence assay to distinguish between live and dead cells, will be used. The live cells are identified by the polyanionic dye calcein that produces an intense green fluorescence, while the dead cells with damaged membranes are identified by ethidium homodimer-1 (EthD-1) that produces a bright red fluorescence (5). For cell proliferation, an MTS colorimetric assay will be used. The most commonly used cell proliferation assay is based on whether one can detect an increase or a decrease in the DNA synthesis of a cell by incorporating tritiated (^{3}H) thymidine incorporation, which is radioactive (6). However, the use of a radioactive material puts limitations on the usage of this assay in some testing laboratories due to regulatory restrictions from many clinical manufacturing facilities. The MTS assay is a nonradioactive, nondestructive tetrazolium dye-based colorimetric assay that measures formazan in metabolically active cells (7). Because of this feature, it is used in determining the number of living cells in proliferation assays. The combination of these assays will give preliminary data as to the biocompatibility and mechanical properties of the scaffold and if any optimization of this scaffold is needed prior to its use in vivo. From a broader perspective, similar approaches may be leveraged to characterize other, related neo-organ product candidates with analogous cell/scaffold composition.

2 Materials

2.1 Scaffold Fabrication

1. 50:50 Poly(dl-lactide-*co*-glycolide) (PLGA) (Lactel Absorbable Polymers, Birmingham, AL, USA, catalog # B6010-2).
2. Methylene Chloride (Spectrum Chemical, New Brunswick, NJ, USA).

3. PGA felt 6×20 cm strips, which will eventually be cut into dogbones and square coupons (strips should be cut perpendicular to the direction of the fibers in the felt) (Biomedical Structures, Warwick, RI, USA) (see Note 1).
4. Glass bottle.
5. Stir bar.
6. Stir plate.
7. Polytetrafluoroethylene (PTFE) trough.
8. Source of forced cool air, e.g., blow dryer.
9. Fume hood.
10. Vacuum desiccator.

2.2 Degradation/Mass Loss Profile of PGA-Based Scaffold

1. Coated PGA 6×20 felt strips and 2×2 cm coupons.
2. Low-humidity desiccator.
3. Dogbone die (throat dimensions—25 mm = length, 12.5 mm = width) (see Note 2).
4. Die press (hydraulic).
5. Balance.
6. Calipers.
7. Thickness gauge.
8. Freezer.
9. Lyophilizer.
10. Instron testing machine or appropriate mechanical testing system.
11. Specimen cups (Fisher Scientific, Pittsburg, PA, USA, catalog # 22-146-530).
12. 6-Well tissue culture plates.
13. Oven (37°C).
14. Ethanol (Aaper Alcohol & Chemical, Detroit, MI, USA, Ethyl Alcohol, 200 proof).

    Phosphate-buffered saline (PBS) 1×.
16. 10× PBS reconstituted to 0.1 M solution with phenol red (see Note 3).

2.3 Cell Culture on PGA-Based Scaffold

1. Coated PGA 1×1 cm coupons.
2. Adipose-derived smooth muscle cells (Ad-SMC).
3. Cell culture media: Dulbecco Modified Essential Medium (DMEM) with 10% Fetal Bovine Serum (FBS).
4. PBS 1×.
5. LIVE/DEAD® viability/cytotoxicity kit (Invitrogen, catalog # L3224).

6. MTS colorimetric assay—Cell Titer 96® AQ_{ueous} Non-Radioactive Cell Proliferation Assay (Promega, Madison, WI, USA, catalog # G5421).
7. Ultra-low cell binding 6-well plate (Corning, catalog # 3471).

3 Methods

3.1 Scaffold Fabrication

1. Place a stir bar in glass bottle and put in the fume hood (see Note 4). Make a 4.25% PLGA solution by putting 4.25 g of PLGA in the glass bottle and adding 100 ml of methylene chloride solution. Let PLGA dissolve for about an hour while stirring on stir plate.
2. After PLGA has dissolved, apply two coatings of PLGA solution to the PGA strips in the fume hood to achieve approximately a 50% increase in weight of the PGA strips. Apply first coating of PLGA solution to the PGA strips in a Teflon trough (see Note 5) and evaporate the methylene chloride using forced air. After first coat has dried, apply second coat to coated PGA strips and dry using forced air. Place in vacuum desiccator for at least 12 h (see Note 6).

3.2 Tensile Testing of PGA-Based Scaffold

1. Cut coated PGA strips into dogbones using dogbone die and die press as per manufacturer's instructions. Total amount of dogbones needed is 50.
2. Test five dogbones for tensile strength using tensile testing machine as per ASTM standard (8) and instrument protocol to represent time $(t) = 0$ (Fig. 1). Briefly,
   (a) Use calipers to measure the length and width of the narrowest part of the dogbone and record measurements.
   (b) Use a thickness gauge to measure the thickness of the dogbone, and record.
   (c) Securely clamp dogbones to the grips of the tensile testing machine and test sample in tension.
   (d) Record maximum stress.
3. Package the remaining dogbones and sterilize via ethylene oxide (EtO).
4. After sterilization, test five dogbones for tensile test to represent $t = 0^*$ (see Note 7).
5. The rest of the dogbones can immediately undergo pre-wetting, which is the next step, or be stored in a low-humidity desiccator until further use (see Notes 8 and 9).

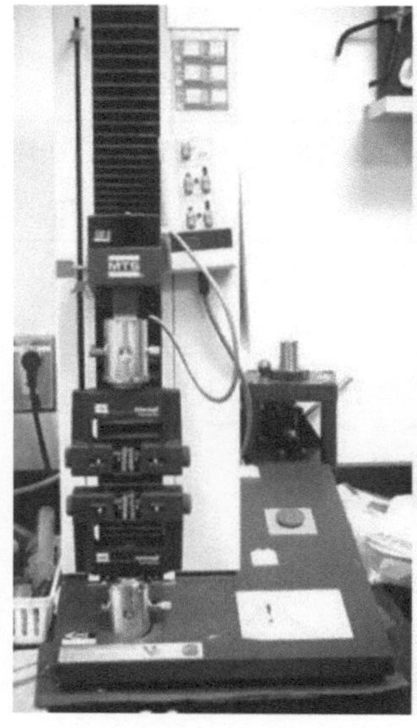
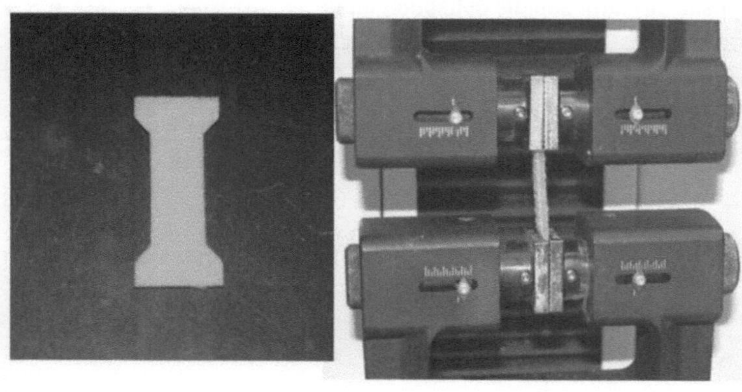

Fig. 1 Image showing a tensile testing instrument (*left panel*), a dogbone (*middle panel*), and a dogbone in between the grips of the tensile testing instrument (*right panel*)

6. Pre-wet the dogbones in a large beaker, using 60% ethanol (EtOH) in PBS for 10 min and afterwards place the dogbones in PBS alone for 5 min.
7. Take five dogbones and place in a sample cup. Repeat for the remaining dogbones.
8. Fill cups with 10× PBS with phenol red, seal, and place in a 37°C oven.
9. Change PBS solution when phenol red changes in color.
10. Remove dogbones from sample cup according to Table 1 and test for tensile strength. The times are represented in days.
11. Generate a degradation profile with the tensile strength (stress—kPa) on the *Y*-axis and time on the *X*-axis (Fig. 2).

3.3 Mass Loss of PGA-Based Scaffold

1. Cut coated PGA strips into (24) 2×2 cm coupons.
2. Package coupons and sterilize via EtO.
3. After sterilization, weigh 2×2 cm coupons on weigh balance and record weights—initial weight (W_{IC}).
4. Pre-wet coupons using the protocol as stated in Subheading 3.2.
5. Place coupons in 6-well tissue culture plates, three coupons to each plate. Mass loss will be monitored on the days according to Table 2.
6. Fill plates with 10× PBS with phenol red.

Table 1
Tensile strength sample collection table

$t=0$	$t=0$*	$t=3$	$t=6$	$t=7$	$t=9$	$t=12$	$t=15$	$t=18$	$t=21$	Total
5	5	5	5	5	5	5	5	5	5	50

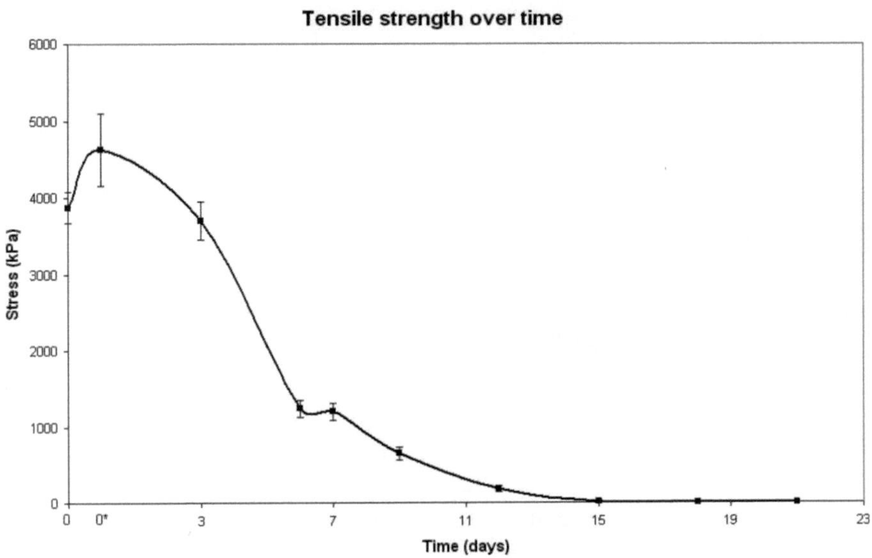

Fig. 2 Sample graph showing tensile strength over time

Table 2
Mass loss sample collection table

$t=3$	$t=6$	$t=7$	$t=9$	$t=12$	$t=15$	$t=18$	$t=21$	Total
3	3	3	3	3	3	3	3	24

7. Change PBS solution when phenol red changes color.
8. On designated days, take coupons out of solution and rinse with DI water.
9. Place coupons in pre-weighed tubes (W_T) and put in a freezer.
10. Lyophilize coupons using standard protocol and weigh coupons plus tubes (W_{CT}).

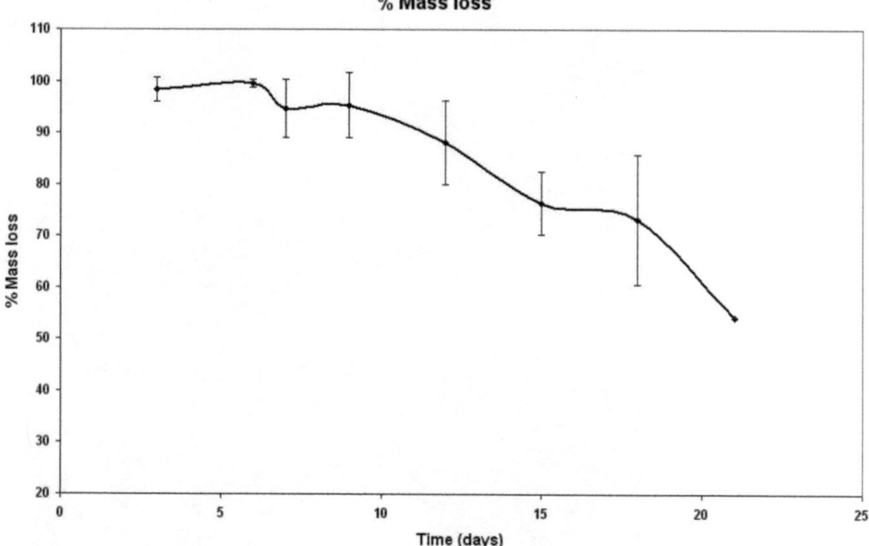

Fig. 3 Sample graph showing % mass loss over time

11. Calculate the final weight of the coupons by subtracting the weight of the tube (W_T) from the weight of the coupons plus tubes (W_{CT}). So $W_{CT} - W_T = W_{FC}$.
12. To calculate mass loss of the samples (W_M), subtract the final weight of the coupons (W_{FC}) from the initial weight of the coupons (W_{IC}). So $W_{IC} - W_{FC} = W_M$.
13. To obtain the percentage mass loss over time, divide the mass loss of the samples (W_M) by the initial weight of the coupons (W_{IC}) and multiply by 100. So $(W_M / W_{IC}) \times 100$. Plot the time on the *X*-axis and % mass loss on the *Y*-axis to generate a % mass loss profile (Fig. 3).

3.4 Cell Viability

1. Cut coated PGA strips into (6) 1×1 cm coupons.
2. Sterilize via exposure to EtO.
3. Divide the coupons into two sets of three and place each set in an ultralow cell binding 6-well plate, one coupon per well (see Note 10). The two plates are to be evaluated at two different time points (24 h and 7 days).
4. Pre-wet coupons using 60% EtOH in PBS and PBS as described in Subheading 3.2.
5. Incubate coupons in cell culture media for 15 min at room temperature (see Note 11).
6. Aspirate as much media as possible from the coupons (see Note 12).

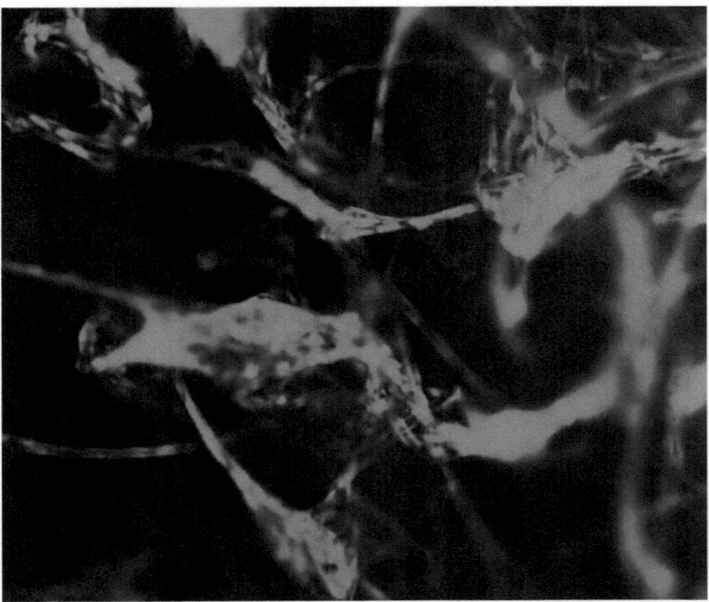

Fig. 4 LIVE/DEAD image of Ad-SMC on a PGA-based scaffold

7. Seed Ad-SMC on PGA-based scaffold at a seeding density of 8,000–13,000 cells/cm^2 (based on the total internal surface area of the scaffold) in about 50 μl of cell culture media.
8. Allow cells to attach to scaffold for 2–4 h in a 37°C, 5% CO_2 incubator.
9. Fill wells with 3 ml of media and culture cells in a 37°C, 5% CO_2 incubator with one media change in between for the scaffolds cultured for 7 days.
10. At 24 h and the seventh day, rinse scaffolds with PBS and apply LIVE/DEAD viability/cytotoxicity kit to the cells on scaffold as per kit instructions (http://tools.invitrogen.com/content/sfs/manuals/mp03224.pdf).
11. View and image cells using a fluorescent microscope (Fig. 4).

3.5 Cell Proliferation

1. Repeat steps 1–9 from Subheading 3.4.
2. At 24 h and the seventh day, rinse scaffolds with PBS and apply MTS solution to the cells on scaffold as per assay instructions (7).
3. Read absorbance at 490 nm.
4. Subtract the absorbance reading from the values obtained on the seventh day from that at 24 h to give an estimate of cell proliferation from the first to the seventh day of the cells in culture (see Note 13).

4 Notes

1. The direction of the tensile test should be taken into consideration when cutting the strips. The strips are cut so that they are tested in the perpendicular (cross machine) direction because that is the weakest direction, thereby looking at worst-case scenario.
2. The use of a dogbone allows for reproducible results and it has a uniform cross-sectional area that is representative of the material undergoing stress. The ASTM standard D638-10 (8) specifies a width of 6.0 mm; however the width for this study was changed to 12.5 mm to account for the intact fiber length in the PGA mesh.
3. As the PLGA/PGA scaffold degrades, it breaks down into lactic and glycolic acids. These degradation products lower the pH and change the phenol red indicator from a red to an orange color, which indicates the need to change the solution to keep it at a more neutral pH. In addition, use of a more concentrated form of PBS (10×) reduces how often the buffer solution needs to be changed.
4. Methylene chloride (also known as dichloromethane) is an organic solvent that gives off toxic fumes and should be handled in a fume hood with appropriate personal protective equipment at all times for safety reasons. Methylene chloride also dissolves many plastics, so use of a glass bottle for handling this solvent is necessary.
5. A PTFE trough is used to contain the PLGA solution while it is being fully absorbed by the scaffold. PTFE was chosen because it is resistant to methylene chloride and will not adhere to the scaffold, which can become tacky during the coating process.
6. A vacuum desiccator for this period ensures removal of methylene chloride to normal safety levels.
7. Because some degradation does occur with EtO sterilization, it is important to see what effect it may have on the degradation profile of the PGA felt being tested.
8. Pre-wetting is the process by which the scaffold material is wetted with ethanol and PBS to overcome the hydrophobic effects of the PLGA coating on the PGA felt to allow for easy cell attachment and proliferation through the scaffold.
9. Degradation of the PGA felt occurs by hydrolysis and hence sensitive to moisture; a low-humidity desiccator prevents premature degradation and keeps the scaffold dry.
10. An ultralow binding plate minimizes the amount of cells that will adhere to the surface of the well, thereby promoting more cells to adhere to the scaffold.

11. Incubating the coupons in media allows adsorption of serum proteins onto the scaffold, thereby promoting cell attachment.
12. Aspirating as much media as possible from the scaffold allows the cell suspension in the next step to be contained within the pores of the scaffold. This increases the chance that the cells will successfully attach to the scaffold.
13. The absorbance readings obtained at 490 nm are directly proportional to the number of living cells in culture (7).

References

1. Standard guide for characterization and testing of biomaterial scaffolds used in tissue-engineered medical products. ASTM Standard: F2150-07
2. Barbucci R (2002) Integrated biomaterials science. Kluwer,
3. Basu J, Jayo MJ, Ilagan RM et al (2012) Regeneration of native-like neo-urinary tissue from nonbladder cell sources. Tissue Eng Part A 18:1025–34
4. Ratner BD, Hoffman AS, Schoen FJ et al (1996) Biomaterials science: an introduction to materials in medicine. Elsevier Science,
5. Maltaris T, Kaya H, Hoffmann I et al (2006) Comparison of xenografting in SCID mice and LIVE/DEAD assay as a predictor of the developmental potential of cryopreserved ovarian tissue. In Vivo 20:11–16
6. Soman G, Yang X, Jiang H et al (2009) MTS dye based colorimetric CTLL-2 cell proliferation assay for product release and stability monitoring of Interleukin-15: Assay qualification, standardization and statistical analysis. J Immunol Methods 348:83–94
7. CellTiter 96 Aqueous Non-Radioactive Cell Proliferation Assay (MTS) - http://www.promega.com/products/cell-health-assays/cell-viability-assays/celltiter-96-aqueous-non_radioactive-cell-proliferation-assay-_mts_/. Accessed on January 22, 2013
8. Standard test method for tensile properties of plastics 2010. ASTM Standard: D638-10

Chapter 16

Migration Assay to Evaluate Cellular Interactions with Biomaterials for Tissue Engineering/Regenerative Medicine Applications

Kelly I. Guthrie, Namrata Sangha, Christopher W. Genheimer, Joydeep Basu, and John W. Ludlow

Abstract

Regenerative medicine and tissue engineering approaches for solving current medical dilemmas such as organ failure, congenital defect, or reconstruction following disease or trauma typically require specific considerations regarding biomaterial selection, identification of key cell types, and applicable surgical techniques (Lanza et al. Principles of tissue engineering, Academic, 2007; Kikuchi, Kanama., Quart Rev 24:51–67, 2007). The ability to evaluate these components in vitro under conditions which simulate relevant in vivo environments can reduce development risks including time and money costs associated with early-stage product development. Similarly, such methods can be useful in making progress in researching features of natural and synthetic biomaterial such as porosity, strength, surface topography, and functionalization, and their singular or collective effects on cell behavior (Kikuchi and Kanama., Quart Rev 24:51–67, 2007; Furth et al. Biomaterials 28:5068–5073, 2007; Mieszawska and Kaplan., BMC Biol 8:59, 2010).

Adhesion, migration, and gene and protein expression are all cell behaviors that can be affected by properties of a chosen biomaterial and vary based upon organ system (Cornwell et al. J Biomater Res 71A:55–62, 2004; David et al. Tissue Eng 8(5):787–798, 2002). Understanding of these properties and their role in combination with biomaterial in remodeling is sought in order to fully harness and direct regeneration (Lanza et al. Principles of tissue engineering, Academic Press, 2007; Mieszawska and Kaplan. BMC Biol 8:59, 2010; Matragotri and Lahann J. Nat Mater 8:15–23, 2009).

Key words Cell migration, Esophagus, Biomaterial

1 Introduction

Understanding cell migration within the organ system to which a regenerative medicine solution is proposed is important in creating successful outcomes (1–7). An in vitro migration assay utilizing either native tissue or, alternatively, material seeded with primary cells or cell lines characteristic of the native tissue is an important tool to identify suitable biomaterials and surgical methods (see Note 1) for application. A simple method for evaluating cellular migration applicable to a wide variety of materials and organ systems

Joydeep Basu and John W. Ludlow (eds.), *Organ Regeneration: Methods and Protocols*, Methods in Molecular Biology, vol. 1001, DOI 10.1007/978-1-62703-363-3_16, © Springer Science+Business Media New York 2013

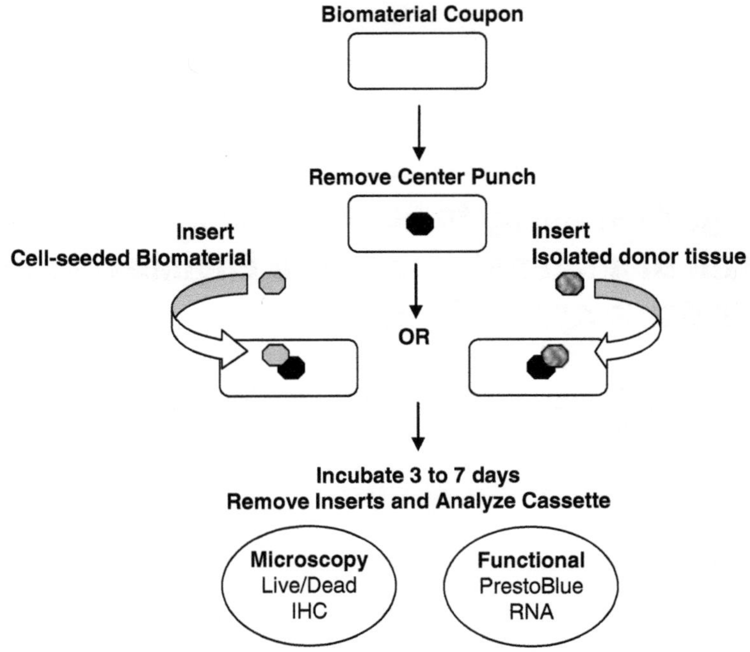

Fig. 1 Overview of migration assay

and allowing for a broad spectrum of post-migration analysis is described (see Fig. 1). The assay was specifically designed for an esophageal application; however by varying donor tissues and relevant cell types the assay can provide information applicable to other organ systems (see Note 5).

2 Materials

2.1 Migration Assay Components

1. Biomaterial cassettes—sample pieces or "coupons" of biomaterials for evaluation with center punch removed, sterilized by method appropriate to material (see Note 2).
2. Native tissue aseptically collected, full thickness or layer-specific piece sized for insert into biomaterial cassette (with esophagus could use epithelial only or submucosal only).
3. Sterile saline.
4. Sterile 6-well low bind culture dish (CoStar 3471) or other appropriate vessel.
5. Sterile 24-well low bind culture dish (CoStar).
6. Cell culture medium of choice containing 1% antibiotic/antimycotic.
7. Sterile forceps, scissors.
8. Biopsy punch tool, 5 mm or other selected size.

9. Cell culture incubator with conditions of choice, typically 5% CO_2 although hypoxic conditions could be of interest.
10. Microscope.
11. Others to be added as needed.

2.2 Components for Characterizing Migration

2.2.1 Microscopy-Related Characterization

1. Inverted microscope with fluorescent capabilities.
2. Antibodies of choice for identification of differing cells types: e.g.: pan-cytokeratin for epithelial cells and smooth muscle alpha actin for smooth muscle cells.
3. Live/Dead/Viability/Cytotoxicity Kit (Invitrogen, L3224).
4. DPBS.
5. 2% PFA.

2.2.2 Metabolic or Cell Proliferation-Based Assays for Cell Quantitation

1. PrestoBlue Cell Viability Reagent (Invitrogen, A13261).
2. 96-well spectroscopy plate.
3. Plate reader with fluorescent capabilities.
4. Or other cell quantitation assays of choice.

2.2.3 RNA/DNA Characterization of Cells

1. DPBS.
2. Trizol reagent.
3. Storage tubes for freezing cell lysate until analysis.

3 Procedure

3.1 Migration Assay

3.1.1 Preparation of Biomaterials to Be Tested

1. Identify biomaterials to be tested. For example, synthetic scaffold materials such as PCL foams of varying pore size and thickness, woven PGA mesh, PGA felts of varying thickness and volume, or natural scaffolds such as SIS.
2. Cut scaffold "coupons" (square or rectangular pieces) from each of the materials such that they will lie flat in each well of a 6-well low binding plate (scaffold coupon approx 3 cm × 1.5 cm). Alternatively, to conserve material or for assaying large number of materials, cut to lie flat in 24-well low bind dish (scaffold coupon approx 1 cm × 1 cm).
3. Using a coring punch biopsy tool (5 mm) or scalpel remove a portion from the center of each biomaterial coupon to create the biomaterial cassette. The removal of material creates the receptacle into which you place either a piece of native tissue or a cell-seeded biomaterial insert as described later.
4. Reserve the excised pieces of biomaterial as these can be used for creating the cell-seeded inserts to be used later (refer to Subheading 3.1.3, step 2).

5. Sterilize biomaterials, if not provided as such, by means appropriate to the nature of the material to be tested. Ethylene oxide, autoclaving, or ethanol sterilization are applicable methods but dependant upon the biomaterial.

3.1.2 Preparation of Donor Tissue

1. Isolate a section of esophagus from donor animal(s) proportional to the number of biomaterials or conditions to be tested. For example, approximately four appropriately sized sections of esophagus can be derived from one rat.
2. Flush esophagus with copious amounts of saline. Tissue may be opened by cutting the full length of the esophagus with sterile scissors for ease of cleaning.
3. Allow tissue to remain in growth medium containing antibiotic/antimycotic until use, up to 24 h from harvest. Store refrigerated.
4. Using a scalpel or punch biopsy tool, aseptically cut full-thickness inserts from the esophagus (epithelial side up) of sufficient size to place in the biomaterial cassette created above.

3.1.3 Preparation of Cell-Seeded Inserts

1. Determine the appropriate cell type(s) of interest to be evaluated. The cells can be either stable cell lines or primary cells isolated from native tissue. For example: Relative to the esophagus, a mixed primary cell culture can be generated from enzymatic digest of donor tissue and seeding onto biomaterial inserts.
2. Retrieve the sterilized punches or inserts removed previously from biomaterials to be tested and while working within a laminar flow hood, pre-wet or soak the biomaterials with serum-containing growth medium in a 24-well low bind dish. Each well should contain one insert.
3. Harvest cells, pellet, and resuspend to create a highly concentrated cell suspension with which to seed the biomaterial inserts. Use approximately 150,000 cells to seed a 5 mm punch.
4. Remove the pre-wetting medium from each well containing an insert and remove excess liquid volume from the biomaterial by aspiration.
5. Pipette the cell suspension onto the insert taking care to cover the surface of the insert well, and evenly, with the seeding material (example: 20 μl volume per 5 mm insert).
6. Place at 37°C/5% CO_2 for 1–2 h to allow cells to attach.
7. Gently pipette sufficient volume of growth medium into each well to cover the insert. Deliver volume against the side of well and away from insert to minimize disturbance of cells.
8. Incubate at 37°C/5% CO_2 for a minimum of 1 day until cells are well established and abundant. Do not incubate for such extended periods that biomaterial integrity is compromised, making handling of the insert difficult.

3.1.4 Establishing Migration Assay

1. Retrieve sterile coupon cassettes, culture dishes, sterile instruments, and the inserts, either native tissue or previously cell-seeded biomaterial, and place in laminar flow hood.
2. Using sterile forceps, distribute biomaterial cassettes to be tested into a low bind 6-well dish, one coupon per well. If using smaller cassettes, use a 24-well low bind dish.
3. Include blank cassettes as negative controls for analytical assays.
4. Pre-wet biomaterial cassettes with complete growth medium by adding 2–3 ml of media per well. Utilizing a pipette placed firmly onto the biomaterial, draw medium into the biomaterial to displace any trapped air.
5. Allow pre-wetting to occur at room temperature for a minimum of 15 min (see Note 3).
6. Aspirate media, and using sterile instruments and technique, insert either the previously prepared portion of native tissue or cell-seeded biomaterial punch into the biomaterial cassette. See Fig. 2.
7. Add media in sufficient volume as to cover cassette and donor tissue, 3–5 ml for a 6-well plate and 0.5–1 ml for a 24-well plate.
8. Incubate cassettes at 37°C/5% CO_2 for 3–7 days (see Note 4).
9. Exchange medium every 2–3 days.
10. Following incubation, remove tissue explant or cell-seeded insert from the cassette.
11. At this point the entire cassette can be analyzed as a whole or sectioned for end point analysis.

Fig. 2 PGA coupons in two wells of a low bind culture dish. The first well contains an empty PLGA coupon cassette (with *central hole*), while the second well is a duplicate cassette into which a cell-seeded insert has been firmly seated

4 Characterization of Migration

4.1 How Much Migration Occurred?

4.1.1 Live/Dead Stain

1. Wash cassette, or alternatively section of cassette, with DPBS 2× to remove excess media and unattached cells or remnants of tissue.
2. Follow manufacturer's instructions for preparation of the staining solution which stains live cells green (calcein-AM) and dead cells red (ethidium homodimer-1).
3. Visualize cells utilizing fluorescent microscope capable of reading both calcein and ethidium.
4. Observations of cells which have migrated onto cassette can be made quantitative with the use of quantitative microscopy software if available.

4.1.2 PrestoBlue Quantitation

1. Wash cassette, or alternatively section of cassette, with DPBS 2× to remove excess media and unattached cells or remnants of tissue.
2. For entire large cassette, place in 1 well of 6-well dish. For sections or small coupons, place in 1 well of 24-well dish.
3. Dilute PrestoBlue reagent 1:10 in growth medium, adding minimal volume to cover cassette (approx 1.5 ml for 6 well) and incubate at 37°C/5% CO_2 for 2 h.
4. Diluted PrestoBlue reagent alone or incubated with a blank coupon section serves as background/negative control.
5. Sample and transfer 200 μl from each well to a 96-well spectroscopy plate (see Note 6).
6. Read using a fluorescent plate reader (Ex 540-570, Em 580-610).
7. Higher RFU is indicative of more metabolic activity which can be related to increased cell numbers having migrated.

5 What Types of Cells Migrated?

5.1 Immunofluorescent Labeling

1. Wash cassette(s), or alternatively section of cassette(s), with DPBS 2× to remove excess media and unattached cells or remnants of tissue.
2. Transfer cassette(s) to a new culture dish well.
3. Add blocking buffer (0.5% BSA in 1% Tween DPBS) to cassette or pieces to be stained.
4. Incubate at room temperature for a minimum of 30 min.
5. Dilute antibody 5 μg/ml (or mfg recommendation) in DPBS. Approximately 1.5 ml is required for intact large cassette, and 0.5 ml for smaller.
6. Remove blocking buffer by aspirating taking care not to contact biomaterial.

Fig. 3 Representative immunohistochemistry panels from migratory assay with esophageal tissue inset. The three panels above are representative of cassettes stained with cytokeratin antibodies in order to visualize the migration of esophageal epithelial cells onto PLGA/PGA cassettes with and without prior cell seeding of the biomaterial. The *light spots* are representative of cytokeratin-positive epithelial cells having migrated from the tissue insert onto the cell-seeded cassette in greater number than the blank cassette. The *right side* of each panel was in proximity to the tissue insert

7. Add primary antibody solution to cover cassette pieces.
8. Incubate for 1 h with gentle rotation.
9. Remove primary antibody solution and wash 3× in DPBS allowing a 5-min (minimum) soak with gentle rotation on a plate mixer.
10. Add fluorescently labeled secondary antibody in DPBS (1–5 μg/ml) and add to each well.
11. Incubate at room temperature with gentle rotation.
12. Remove secondary antibody solution and wash 3× in DPBS allowing a 5-min (minimum) soak with gentle rotation on a plate mixer.
13. Add 2% PFA as a fixative and incubate at room temperature with gentle rotation for 30 min.
14. Wash as before and store wet in DPBS until microscopy can be performed (Fig. 3).

5.2 PCR Analysis of Migrated Cells

1. Wash cassette, or alternatively section of cassette, with DPBS 2× to remove excess media and unattached cells or remnants of tissue.
2. Lift cassette, or section of cassette, with forceps and touch edge of the biomaterial to an absorbent wipe to wick excess moisture out of the biomaterial.
3. Transfer intact cassette to a new culture dish well.
4. Add Trizol reagent directly to the coupon to lyse cells.
5. Collect resultant lysate and follow standard protocols for preparation of RNA or DNA for PCR-based analysis targeting organ- or cell type-specific markers.

6 Notes

1. Elements of different surgical approaches, for example the use of fibrin glue, can be applied in this assay to determine if there is any effect on cellular migration.
2. Two punches can be created in a single cassette to determine if a functionalized or cell-seeded insert would influence the migration of cells from native tissue within the biomaterial.
3. The cassette itself may be seeded to represent the potential application of a combination product (biomaterial plus cells). Post-migratory analysis would need to be adjusted to determine differences between seeded cells vs. migrating cells.
4. Cassettes may be incubated in a low-oxygen environment to more closely mimic in vivo environments.
5. A modification of this technique has been used to examine interaction of small intestine when anastomosed with tubular biomaterial.
6. The PrestoBlue reagent is nontoxic and the assay is not terminal. Cassettes, or sections of cassettes, utilized for PrestoBlue analysis can be repurposed for additional analysis.

References

1. Lanza R, Langer R, Vacanti J (2007) Principles of tissue engineering. Elsevier Academic, Amsterdam
2. Kikuchi M, Kanama D (2007) Current status of biomaterial research focused on regenerative medicine. Quart Rev 24:51–67
3. Furth ME, Atala A, Van Dyke ME (2007) Smart biomaterial design for tissue engineering and regenerative medicine. Biomaterials 28:5068–5073
4. Mieszawska A, Kaplan DL (2010) Smart biomaterials-regulating cell behavior through signaling molecules. BMC Biol 8:59
5. Cornwell KG, Downing BR, Pins GD (2004) Characterizing fibroblast migration on discrete collagen threads for applications in tissue regeneration. J Biomater Res 71A:55–62
6. Geer DJ, Swartz DD, Andreadis ST (2002) Fibronectin promotes migration in a three-dimensional in vitro model of wound regeneration. Tissue Eng 8(5):787–798
7. Matragotri S, Lahann J (2009) Physical approaches to biomaterial design. Nat Mater 8:15–23

Chapter 17

Care of Rodent Models Used for Preclinical Evaluation of Tissue-Engineered/Regenerative Medicine Product Candidates

Kim L. Mihalko

Abstract

The pre-, peri-, and postoperative care of animal surgical models used for testing tissue engineering/ regenerative medicine product candidates includes the thoughtful consideration of several important factors. It must ensure the health and comfort of the animals and the success and reproducibility of the model. In order to reduce the number of animals needed in creating the model and to reduce costs, a preliminary evaluation of surgical procedures and instruments should be performed on cadavers. Once a minimal level of proficiency has been acquired, non-survival surgeries should be executed successfully before attempting survival surgeries. Planning ahead is crucial and will involve all aspects of the animal's care such as allowing the animal to become accustomed to soft foods (as in the case of gastrointestinal surgeries), planning appropriate pain management, and the use of positive reinforcement. We present specific examples of pre-, peri- and post-operative care of rodents using our experiences in developing tissue engineering products for kidney, esophagus, small intestine and lung.

Key words Tissue engineering, Regenerative medicine, Rodent model, Rodent care, Veterinary care, Tissue engineered kidney, Esophagus, Small intestine, Lung

1 Introduction

It is unfortunate that many published studies using research animals show deficiencies in describing pertinent details. Information such as the source of the animals, the strain, age, sex, pathogen status, diet, type of housing, anesthetics used, and antibiotics and analgesics administered may not be reported. The absence of these may affect the reproducibility of the study and, more significantly, may cause the reader to misinterpret the study and lead to unnecessary use of animals.

To this end, we have attempted to correct these inadequacies in our own work by standardizing surgical approaches wherever possible. The protocols below summarize our learning from work on the evaluation of tissue-engineered constructs for regeneration

Joydeep Basu and John W. Ludlow (eds.), *Organ Regeneration: Methods and Protocols*, Methods in Molecular Biology, vol. 1001, DOI 10.1007/978-1-62703-363-3_17, © Springer Science+Business Media New York 2013

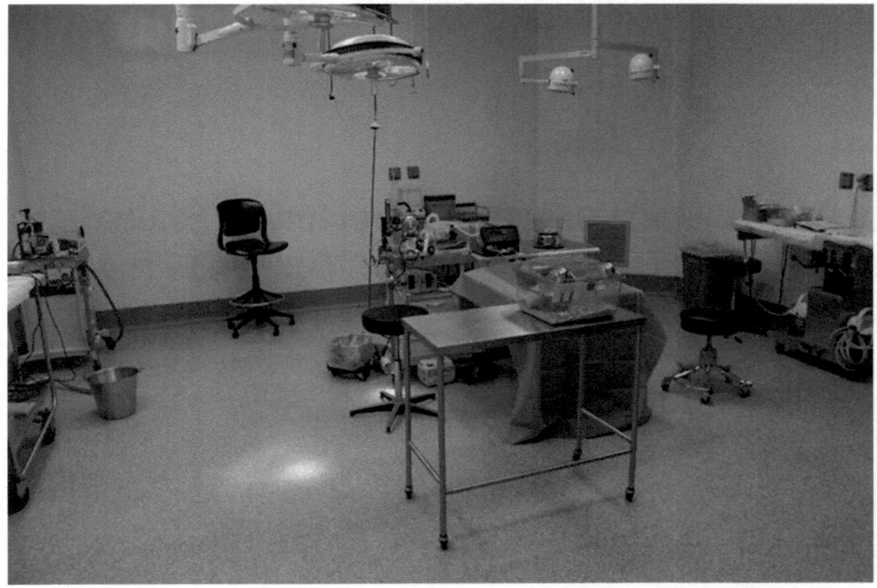

Fig. 1 Overall view of surgical suite appropriate for rodent work

of esophagus, small intestine, lung, and kidney. For additional details, the reader is referred to our relevant recent publications (see refs. 1–3). Commonalities of methodology emerge as well as idiosyncrasies peculiar to certain organ or tissue types. Both of these are summarized below.

2 Materials

1. For the procedures described here all animals used were female Lewis rats obtained from Charles River Labs, MA (see Note 1).
2. The animals were pair housed in Tecniplast ventilated cages (www.tecniplast.it/usa/index.php). The room lighting was set on a 12-h cycle.
3. Diet was Harlan Rodent Diet and included a wide range of treats from Bio-Serv (One 8th Street, Suite One, Frenchtown, NJ 08825) (see Note 2).
4. Surgical suite appropriate for small animal work (see Fig. 1).
5. Sterile gauze, surgical tools suitable for small animals (scissors, tweezers, etc., see Figs. 1 and 2).
6. Surgical scalpels, #10, #15. Bard-Parker, rib back carbon steel surgical blade.
7. Ethicon skin stapler PXR35.

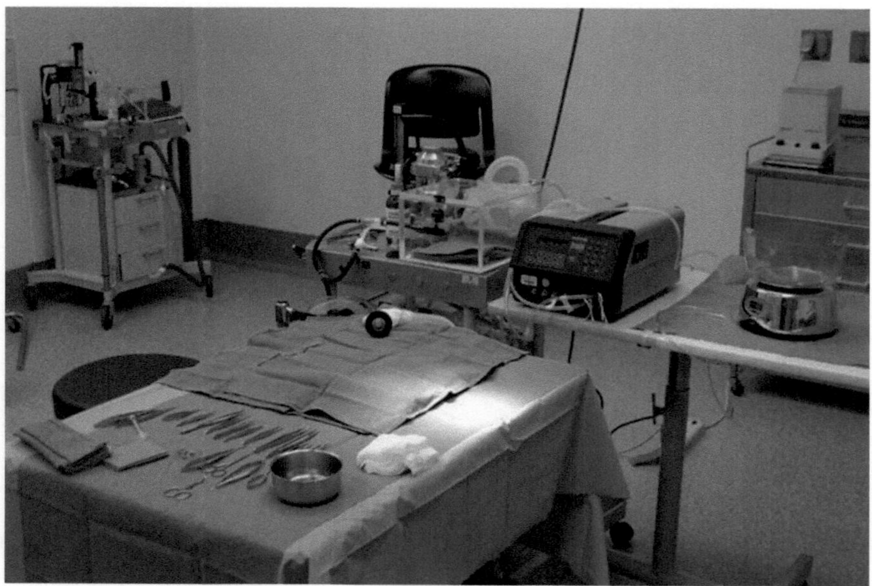

Fig. 2 View of surgical table showing instruments and isoflurane chamber

3 Pre-surgical Preparation

1. Prior to initiation of any studies involving live animals, institutional approval for all experimental protocols must be secured through application to the institution's IACUC committee. As part of the application process, justification for the use of live animals must be given and a thorough search of the academic literature executed to establish that the study in question is not needlessly repeating already published experiments.
2. First, the needs and expectations are discussed with the research group. Cadaveric rodents (euthanized from other studies) are used to become familiar with anatomical landmarks, different surgical approaches, sutures, and instruments.
3. Next, non-survival surgery is performed on anesthetized rodents. At the end of surgery, rodents are euthanized while still under general anesthesia. When proficient, survival surgeries are scheduled (see Notes 3–6).
4. Prior to surgery, all rodents are given buprenorphine (supplier: Webster Veterinary), meloxicam (Webster Veterinary), and baytril (Webster Veterinary) subcutaneous injections (see dosage information below). Consideration is given to how long the animals would be anesthetized and if this is enough time for the analgesics to be on board when the animal awakes.

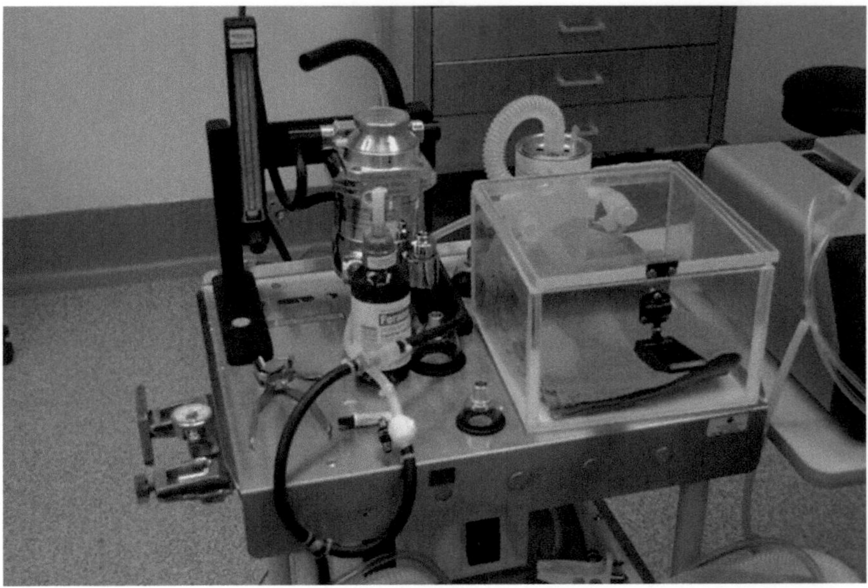

Fig. 3 Closer view of isoflurane chamber used for rodent anesthesia

5. Rodent surgeries are performed in a dedicated surgical suite with a separate scrub room. All participants are required to wear shoe covers, hair bonnets, masks, sterile gowns, and sterile gloves.
6. Rodents are induced with 5% isoflurane (Baxter) in an induction box (Fig. 3). Their eyes are covered with Paralube Vet Ointment (Dechra Veterinary Products, Overland Park, KS 66211).
7. The surgical site is shaved and scrubbed in a separate area of the surgical suite. An alcohol and chlorhexidine scrub (Chloraprep, cat# 260715, Carefusion, KS) is performed over the surgery site.
8. Rodent is placed on the surgery table and a surgical plane of anesthesia is maintained at 0.5–2% isoflurane. The animals' respiration rate is monitored carefully, as the buprenex, combined with the isoflurane (and the decreased kidney function within rodents with hemi- or 5/6-nephrectomy), can quickly cause death at the beginning of surgery (see Note 7).

4 Surgery Suite Setup

1. The dedicated surgery suite (see Fig. 1) contains a rodent anesthesia machine with induction box and Mapleson E nonrebreathing tube and mask. A warm water recirculating plate

(VetEquip DSx Vented Warming Table connected to a Gaymar T/Pump) is used to maintain the body temperature of the animals.

2. The table and plate are draped with a sterile table cover and sterile absorbant towels are placed on top of the cover. The non-rebreathing tube is secured over the towels with sterile tape. The rat is positioned as appropriate for the current surgical procedure and draped with sterile towels or sterile drape.

5 Analgesics, Anesthetics, and Antibiotics

1. All rodents are administered subcutaneous injections of Meloxicam (Boehringer Ingelheim, MO), (2 mg/kg), buprenex (0.01–0.05 mg/kg), and baytril (5 mg/kg) (Webster Veterinary) before surgery. Meloxicam is an NSAID and may last up to 24 h. Buprenex is an opioid that must be re-administered no less than every 12 h. Using both drugs helps to insure a more level plane of analgesia. However, the animals must still be monitored for breakthrough pain. Baytril must be diluted to 10 mg/ml before administering to rodents or skin necrosis will occur.
2. At the end of surgery, the incisions are injected with a mixture of 1% lidocaine and 0.125% bupivacaine (Hospira, Inc., IL). This mixture immediately numbs the area, which stays numb for several hours.

6 Postsurgical Care

1. After surgery, rodent is placed in a towel-lined cage on top of a warm pad. When the rodent becomes mobile it is placed in a clean cage with food, water, treats, and enrichment.
2. Depending on the invasiveness of the surgery, rodents are placed on a regimen of meloxicam once or twice a day for 2 or 3 days and buprenex is given a minimum of twice a day for 2–3 days.
3. Rodents are monitored at least twice daily for signs of pain or distress: Porphyrin around the eyes and nose, hunched posture, staying near the edge of cage and not moving, rapid breathing, and little or no interest in treats (4–6). These animals are immediately given an injection of buprenex with or without meloxicam and observed again in 20 min. If the pain continues the animal is euthanized and necropsied. The animal will be examined for obstructions and adhesions, especially if the rat had esophageal or intestinal surgery.
4. Soft food is offered for animals having gastrointestinal surgery (esophagus or small intestine).

7 Harvesting the Construct

1. At specified temporal endpoints, rodents are euthanized by administration of CO_2.
2. CO_2 euthanasia is performed by following the latest recommendations as defined by the AVMA (7). The rodent is placed in a euthanasia chamber within its enrichment tube. The CO_2 is set at a flow rate to displace the air in the container at 20% per minute. When respiration ceases, the flow is continued for several more minutes.
3. Alternatively, euthanasia may be performed by severing a major vessel while under a surgical plane of anesthesia using isoflurane, or by intracardiac injection of 0.5 ml Euthasol (Virbac Animal Health, Inc., TX).
4. During necropsy, care is taken not to harm the construct:
5. Kidney constructs are examined grossly for size, shape, and the presence of urine-filled blisters or nephrosis.
6. Intestinal constructs are sometimes difficult to find and involve the careful removal of the entire intestinal tract. Often adhesions grow around the area of the construct and require careful dissection to find the surgical area. Older animals may show an increase in fatty deposits throughout the abdominal cavity covering the construct.
7. Lung constructs require the removal of the entire lung and trachea. Care is taken not to puncture the lung tissue. After removal, a catheter is tied into the trachea and air introduced with a syringe. All lobes are expanded with air, the trachea is tied off, and the air-filled lung placed in 10% neutral buffered formalin.

8 Lessons Learned

1. Certain problems occurred at the beginning of the esophageal and small intestine surgeries due to the natural behavior of rats. Several rodents were euthanized soon after surgery due to not eating. Necropsy revealed pieces of green towel lodged within the esophageal construct of one animal and the small intestine construct of the other. Thereafter rodents were immediately removed from their post-op cage which was lined with green towels and placed in a clean cage when they became mobile.
2. The regular animal bedding posed a problem at one point in the small intestine study. An animal was euthanized when it stopped eating and bedding was found lodged in the construct. Hair was also found obstructing esophageal constructs.
3. Abdominal surgery posed an additional problem with chewing on the incision. Incisions made on the chest or side and closed

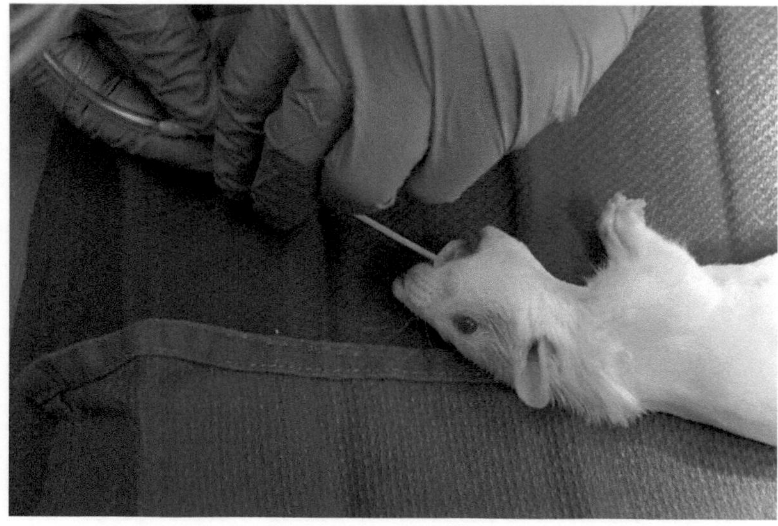

Fig. 4 Intubation of rat for delivery of biomaterials to rodent lung

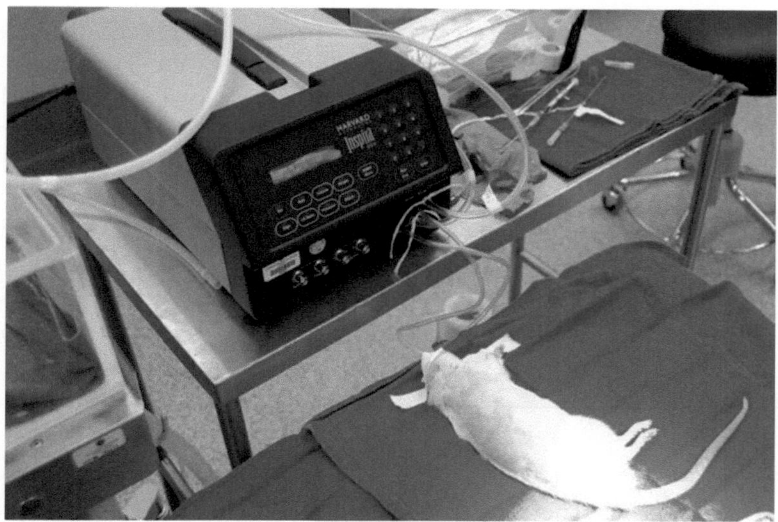

Fig. 5 Intubation of rodent lung

with staples were typically left alone by the rat. Skin incisions on the abdomen may be opened up by the rat removing the staples. When this occurred, the rat would be re-anesthetized and the incision cleaned and sutured with subcutaneous sutures. Lidocaine and bupivacaine would be injected at the site. The rats would then leave it alone. It was also found that the rats would not chew at skin incision glue.

4. During the lung study several surgical models were tested. At the beginning of the study the rats were intubated (Fig. 4) and placed on a Harvard respiratory machine which was connected to an isoflurane anesthesia machine (Fig. 5). This allowed open

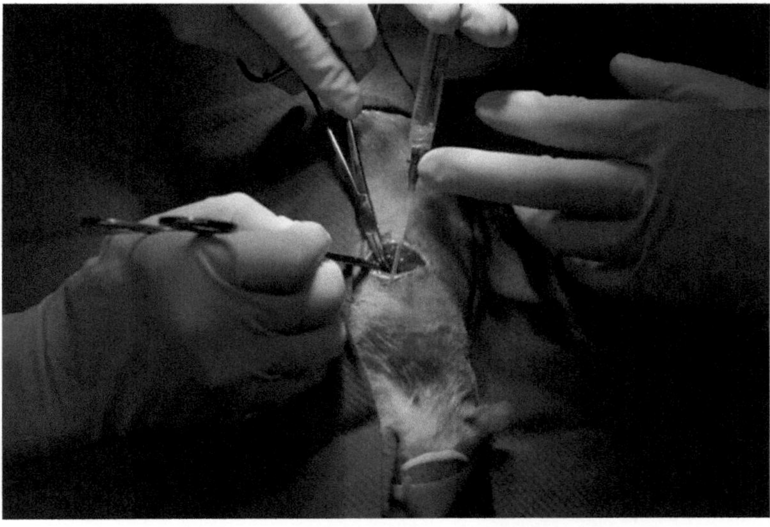

Fig. 6 Open thoracic surgery showing direct delivery of biomaterial to rodent lung

thoracic surgery to be performed (Fig. 6). The respiration rate of the Harvard apparatus was slowed. The tubing was manipulated to create a full inhalation allowing a lobe of the lung to be placed outside the chest. The construct was embedded and sutured into the lung lobe. The lobe was replaced into the chest and the incision closed. However, at necropsy the construct often had disappeared. A new approach was needed and the suggestion was made to inject the construct into the trachea where it would flow into the lobes of the lungs (Fig. 7). This proved to be a very simple surgery with no postoperative problems.

9 Notes

1. These are inbred rats, MHC Haplotype RT1, typically used for transplantation studies. The rats are raised in barrier production rooms or isolators and are routinely tested as free for an extensive list of pathogens. They were originally developed by Dr. Lewis from Wistar stock in the early 1950s and came to Charles River from Tulane in 1970. The rats we used were approximately 2–4 months old. The animals were typically between 175 and 250 g. Age was important as animals older than one year had more complications that required euthanasia before the end point.
2. The treats were given when the cages were changed once a week. This was important to decrease fighting and injuries when animals were placed in clean cages. Treats were also given after injections, staple removal, and after urine and blood

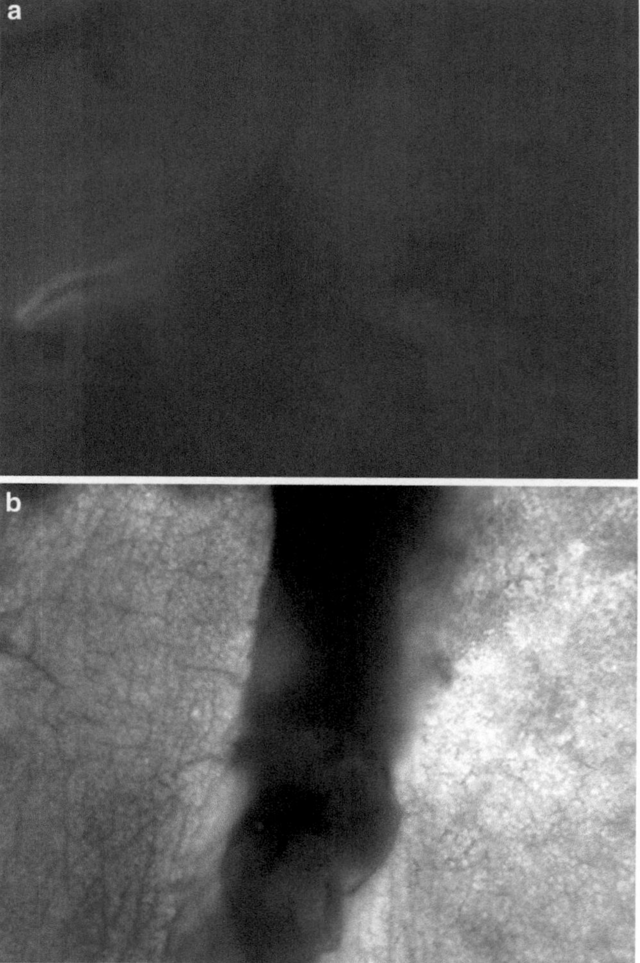

Fig. 7 Delivery of FITC–gelatin particles to rat lung by intra-tracheal injection (whole lungs removed and examined by inverted fluorescence): (**a**) negative control, whole rat lungs without biomaterial (**b**) experimental, whole rat lungs removed after delivery of FITC-gelatin particles showing extensive fluorescence throughout all lobular compartments

collection. All cages contained enrichment including rat tubes and chew bones.

3. Animal surgeries were most often performed in the morning at least 2 days before the weekend so that there was a full staff to monitor the animals after surgery. The Vivarium staff could easily tell if an animal did not feel well, as the healthy animals would jump to the front of the cage when approached, expecting a treat.
4. Rodents requiring a special soft diet are introduced to the soft food a week before surgery. Rodents themselves choose what type of soft food they prefer (pureed pumpkin, squash, and jello were favorites). It is important to offer soft food in the

days before the surgery so that the rodent is familiar with it. Rodents would not touch the commercial rodent gel food (Harlan).

5. Rodents to undergo esophageal or intestinal surgery are fasted for 4 h; however, this was changed to no fasting when it was found to not affect the outcome of the study.
6. Pre-surgical blood draws and urine collections are required in some studies. The blood draws are performed a week before the scheduled surgeries. This establishes a baseline for evaluation of key hematologic parameters such as hematocrit, serum creatine, blood urea nitrogen (BUN), etc. This is especially important in studies of renal therapeutics in rodents with compromised kidney function due to prior kidney removal (hemi- or 5/6 nephrectomy) (8).
7. Rodents with hemi- or 5/6 nephrectomy will typically require 0.5% isoflurane. Blink reflex is continuously monitored at this percentage isoflurane and adjustments to isoflurane concentration made as needed.

Acknowledgements

I wish to acknowledge the Vivarium Surgical Staff of the Cannon Research Center, Carolinas Healthcare System, whose expert assistance made this chapter possible.

References

1. Basu J, Mihalko KL, Payne R et al (2012) Extension of bladder-based organ regeneration platform for tissue engineering of esophagus. Med Hypotheses 78:231–234
2. Basu J, Mihalko KL, Payne R et al (2011) Regeneration of rodent small intestine tissue following implantation of scaffolds seeded with a novel source of smooth muscle cells. Regen Med 6:721–731
3. Basu J, Genheimer CW, Rivera EA et al (2011) Functional evaluation of primary renal cell/biomaterial Neo-Kidney Augment prototypes for renal tissue engineering. Cell Transplant 20:1771–1790
4. Guide for the care and use of laboratory animals, 8th edn. National Research Council of the National Academies, 2011. www.nap.edu/catalog.php?record_id=12910. Accessed on 22 Feb 2013
5. Recognitions and alleviation of pain in laboratory animals, National Research Council of the National Academies, 2009. www.ncbi.nlm.nih.gov/books/NBK32658/pdf/TOC.pdf. Accessed on 22 Feb 2013
6. Recognition and alleviation of distress in laboratory animals, National Research Council of the National Academies, 2008. www.nap.edu/catalog.php? record_id=11931. Accessed on 22 Feb 2013
7. 2011 AVMA Euthanasia Guidelines. www.avma.org/issues/animal_welfare/euthanasia.pdf. Accessed on 22 Feb 2013
8. Genheimer CW, Ilagan RM, Spencer T, Kelley RW, Werdin E, Choudhury S, Jain D, Ludlow JW, Basu J (2012) Molecular characterization of the regenerative response induced by intrarenal transplantation of selected renal cells in a rodent model of chronic kidney disease. Cells Tissues Organs. doi:10.1159/000336028

Chapter 18

Bioreactor Design Considerations for Hollow Organs

Jeff Fish, Craig Halberstadt, Darell W. McCoy, and Neil Robbins

Abstract

There are many important considerations in the design, construction, and use of a bioreactor for growing hollow organs such as vessels, gastrointestinal tissue, esophagus, and others. The growth of new organs requires a specialized container that provides sterility and an environment conducive to cell-seeding and attachment onto a three-dimensional bioabsorbable porous scaffold, incubation, maturation, and shipping for implantation. The materials' selection, dimensions, manufacturing, testing, and use of the bioreactor are all factors that should be considered in designing a bioreactor for the development of hollow organs.

Key words Bioreactor, Organ, Incubation, Transport, Medium

1 Introduction

A bioreactor for the development of a hollow organ consisting of cells and a biomaterial scaffold must provide many functions. The neo-organ requires an environment of controlled temperature and gas mixture, and there needs to be a means for cell-seeding and attachment, media exchange, and sampling. An optimum design supports the ability to sterilize the scaffold within the bioreactor; supports well-distributed cell-seeding, maturation, and sampling of contents by designing a functionally closed system; and supports long-term stability of the final product with eventual shipping to a clinical site. Sterility must be maintained throughout the process, and the bioreactor should be opened easily upon delivery of the organ to the surgical suite, but never have accidental breaches of sterility or leak of medium.

A scaffold on which to seed the cells may be suspended in a frame which supports the structure in the center of the bioreactor. It is important to minimize contact of the scaffold with the walls of the bioreactor. Cell damage could occur at contact points, and cell attachment distribution will be uneven if the scaffold is not centrally located.

Joydeep Basu and John W. Ludlow (eds.), *Organ Regeneration: Methods and Protocols*, Methods in Molecular Biology, vol. 1001, DOI 10.1007/978-1-62703-363-3_18, © Springer Science+Business Media New York 2013

Cell-seeding may be accomplished in an open bioreactor within a biological safety cabinet (BSC), but a much-preferred process is using a functionally closed system. To accomplish this, a scaffold is suspended in a frame using suture material; the frame is inserted into the bioreactor, which is subsequently sealed; and cells are seeded onto the scaffold using an agitation system such as continuous rotation or perfusion. Submicron filters permit gas exchange, yet prevent contaminants from entering the bioreactor. This assembled bioreactor with a bioabsorbable scaffold undergoes a sterilization cycle of ethylene oxide gas. Other methods of sterilization, such as autoclaving or gamma radiation, may have negative and uncontrolled effects on the bioabsorbable scaffold material.

This chapter describes a method for designing, building, and using a bioreactor for a Neo-Urinary Conduit™ (NUC) tissue used to support the transport of urine from the ureters following a cystectomy which consists of autologous smooth muscle cells seeded onto a bioabsorbable scaffold. The dimensions, volumes, and rotation rates may be scaled to smaller and larger than those mentioned. It should be noted, though, that media volume, thermal exchange, surface areas, and other aspects do not scale linearly.

2 Materials

2.1 Component List (Fig. 1)

1. Bioreactor body (polycarbonate tubing, 2.75 in. OD, 0.125 in. walls, 8.5 in. length).
2. Fixed end-cap (polycarbonate sheet, 0.75 in. thickness, 5 in. diameter round, with O-ring groove, secured with socket head cap screws; integral barbs machined in for 0.125 in. ID tubing connection).
3. Removable end-cap (polycarbonate sheet, 0.5 in. thickness, 5 in. diameter, with O-ring groove, secured with stainless steel knobs).
4. Knobs or clasps to close/open lid (stainless steel).
5. Shroud to protect tubing/filters/connections (polycarbonate).
6. Cage structure to support scaffold (polycarbonate sheet and rod).
7. Tubing (quantity 4, 12 in. length).
8. Clamps (quantity 4, polypropylene ratcheting pinch).
9. In-line submicron filters (quantity 4, PTFE filter material in polypropylene housing).

2.2 Bioreactor Construction Components

There are a variety of materials appropriate for the construction of a bioreactor, and the selection depends on individual project requirements. Factors include cost, single or multiple use, transparency needs, and quantities. For example, 316L grade stainless steel makes a very durable bioreactor, but could be cost prohibitive for a single-use

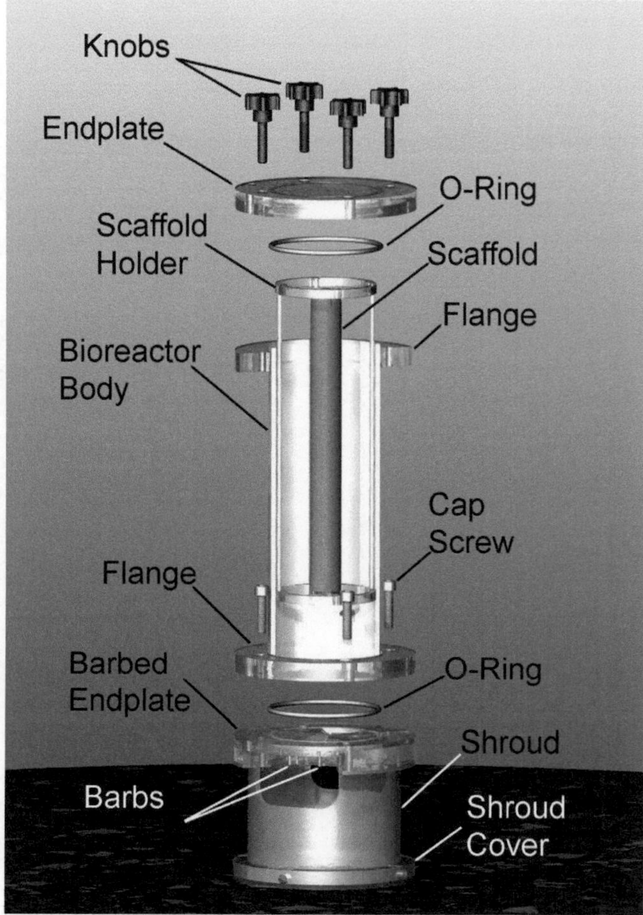

Fig. 1 Diagram of bioreactor containing a scaffold, showing components for securing the scaffold, protecting the barb fittings, maintaining the hermetic seal through incubation and shipping, and a means of opening in the clinical setting. The bioreactor body is constructed of transparent material to facilitate filling and draining during medium exchanges

product and it is opaque. Many polymers are available which are inexpensive, can withstand the sterilization process, and are appropriate for single-use applications. For quantities that can justify the initial capital cost of mold-making, injection molding is a good approach to manufacturing. For smaller quantities, mechanical machining is an option. The cost per piece is higher than injection molding, but there is no capital investment in molds up-front. For the initial clinical trials for the NUC, polycarbonate was used as a material for the bioreactor body, ends, and support cage, chosen for its ability to be machined, transparency, durability, and its ability to be sterilized using ethylene oxide. Other components that were used included silicone gasket material, stainless steel fasteners, Tygon® tubing, PTFE filters with polypropylene housing, and polypropylene tubing clamps.

Fig. 2 Bioreactor with scaffold prepared for shipping. Tubing on barb fittings protected from physical damage by shroud. The vessel is completely filled with fluid medium

Polycarbonate is an example of a very durable, machinable, transparent polymer that can function well as a bioreactor material. It is available in USP class VI and has been shown to have low levels of extractable and leachable components. Some care should be taken to desiccate the material prior to use, as water can dissolve into the polymer, and the subsequent humidity within the bioreactor can begin to degrade in an uncontrolled manner the scaffold prior to seeding. Advantages of transparency in bioreactor construction materials include the ability to visually monitor fill levels, scaffold attachment security, and turbidity or other signs of contamination.

A silicone or EPDM seal, either O-ring or gasket, may be used to provide a leakproof hermetic seal. These are available in USP Class VI grades. Closure mechanisms may be knobs with threaded stainless steel rods or hinged clasps. It is important that the bioreactor is easily opened at the point of use, but not accidentally opened prior (Fig. 2).

2.3 Cleaning, Packaging, and Desiccation

The bioreactor materials of construction should not have an affinity for cell attachment, but should be inert to cells. In an ideal situation, the cells should have an opportunity to land on the scaffold and attach, but not attach to the bioreactor walls.

Ideally, the bioreactor should be able to be opened without the use of tools. Under certain conditions, there can be a pressure differential between the bioreactor interior and the atmospheric pressure at the location where it is being opened. In a bioreactor with a large surface area to be opened, it may be necessary to include a means of breaking the seal and equalizing pressure. This may be done through the sealed-off tubing stubs, or by mechanically forcing the lid open with a threaded fastener acting against a fixed part of the bioreactor body.

2.4 Independent Testing

To qualify the bioreactor for clinical use, several tests need to be run. These include leachables and extractables from the bioreactor materials, container closure of the bioreactor to support culture and shipping of the product, and sterilization validation. NAMSA is an independent laboratory that can provide testing for medical devices, and can evaluate for biocompatibility and histopathological services. The bioreactor components that come in contact with the medium were tested in the following ways: Cytotoxicity Study Using the ISO Elution Method; USP Physicochemical Tests for Plastics (Aqueous) Complete, Purified Water Extract; USP Physicochemical Tests for Plastics (Nonaqueous) Complete, Isopropyl Alcohol Extract. Container closure testing may be done by immersing sterile bioreactors filled with tryptic soy broth (TSB) into a chamber of TSB inoculated with 1×10^6 CFU/mL of test organism (Brevundimonas diminuta). After a 24-h period, the bioreactors are removed, and incubated at 30–35°C for 7 days or more. Their external surfaces are washed, and the contained TSB tested for observed growth. This testing was performed by Catalent Pharma Solutions.

2.5 Details

In addition to an opening access lid or port, there are fittings or tubing connections through which liquid may be pumped and gas exchanged. For media exchange, Tygon® tubing (0.25 in. OD, 0.125 in. ID) is connected to the bioreactor through barbed fittings enclosed with ties. The tubing can be welded using a tubing welder such as the Terumo tube welder to media bags. The tubes can also be sealed to close the lines prior to shipping of the final product. Gas is exchanged through filters to maintain sterility.

The scaffold may be a material on which cells can attach, but will be absorbed over time after implantation as the cellular material matures. The scaffold may be a woven material, or a nonwoven material or felt. Polyglycolic acid (PGA) in fibrous form has been shown to be a suitable material.

3 Methods

3.1 Assembly and Sterilization

1. The bioreactor is assembled with scaffold held within a suspension cage.
2. It is inserted into sterilization pouches, which are sealed into a polymer bag with desiccant.
3. At the time of sterilization, the bioreactor is removed from the polymer bag and sterilized using an ethylene oxide gas cycle.
4. After sterilization, it is replaced into the polymer bag with desiccant until seeding.

3.2 Seeding of Cells

1. Cells are seeded onto the scaffold and are given time to attach. The bioreactor is rotated to establish uniform attachment.
2. If an open system is used for seeding, the bioreactor is opened and a known number of cells are applied via pipette at a known concentration, distributed evenly over the scaffold surface.
3. Filter tubes should be clamped closed during rotation. In a closed system, the distribution of cells may be accomplished by partially filling the bioreactor with a medium cell suspension (making sure that the scaffold is immersed in the media), and then rotating the bioreactor about a horizontal axis for 24–48 h at 0.25–10 revolutions per minute in an incubator or a warm room. The time and rotation speed are dependent on the dimensions of the bioreactor and scaffold, medium viscosity and density, and the size and density of the cells. An alternative to rotating the bioreactor as a closed system is to design a perfusion bioreactor in which medium is continuously circulated through the system.
4. The rotation speed of the bioreactor should be matched to the settling rate of the cells within the medium such that cells move toward the center of the bioreactor where the scaffold is located. Wall velocities of the bioreactor cylinder wall have been tested between 0.3 and 1.3 cm/s, with the best attachment toward the high end of that range.
5. The larger the bioreactor cylinder diameter, the lower the rotational speed required to achieve these velocities. This may be optimized with each bioreactor diameter, after cell type and medium are selected.
6. The length of the bioreactor is not a significant factor in this, so for hollow organs of two different lengths, bioreactors of two different lengths but the same diameter should function equivalently.
7. Aspects of the bioreactor seeding process that affect cell distribution are the medium volume and the headspace. Air in the headspace helps with the mixing of the medium and liquid gas

interface area for gas exchange to the medium, and prevents cell settling.

8. The medium volume should be matched with the scaffold mass and cell count such that the medium is not expended during the cell attachment period.
9. The filters need to stay dry to be functional, so the tubes leading to the filters are clamped during the rotation phase of cell attachment.
10. In a cylindrical bioreactor, medium fill volumes between 60 and 100% of total bioreactor volume were explored, with best results in the 80% range.
11. In designing dimensions, the long axis of the organ should be along the rotational axis of a cylindrical bioreactor. The length of the bioreactor should be long enough to accommodate the scaffold and supporting structure, but not so long as to have appreciable volume at the ends which could be inefficient for cell attachment.
12. The diameter of the bioreactor should be large enough to keep the scaffold submerged during rotation, even with the headspace of an 80% volume-filled cylinder.

3.3 Incubation and Maturation

1. Subsequent to seeding and attachment, the bioreactor and its contents are put into an incubator with controlled temperature, humidity, and CO_2.
2. The clamps on the filter tubes are unclamped during incubation, and the bioreactor is positioned such that the filter tube connection to the bioreactor is in the gas headspace, not immersed in medium.
3. The growth medium will need to be replaced occasionally, with the frequency of replacement depending on cell metabolism, medium volume, and cell numbers.

3.4 Shipping

1. After a cell maturation phase, the bioreactor may be shipped to the clinical site for implantation.
2. On the final fill prior to shipping, the bioreactor is filled with the shipping medium to as close as possible to 100% capacity, for the purposes of minimizing gas expansion during pressure changes, minimizing fluid turbulence during shipping, and maximizing medium volume for the cells.
3. The tubes with filters are sealed/cut in a sterile fashion using a heat-sealing device (Terumo, SCD IIB).The design of the bioreactor should include wall strength and sealing ring/gasket design that can withstand pressure differentials during transport by air. These pressure differentials are ~3.5 psi (24 kPa) in normal flight, but can be as high as ~12 psi (83 kPa) if the

aircraft depressurizes at cruising altitude. The temperature of the bioreactor should be maintained within a range during shipping. The range chosen depends somewhat on the nature of the particular product. Common ranges used are 5°C (±3°C) and 22°C (±5°C).

4. This may be accomplished through the use of an insulated container which may contain foam insulation, vacuum-insulated panels (VIP), or a combination of the two. A mass of phase change material (PCM) may be preconditioned to a temperature near the center of the target temperature range to prevent an excursion of the temperature profile out of range by using heat of crystallization and heat of fusion.
5. There are testing guidelines by the International Safe Transit Association (ISTA) which use a sequence of temperature extremes for Summer and Winter tests for specified time periods. A test with duration of 48 h is often used, though for international transit, or during a time of extreme storm systems that may affect air traffic for an extended time, there are shippers that can pass a much longer test. It may be preferred to use a robust enough system that the same shipper may be used for different expected ambient extremes, so the only difference needed is preconditioning temperatures for the PCM.
6. Temperature monitors that record data may be included in the payload compartment to provide assurance upon arrival at the implantation site that the package remained within profile, and to provide a record of the profile.
7. The bioreactor should be well padded in the payload compartment, and an outer box of corrugated fiber board or polymeric sheets protects the more fragile insulating layers from the rigors of transport.

4 Notes

1. It is preferred to have no small parts that could fall out upon opening the bioreactor.
2. The bioreactor was designed to have as few potential leak points as possible. For example the barbed fittings for tubing attachment were machined into a monolith of ¾ in. polypropylene sheet, eliminating threaded or glued fittings and small parts.

Chapter 19

Construction of a Multicoaxial Hollow Fiber Bioreactor

Randall McClelland, Katherine Tech, and Jeffrey M. Macdonald

Abstract

Bioreactors are assembled tools conceived to exploit engineering principles with inbuilt biological relevance. Such reactors are created as in vitro models to better replicate natural in vivo organs. These biotools are subsets within the interdisciplinary tissue engineering field and are established as inert devices to improve upon biological stimuli while simultaneously allowing tissue functional properties to be nondestructively measured. Design and fabrication efforts are focused on two-dimensional (2D) and three-dimensional (3D) physical constructs while linking environment–cell relations, the microenvironment. Product proficiencies generally involve material scaffolds, nutrient dispersion, compartmentalized units, passive and kinetic flow channels, temperature regulation, pressure management, and cell line or primary cells from assorted organs as tissues. Bioreactor advancements continue with interdisciplinary principles such as energy conservation, cell ecosystems, system-biological approaches, and viable-cell design innovation. Herein, we describe the design and construction of a hollow fiber multicoaxial bioreactor with integral oxygenation (i.e., oxygenation within the bioreactor proper) for use with liver cells, but it could be used with any anchorage-dependent cell type.

Key words Multicoaxial bioreactor, Hollow fiber, Membrane-type bioreactor, Integral oxygenation, Bioartificial liver, Hepatocytes

1 Introduction

A core task in bioreactor engineering is to recreate tissue function by constructing a viable and functional end product, a tissue. One common way to approach this goal is to recreate or replicate the native physiological in vivo environment. This task involves mimicking a variety of features of the microenvironment to include cell–cell and cell–matrix communications along with the biochemical and mechanical settings between tissues. Achieving this manufacturing goal is challenging and is strictly based on fundamental axioms of living cells. For liver tissue, the coaxial design mimics the architecture of the smallest unit of the liver (1), the liver lobule, or acinus, and if the hollow fiber diameters are chosen appropriately, the distance across the liver acinus, 0.5 mm, can be replicated. Figure 1 is a drawing of the liver acinus and a schematic

Joydeep Basu and John W. Ludlow (eds.), *Organ Regeneration: Methods and Protocols*, Methods in Molecular Biology, vol. 1001, DOI 10.1007/978-1-62703-363-3_19,

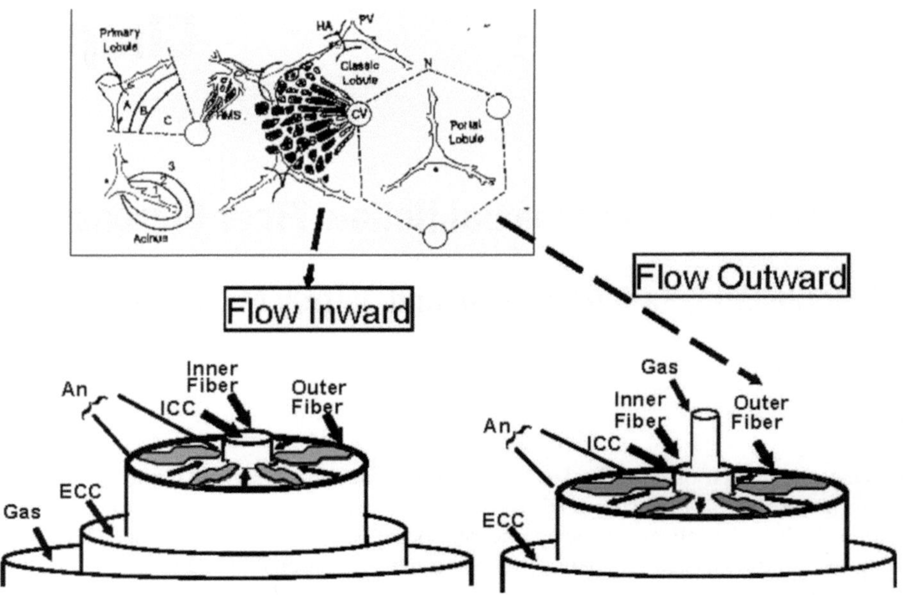

Fig. 1 Drawing of the liver acinus and a schematic representation of the hollow fiber coaxial bioreactor demonstrating the similarity of scale and architecture

representation of the hollow fiber coaxial bioreactor demonstrating the similarity of scale and architecture. The multicoaxial bioreactor (MCB) has a third oxygenation tube creating a fourth compartment between the housing and the tube for gas exchange, permitting integral oxygenation, which maintains oxygen concentrations axially (2). This compartment oxygenates the media compartment, which is adjacent to the compartment containing cells. The innermost compartment at the center of the MCB also contains media, thereby sandwiching the cell compartment between media supplies (3). The MCB has been shown to maintain viability and function of rat hepatocytes for 7 days (4), and healthy human hepatocytes for 30 days (5).

Oxygen is generally the limiting factor in hollow fiber bioartificial livers (5–8), primarily because hepatocytes are highly aerobic, but also because designs often do not have optimal diffusion distances of hepatocytes to an oxygen source. For tissue densities, this distance is typically accepted to be less than 200 μm (1, 5, 9). Since the MCB does not include hemoglobin-like blood, mass transfer of oxygen is dominated by diffusion from the media and into the cell compartment. This is plausible since the cells are sandwiched by two media compartments and result in 250 μm diffusion distances if hollow fibers are coaxially configured with an interfiber space of 500 μm, as shown in Fig. 1. For the MCB described in this chapter, the outer diameter (o.d.) of the inner hollow fiber is 1.3 mm and the inner diameter (i.d.) of the outer hollow fiber is 2 mm, resulting in an interfiber space of 350 μm, or a maximum diffusion distance of 175 μm. The rationale of this chapter is to give the reader specific step-by-step instructions on how to construct the MCB. Since the initial construction of the MCB, there have been some modifications that will be described throughout, but not specifically shown in parts depicted in the step-by-step images.

2 Materials

1. 5 cm lengths of polycarbonate tubing used for the housing, 13 mm o.d. (10 mm i.d.) (Piedmont Plastics Inc., Raleigh, NC, USA).
2. 6 machined manifolds, 1 in. diameter polypropylene rod, per bioreactor (Piedmont Plastics Inc., Raleigh, NC, USA).
3. 12 cm lengths of 1.3 mm o.d., 0.9 mm i.d., 0.65 μm pore size inner polysulfone hollow fibers (AG/Technology, Needham, MA, USA).
4. 9 cm lengths of 2.8 mm o.d., 2.0 i.d., 0.2 μm pore size middle polypropylene hollow fiber (Akzo-Nobel, Wuppertal, Germany).
5. 7 cm lengths of 8 mm o.d. Silastic™ gas exchange tubing and silicone glue (Dow Corning, Midland, MI, USA).
6. 0.394 in. by 0.079 in. thick Viton® O-rings (Apple Rubber Products, Lancaster, NY, USA).
7. Medical-grade polyurethane (Caschem Inc., Bayonne, NJ, USA).
8. A razor blade of an X-Acto™ knife is required for slicing the potted fiber flush for fitting with subsequent manifolds in Subheadings 3.1–3.3 below.
9. Polypropylene barbed fittings.
10. Perfusion loop consisted of a 1–60 rpm variable pump, Pharmed™ tubing, and plastic tubing clamps, all purchased from Cole Parmer (Chicago, IL, USA).
11. Vacuum apparatus for degassing.
12. Epoxy.
13. Wooden toothpicks.
14. Isothiocyanate.

3 Methods

For each step, the two parts of epoxy must be thoroughly mixed with a plastic rod using one to one ratios, typically in a 10–50 ml disposable container depending on the number of bioreactors

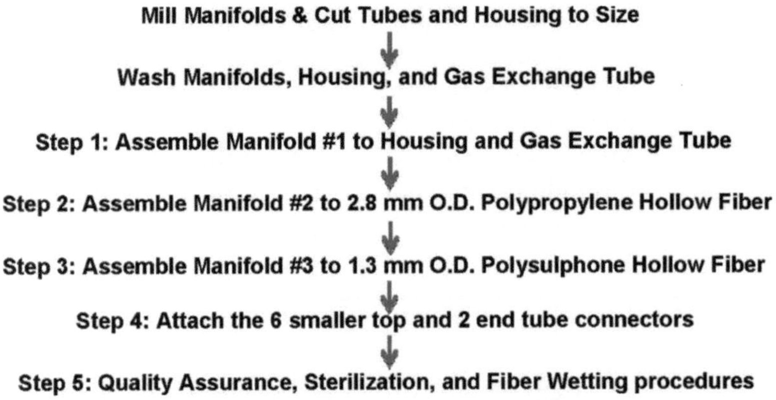

Mill Manifolds & Cut Tubes and Housing to Size

↓

Wash Manifolds, Housing, and Gas Exchange Tube

↓

Step 1: Assemble Manifold #1 to Housing and Gas Exchange Tube

↓

Step 2: Assemble Manifold #2 to 2.8 mm O.D. Polypropylene Hollow Fiber

↓

Step 3: Assemble Manifold #3 to 1.3 mm O.D. Polysulphone Hollow Fiber

↓

Step 4: Attach the 6 smaller top and 2 end tube connectors

↓

Step 5: Quality Assurance, Sterilization, and Fiber Wetting procedures

Fig. 2 Flow diagram of the process and timing of each step of construction

being constructed. The top of the disposable container must be attached to vacuum to degas the epoxy so bubbles do not form in the cured epoxy, once hardened. Care must be taken to account for the timing of the curing process, typically about 30 min. The isothiocyanate expires within a year and once opened, must be used within a few months. Most medical grade polyurethanes are polyols and composed of castor bean oil (Part A: Polycin® 936) with a catalyst cross-linking agent composed of 4,4′-methylene diphenyl diisocyanate (Part B: Vorite® 689) *without* metals included as a catalyst (see Note 1).

Figure 2 is a flow diagram of the process and timing of each step of construction described below. For assembly, the first three steps are used to align three manifold pairs. Prior to use, the machined Manifold 1–3, the polycarbonate housing, and the Silastic™ tubing should be thoroughly washed in 70% ethanol and set in a laminar flow hood to dry. A supporting rack is required, which is essentially two rods spaced the distance of the housing so Manifold #1 can rest in the slot and epoxy is placed in the well created around the gas exchange tube and the hollow fibers for potting and curing of epoxy. The fourth step attaches tube connectors to make a final MCB useable product, and then the fifth step is quality assurance and product preparation.

Figure 3 shows the various parts of the bioreactor: three sets of machined manifolds, the polycarbonate housing, the silicone gas exchange tube, the two macroporous hollow fibers, and numerous connectors.

3.1 Assembly of Manifold #1

Figure 4 shows the various parts of the bioreactor for the first step and includes two Manifold #1, polycarbonate tube (housing), Silastic™ gas exchange tube, and two tube connectors. The three sets of manifolds are machined from 1-in. polypropylene rods.

Step 1 - Assemble Manifolds #1

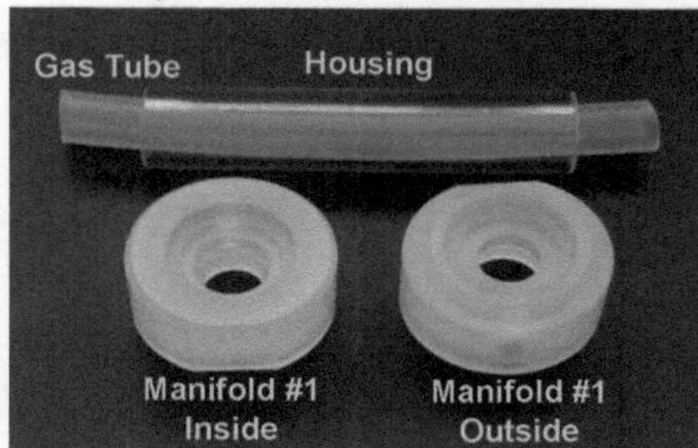

Step 2 - Assemble Manifolds #2

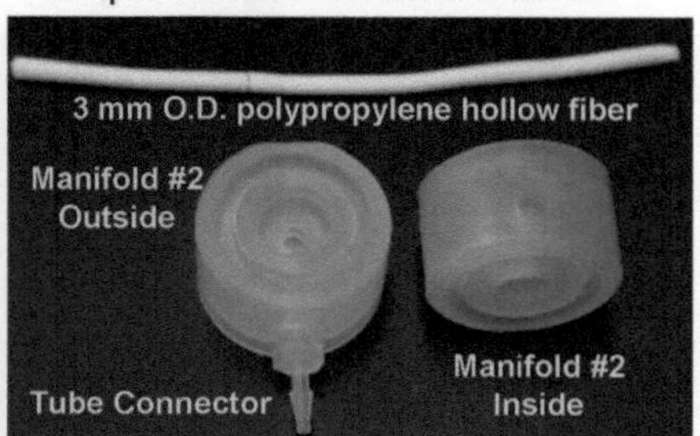

Step 3 – Assemble Manifolds #3

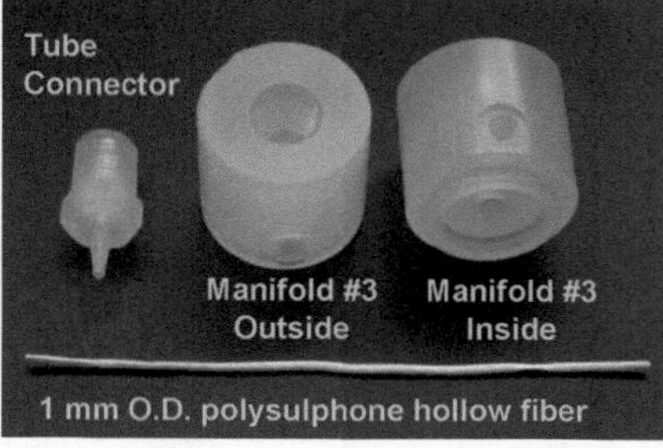

Step 4 – Attach Connectors

Fig. 3 Enlarged parts view and assembly

1. To pot Manifold #1 with the housing and gas exchange tube via epoxy, the two manifolds pop into the housing and the Silastic™ tube is pushed through the middle hole of each manifold (Fig. 4) (see Note 2).
2. The assembled component is placed on the rack and epoxy is applied with a small plastic or wood rod or toothpick in the two wells created by the Silastic™ tube and housing.
3. After an hour or once the epoxy has cured the assembled component is inverted (Fig. 4), and placed on the rack and the same procedure of epoxy placement is performed (see Note 3).

3.2 Assembly of Manifold #2

Figure 5 shows the various parts of the bioreactor for the second assembly step, and includes two Manifold #2 parts, the outer 2.8 mm outer diameter polypropylene hollow fiber, and the two associated tube connectors.

1. For the assembly of the two Manifold #2 parts with the 2.8 mm o.d. polypropylene hollow fiber, one side at a time is epoxied. First a small amount of epoxy is placed on the male portion of the manifold and squeezed together a complete symmetric fit is obtained.
2. Then the 2.8 mm polypropylene hollow fiber is pushed through the hole in Manifold #2. It is important to no epoxy seeps into

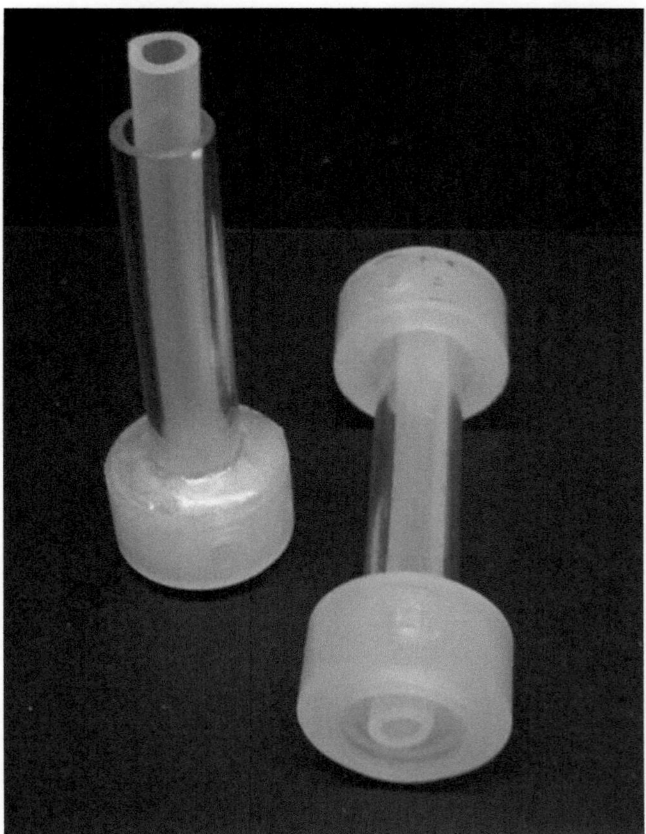

Fig. 4 Intermediate step in gluing parts together

the inner part of this fitting; otherwise the epoxy will coat the length of the hollow fiber as it is pushed through the hole in Manifold #2 (see Note 4).

3. The end of the hollow fiber is pushed down to be flushed with the end of the manifold.
4. The well created between the hollow fiber and the manifold is filled with epoxy.
5. After an hour of curing, the assembled component is inverted and the other side is assembled in a similar fashion (see Note 5).
6. Once cured, a razor blade or X-Acto™ knife is used to cut the epoxy and hollow fiber flushed with the end of the manifold so that Manifold #3 can be placed in step 3 (Fig. 5).

3.3 Assembly of Manifold #3

Figure 6 shows the various parts of the bioreactor for the third step and final epoxy stage for the MCB. The parts include two Manifold #3, the inner 1.3 mm polysulfone hollow fiber, and the end piece tube connectors.

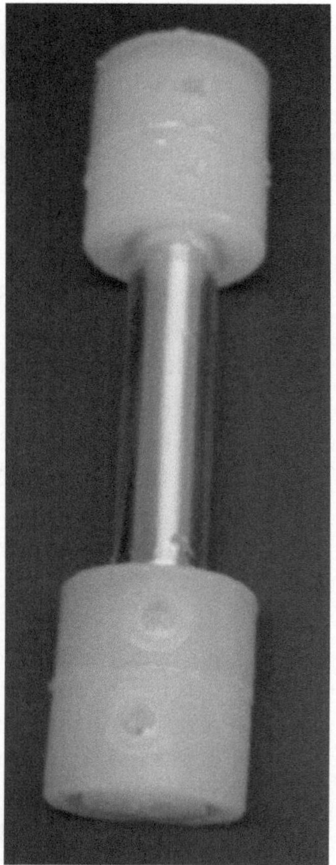

Fig. 5 Completed glued assembly

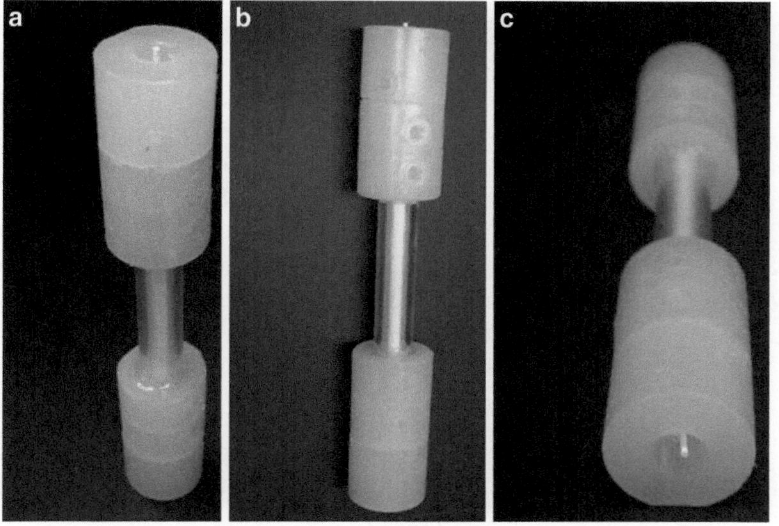

Fig. 6 Positions of top end (**a**), side end (**b**), and bottom end (**c**) holes for attaching barbed fittings

1. One side at a time is epoxied for the assembly of the two Manifold #3 parts with the 1.3 o.d. polysulfone hollow fiber. First a small amount of epoxy is placed on the male portion of the manifold and squeezed together to make sure a complete symmetric fit is made. It is important not to put epoxy on the inside area of the male part of the manifold. Even a small amount of epoxy on this inside portion can squeeze inward and clog the outlet/inlet ports.
2. Place Manifold #3 on the outside face on the table (Fig. 6), while the inner 1.3 mm hollow fiber is inserted through the middle of the assembled component of the MCB.
3. Next thread the 1.3 mm polypropylene hollow fiber through the hole on the inside face (i.e., the side with the epoxy) of Manifold #3 (see Note 6).
4. Squeeze the inside face Manifold #3 (Fig. 6) against the outside face of Manifold #2 (Fig. 5), and set on the rack. Make sure all of the epoxy has squeezed out the sides of the contact between Manifold #2 and Manifold #3, and Manifold #3 is flushed. Ineffectively, if there is too much epoxy, Manifold #3 will lift up away from Manifold #2, and this should be monitored for a minute or two to ensure this does not occur. This effect is visible in the middle photo of the assembled MCB in Fig. 6 (see Note 7).
5. The end of the 1.3 mm polysulfone hollow fiber should extend from the end of Manifold #3 at least an eighth of an inch.
6. The well created between the hollow fiber and the Manifold #3 is the trickiest to fill with epoxy, because there is a small space between the hollow fiber and the threads. If the threads get epoxy coating on them, the final end-port cannot be screwed in. The epoxy is applied with a small plastic stick, similar to a toothpick, and the hollow fiber is coated until enough has made it to the well at the contact of Manifold #3 and the hollow fiber. After an hour of curing, the assembled component is inverted and the other side is assembled in a similar fashion (see Note 8).

3.4 Attachment of Tube Connectors

Figure 7 shows the final MCB for the fourth step with the three manifold sections for each of Subheadings 3.1–3.3 demarcated and the tubing connectors attached. Each of the connectors is shown in Fig. 3, but consists of six smaller top tube connectors, and two end tube connectors.

1. After epoxy curing, the excess epoxy is shaved off with a razor along the demarcation lines for each of the manifolds shown in Fig. 7.
2. The inside 1.3 mm hollow fiber is cut near the epoxy–fiber interface in the wells of Manifold #3. Each of the connectors,

Fig. 7 Final assembled bioreactor

the six top connectors and the two end connectors, is then screwed in to create final MCB embodiment shown in Fig. 7.

3.5 Quality Assurance and Sterilization

1. Once the bioreactor is complete as shown in Fig. 7, a 5–10 cc disposable syringe is filled with air, a tube is used to connect the syringe to each of the ports, and the air is pushed through each port to see if it is open and not clogged with epoxy.
2. The MCB is then placed in an autoclavable bag and subjected to ethylene oxide treatment for sterilization.

3.6 Fiber Wetting

Before assembling the bioreactors into a perfusion loop, the middle polypropylene fiber must be wetted with ethanol to make it permeable to water, and this must be done uniformly across the length of the fiber so that the entire fiber is properly wetted. The process of fiber wetting is performed in a cell culture hood.

1. The two top ports on Manifold #3 (Fig. 7) accessing the cell compartment, or annular space (Fig. 1), should be attached with a piece of tubing.
2. The extracapillary compartment (ECC) ports on Manifold #2 should be attached with two pieces of tubing connected to a Tee connector (Fig. 1).
3. Attach a 5–10 cc syringe containing 95–100% ethanol via tubing to the end port of Manifold #3 and place the MCB upright on its end.
4. Fill up the annular and intracapillary compartments (ICC) by pushing in 1 cc of ethanol until ethanol flows out of the tube at the top of Manifold #3.
5. Disconnect the top tube from Manifold #3, and hold tube exit and top Manifold #2 port so they are at the same elevated level.
6. Place a sterile gloved finger over the top exit port of Manifold #3 of the ICC, and force the ethanol laterally out of annular compartment port. Ethanol should flow from both ports if the tube attached to the lower port on Manifold #3 is at the same level as the top port of Manifold #3.
7. Tap the MCB to dislodge bubbles and push more ethanol out while holding the exit port of Manifold #3.

8. Replace top tube to the top of Manifold #3 port; the annular compartment and ICC are now full of ethanol.
9. Pressurize the ethanol-containing compartments by compressing the syringe until it flows out through the ECC ports.
10. Quickly switch the syringe and flow through four washes of distilled water, washes of phosphate-buffered saline.
11. Minimize exposure of the MCB to ethanol as it will dissolve the epoxy and may clog the pores of the hollow fibers. The MCB is now ready for inoculation and integration into perfusion loop as previously shown (see Note 9) (10).

4 Notes

1. Remember to purchase polyurethane that cures in about 1 h and degas the polyurethane prior to use, so bubbles will not form which will form breaches in the integrity of the bioreactor. Also, for mass production of manifolds, an injection mold for each manifold would be more economical than machining each one.
2. Instead of using epoxy, compression fitting would speed up this process using o-rings described in Subheading 2, whereby the housing and Silastic™ tubing would create the seal by either popping or screwing into place as described previously (3).
3. Each side is epoxied separately and a separate batch of epoxy will need to be mixed. If timed properly, the epoxy can be mixed and degassed about 45 min into the mixing of the first epoxy mixture.
4. Make sure not to insert the entire length of the hollow fiber through the hole in Manifold #2 as it may score the full length of the fiber if there is a shard or the hole is not perfectly smooth.
5. Each side is epoxied separately and a separate batch of epoxy will need to be mixed.
6. Again, make sure not to insert the entire length of the hollow fiber through the hole in Manifold #3 as it may score the full length of the fiber if there is a shard or the hole is not perfectly smooth.
7. The hole in Manifold #3 has more clearance the one in Manifold #2. The clearance for the port, however, is significantly smaller and one must carefully spread epoxy only on the male portion of the manifold outward, as not to squish excess epoxy into this cavity upon assembly of Manifold #3 to Manifold #2. If epoxy has seeped into the inner part of this fitting, it is likely the inlet/outlet port will be clogged, and at this point, it can only be checked during the quality assurance step at the end of the MCB assembly process.

8. A small syringe with a needle was initially used to apply the epoxy; however when the needle was extricated after application, the epoxy had a good chance of smearing onto the threads. It is also important that the consistency of the epoxy is similar to cold honey, so it does not seep through the small space between the hollow fiber and Manifold #3 and clog the port. Typically, and if the epoxy cures in about 60 min, this epoxy step should be done about 30–40 min after the epoxy was mixed. Using this process, a support stand that holds 4–5 bioreactors at one time is the maximum number of MCBs that can be simultaneously constructed, if one is concerned about efficient use of epoxy. Each side is epoxied separately and a separate batch of epoxy will need to be mixed.
9. The five steps described in Subheading 3 describe the construction of the MCB, through sterilization, wetting of the polypropylene fibers, and just prior to insertion into a perfusion loop. For a description of inoculation and a picture of the perfusion loop used in our laboratories, the reader is directed to McClelland and Coger (4), Seagle and others (10), and Jeffries and Macdonald (5). Briefly, the MCB is best inoculated with the axis of the multicoaxial tubes in the vertical position, with ECM mixed with cells in a 1 cc syringe attached to the bottom port of the annular compartment (Figs. 1 and 3). The MCB is perfused in this same position and a picture of the perfusion loop inside a conventional incubator is shown in Seagle and others (10), and the parts for the perfusion loop are listed in the Materials section of this publication. There are several flow configurations one can use, but a cross-flow configuration minimizes membrane fouling (11).

Acknowledgments

The authors would like to acknowledge NIH grant R43ES020638 for partial salary support during the writing of this manuscript.

References

1. Macdonald JM, Grillo M, Schmidlin O et al (1998) NMR spectroscopy and MRI investigation of a potential bioartificial liver. NMR Biomed 11:55–66
2. Wolfe SP, Hsu E, Reid LM et al (2002) A novel multi-coaxial hollow fiber bioreactor for adherent cell types. Part 1: hydrodynamic studies. Biotechnol Bioeng 77:83–90
3. Macdonald JM, Wolfe SP (2005) Bioreactor design and process for engineering tissue from cells. USA. pp 1–45
4. McClelland RE, Coger RN (2004) Effects of enhanced O(2) transport on hepatocytes packed within a bioartificial liver device. Tissue Eng 10:253–266
5. Jeffries RE, Macdonald JM (2012) New advances in MR-compatible bioartificial liver. NMR Biomed 25:427–442
6. Macdonald JM et al (1999) Bioartificial livers. In: Kuhtreiber W, Lanza RP, Chick WL (eds) Cell encapsulation technology and therapeutics. Birkhauser Boston, Cambridge

7. Macdonald JM et al (2001) Epithelial cell culture: liver. In: Lanza WL, Langer R, Vacanti J (eds) Methods of tissue engineering. Academic, San Diego, CA
8. McClelland RE, MacDonald JM, Coger RN (2003) Modeling O2 transport within engineered hepatic devices. Biotechnol Bioeng 82:12–27
9. McClelland RE, Coger RN (2000) Use of micropathways to improve oxygen transport in a hepatic system. J Biomech Eng 122:268–273
10. Seagle C, Christie MA, Winnike JH et al (2008) High-throughput nuclear magnetic resonance metabolomic footprinting for tissue engineering. Tissue Eng Part C Methods 14:107–118
11. Macdonald JM, Wolfe SP, Roy-Chowdhury I, Hiroshi K, Reid LM (2001) Effect of flow configuration and membrane characteristics on membrane fouling in a novel multicoaxial hollow fiber bioartificial liver. Ann N Y Acad Sci 944:334–343

Chapter 20

Isolation of Human Cadaveric Pancreatic Islets for Clinical Transplantation

Craig Halberstadt, Deana Williams, and Paul Gores

Abstract

Diabetes is a debilitating condition which can lead to chronic vascular, renal, and ophthalmic disease. Type I or Juvenile Diabetes is caused by the destruction of beta cells within the islets of Langerhans within the pancreas. The beta cells are able to maintain tight control of blood glucose levels by virtue of their ability to secrete insulin in response to small increases in blood glucose concentration. In the absence of beta cells patients with Type I diabetes are dependent on the exogenous administration of insulin. This results in imperfect control of blood glucose levels. In early animal and human studies, it was shown that the transplantation of allogeneic pancreatic islets into the liver via the portal vein, coupled with low-dose immunosuppression, could lead to insulin independence and tight blood sugar control. Since these seminal studies, it has been clinically demonstrated that islets isolated from cadaveric pancreases and transplanted into the portal vein of immunosuppressed patients can maintain a state of insulin independence for upwards of 5 years. This chapter describes a method of isolating and formulating pancreatic islets from the human cadaveric pancreas.

Key words Diabetes, Human islet isolation, Transplantation

1 Introduction

The successful treatment of patients with Type I diabetes by intra-portal allogeneic islet transplantation has been demonstrated by us and several groups (1–7). An immunosuppression protocol using tacrolimus and sirolimus coupled with a refined process for isolating the fragile islets from cadaveric pancreases (called the Edmonton Protocol) has demonstrated greater than 5-year engraftment of islets into Type I diabetic patients with subsequent reduction or elimination of the need for exogenous insulin (2–7). The following protocols provide a step-by-step method for isolating islets from cadaveric human pancreas by describing methods for the processing of the cadaveric pancreas,

Joydeep Basu and John W. Ludlow (eds.), *Organ Regeneration: Methods and Protocols*, Methods in Molecular Biology, vol. 1001, DOI 10.1007/978-1-62703-363-3_20, © Springer Science+Business Media New York 2013

the digestion and liberation of the islets from the tissue, the separation of the islets from the contaminating acinar tissue, and the preparation of the islets for transplantation.

2 Materials

See Tables 1 and 2 for a list of equipment, supplies, reagents, and materials for the entire process.

2.1 Specific Materials for Steps of the Isolation of Pancreatic Islets

2.1.1 Decontamination of the Pancreas

1. Betadine (150 ml).
2. Ancef (cefazolin sodium, 1 g).
3. Amphotericin B × 2 (50 mg/vial).
4. Hanks buffered salt solution (HBSS), no Phenol Red (670 ml).
5. 70 % Ethanol.
6. Biological safety cabinet.
7. Sterile 35 cc syringe.
8. Sterile 18-gauge needle.
9. Sterile 1,000 ml beaker.
10. Sterile 600 ml beaker × 2.
11. Sterile 100 ml cylinder.
12. Back table cover.
13. EuroCollins (1 l) + Eurocollins additive (20 ml).
14. HBSS.
15. 70 % isopropyl alcohol.
16. Sterile crushed ice (1 bag).
17. Weigh scale.
18. Sterile Saf-T basin.
19. Sterile small stainless steel tray.
20. Sterile large stainless steel tray.
21. Sterile back table cover.
22. Hammer.
23. Sterile towel.
24. Sterile scissors.
25. Sterile pickups × 2.
26. Sterile mosquitos.
27. Sterile needle driver.
28. 15 ml conical tube.
29. 5 ml pipette.

Table 1
Equipment and supplies for entire process

Equipment and supplies	Quantity
1,000 ml Beaker	7
250 ml Graduated cylinder	5
0.45 μm Filter	5
0.5–10 μl Eppendorf pipette tips	1 Box
0.5–10 μl Eppendorf pipetter	1
10–100 μl Eppendorf pipetter	2
10–100 μl Eppendorf pipette tips	1 Box
10 gal Water bath	1
10 × 15 Heat seal pouches	1 Box
100–1,000 μl Eppendorf pipette tips	1 Box
100–100 μl Eppendorf pipetter	1
100 × 20 mm Petri dishes	10
1,000 ml Bottle with a lid	2
100 ml Bottle with a lid	10
100 ml Graduated cylinder	3
12 × 75 mm Glass tubes with caps	9
125 ml Bottle with a lid	2
12 cc Syringe	10
15 ml Conical tube rack	4
2.0 Microcentrifuge tubes	10
−20 °C Freezer	1
24-Well plate	1
24 × 24 Sterilization wrap	1 Case
250 ml Beaker	2
250 ml Conical tube rack	7
3.5 × 8.5 Heat seal pouches	1 Box
37 °C Incubator	1
4 °C Cold room	1
4 °C Refrigerator	2
4-Well chamber coverslip	1
45 × 45 Sterilization wrap	1 Case

(continued)

Table 1
(continued)

Equipment and supplies	Quantity
5 gal Bucket	2
5 gal Water bath	1
500 ml Bottle with a lid	2
50 ml Conical tube rack	3
6-Well plate	2/COBE run
6 × 15 Heat seal pouches	1 Box
6 × 10 Heat seal pouches	1 Box
60 cc Syringe	1
600 ml Beaker	4
7.5 × 15 Heat seal pouches	1 Box
96-Well black-sided, clear bottom plate	1
96-Well flat-bottomed plate	1
Airborne Express Shipment form	1
Aluminum foil	1 Roll
Autoclave	
Autoclaved 1.5 ml tubes	28
Base supports	2
Biohazard safety bag	2
Biological safety cabinet (laminar flow), 1 with vacuum	2
Blue clamps	9
Bottle top filter	4
Caps for 12 × 75 mm tubes	49
Cart with wheels	1
Chamber stand	1
COBE 2991 Blood cell processer	1
COBE bag	1/COBE run
Connectors	4
Digestion chamber with a lid	1
Drummond pipette aid	3
EL800 plate reader	1
Extension clamps	4

(continued)

Table 1 (continued)

Equipment and supplies	Quantity
Florescent microscope	1
Gamma counter	
Glass box	1
Glass slides	1 Box
Glass tube rack	2
Glass tubes	1 Box
Gradient maker holder	1
Green towels	21
Grid eyepiece reticle for microscope	1
Hair bouffant	1/person
Hammer	1
Hollow glass tubes (Candy cane shaped ends, 35.5 cm, 0.64 I.D.)	2
Ice bucket	6
Indicator tape	1 Roll
Iris scissors	2
Isolation mask	1/person
Kimwipes	2 Boxes
Lab counter	1
Large orifice tip	5
Large stainless steel tray	2
Large sterilizing tray	3
Luer lock syringe caps	10
LVC guardian jug (12 l)	2
Stainless steel marbles	9
Masterflex tubing, size 16	200 in.
Masterflex tubing, size 17	80 in.
Mayo scissors	3
Metal mesh screen (500 μm, 9.7 mm diameter, round)	1
Metzenbaum scissors	2
Microcentrifuge	1

(continued)

Table 1 (continued)

Equipment and supplies	Quantity
Microcentrifuge rack	2
Microscope	
Mosquitoes	9
Oven	
Parafilm	1 Box
Peristaltic pump	2
Peristaltic pump head	2
pH meter	1
Pickups	4
Pipette extender	5
Plastic container with lid/small	1
Portal venous access catheter	1
Protective eye glasses	1/person
Red biohazard bags	4
Red biohazard bin	1
Refrigerated centrifuge	1
Rods	2
Scalpel handle	2
Scrubs of appropriate sizes	1/person
Sharps container	
Shoe covers	2/person
Small stainless steel tray	2
Soft bristled brush	2
Sonicator	1
Spray bottles for ethanol	4
Stainless steel heat exchange coil	1
Steam indicator strips	1 Bag
Sterile # 11 scalpel blade	2
Sterile 10 ml pipette	38
Sterile 14-gauge cannulas	2
Sterile 15 ml conical tubes	46

(continued)

Table 1
(continued)

Equipment and supplies	Quantity
Sterile 16-gauge cannulas	2
Sterile 18-gauge cannulas	2
Sterile 18-gauge needle	1
Sterile 20-gauge cannulas	2
Sterile 22-gauge cannulas	2
Sterile 250 ml conical tubes	38 + 8/ COBE run
Sterile 25 ml pipette	12
Sterile 2 ml pipette	3
Sterile 35 cc syringe	3
Sterile 3 cc syringes	1 Box
Sterile 4-0 silk	1
Sterile 50 ml conical tube	17
Sterile 5 ml pipette	6
Sterile back table cover	3
Sterile bag	2
Sterile flip filter	3
Sterile one-way stop cock	1
Sterile Saf-T basin	1
Sterile surgical gloves size 6	2/person
Sterile surgical gloves size 7	2/person
Sterile surgical gloves size 7.5	2/person
Sterile surgical gown with hand towels—XL	1/person
Sterile surgical gown with hand towels—L	1/person
Sterile temperature probe	1
Sterile tubing	2
Stir bar Stir plate	3 1
Styrofoam container	1
Surge protector	2

(continued)

Table 1 (continued)

Equipment and supplies	Quantity
Surgical cap	1/person
T connector (O.D.1/4)	1
T connector (O.D.5/16)	1
T25 Non-tissue culture treated culture flask	2
T75 Non-tissue culture treated culture flask	1
Temperature monitor (+18 to +50 °C)	1
Temperature monitor (−30 to 0 °C)	2
Temperature monitor (−5 to −15 °C)	3
Thermocouple monitor	1
Timer	3
Vortex	1
Waste beaker	3
Wedge—giant	1
Wedge—regular	1
Weigh boat	5
Weigh paper	10
Weigh scale	1
Weighing spatula	10
Y connector (O.D.1/4)	1

30. Glass straw.
31. 4-0 silk.
32. LVC jug.
33. Sterile tubing.
34. Parafilm.

2.1.2 Making Liberase Enzyme

1. Liberase, HI purified enzyme blend (0.5 g).
2. HEPES 1 M (8.5 ml).
3. 1× HBSS, no phenol red (350 ml).
4. Calcium chloride dihydrate ($CaCl_2$) (0.16 g).
5. Sodium hydroxide (NaOH) (endotoxin free) (0.1 N).
6. Hydrochloric acid (HCl) (endotoxin free) (0.1 N).

Table 2
List of reagents and supplies for islet isolation and performing QC assays

Materials	Quantity
125 I-Human insulin w/label hydrating buffer	
10 % Betadine	150 ml
10× Hanks balanced salt solution (HBSS)	24.88 ml/COBE run
1 M HEPES	77.7 ml + 10.08/COBE run
1 N HCl	1 ml
1 N HCl (endotoxin free)	0.5 ml
1 N NaOH	1 ml
1 N NaOH (endotoxin free)	0.5 ml + 0.582/COBE run
1× HBSS	1,200 ml
25 % Human albumin	900 ml
95 % Ethyl alcohol	1 gal
Amphotericin B (50 ml/vial)	1
Ancef (50 mg/vial, Cefazolin sodium)	2
Antibacterial soap	1 Pack
Assay buffer	40 ml
Autoclaved nanopure water	1 l
Calcium chloride	0.748 g
Calf thymus DNA	5 mg
CMRL 1066 media	500 ml
Conflict	1 Bottle
Dimethyl sulfoxide (DMSO)	20 ml
Distilled water	
Dithizone powder	200 mg
Eurocollins	1 l
Eurocollins additive	20 ml
Ficoll density gradient 1.077	130 ml/COBE run
Ficoll density gradient 1.100	290 ml/COBE run
Final wash solution	1 l
Glucose	3.9 g
Heat-inactivated fetal calf serum	26.25 ml
Human insulin specific RIA kit	26 ml
Ice	

(continued)

Table 2
(continued)

Materials	Quantity
LAL assay kit	1 Kit
l-Glutamine	15.25 ml
Liberase	0.5 g
Live/Dead Assay kit	1 Kit
Magnesium sulfate	0.494 g
Minimal essential medium (MEM)	9 l
Nanopure water	300 ml
Nicatinomide	1.22 g
No count radioactive decontaminate	
pH buffer 4	1 Bottle
pH buffer 7	1 Bottle
Phosphate buffered solution (PBS)	1.5 l
Pico green reagent	50 μl
Potassium chloride	0.744 g
Prezyme	10 oz
Precipitating reagent	260 ml
Professional strength lysol	1 Bottle
Scrub brushes with hibiclens	2/person
Sodium bicarbonate	0.704 g
Sodium chloride	15.4 g
Sodium dodecyl sulfate	10 g
Sodium phosphate—dibasic heptahydrate	0.162 g
Sparkleen	1 Box
Sterile ice	2 Bags
Sterile water	11.5 l
Theophylline	0.018 g
Transplant media	500 ml
Tris-EDTA 100×	1 ml
University of Wisconsin solution (UW)	4 l + 80 ml/COBE run
Wash solution	4.5 l + 2 l/COBE run
Z-fix solution	10 ml

7. Sterile ice.
8. pH meter.
9. 37 °C water bath.
10. Weigh scale.
11. Weigh paper.
12. Scoopula.
13. Ice bucket.
14. Sterile bottle top filter.
15. Sterile 100 ml cylinder.
16. Sterile 500 ml bottle with a lid.
17. Sterile 600 ml beaker.
18. 25 ml pipette × 2.
19. 10 ml pipette.
20. Drummond pipette aid.
21. –20 °C freezer.

2.1.3 Digestion System Priming System

1. MEM (minimum essential medium) (1 l).
2. HEPES 1 M (25 ml).
3. 25 ml pipette.
4. Drummond pipette aid.

2.1.4 Digesting the Human Pancreas

1. MEM (8 l).
2. Wash solution (4 l) (MEM + 25 % human albumin).
3. University of Wisconsin solution (100 ml per COBE gradient).
4. 70 % isopropyl alcohol.
5. Dithizone.
6. Sterile ice.
7. Refrigerated centrifuge.
8. Digestion chamber setup (Ricordi Chamber with 7–9 stainless steel marbles) (Figs. 1–3).
9. Timers.
10. Weigh scale.
11. 250 ml conical tubes.
12. Sterile 250 ml conical tube racks.
13. Sterile glass straw.
14. Sterile 1,000 ml beaker.
15. LVC guardian jug.
16. Sterile tubing × 2.

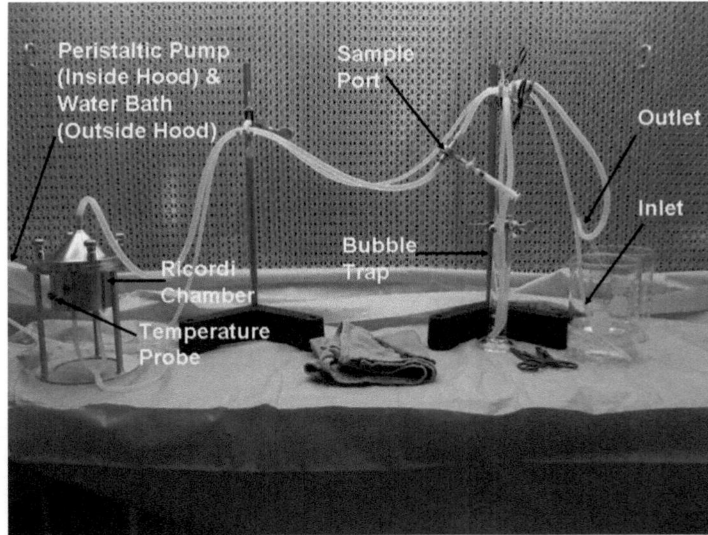

Fig. 1 Ricordi chamber setup in a horizontal laminar flow hood

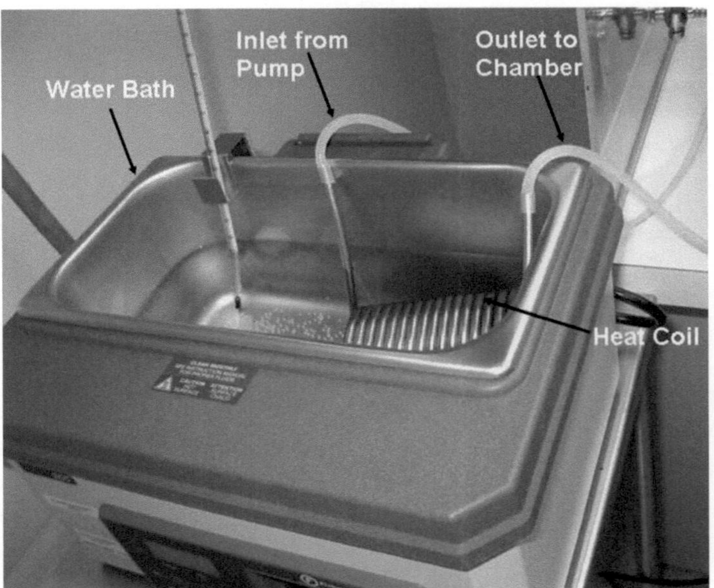

Fig. 2 Heat coil setup for maintaining temperature of the system at around 37 °C. Note temperature of bath was set at 41 °C due to reduction in temperature during flow through the system. Occasionally the heat-exchanger coil is removed from the water bath if the temperature in the Ricordi Chamber went above 38 °C

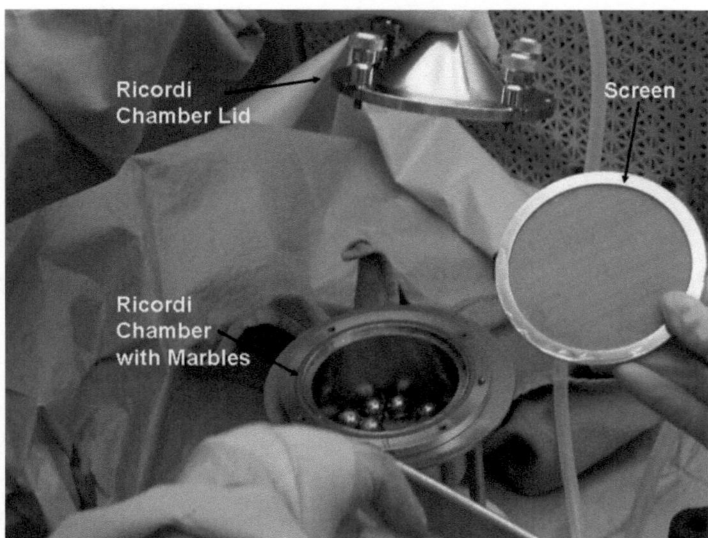

Fig. 3 Ricordi chamber with stainless steel marbles and screen for tissue retention

17. Ice buckets.
18. 60 × 15 mm petri dishes.
19. Microscope.
20. Sterile tip extender with large orifice tip.
21. 10–100 μl Eppendorf pipetter.
22. Drummond pipette aid.
23. 15 ml conical tube
24. 5 ml pipette.
25. Parafilm.

2.1.5 Counting Islets

1. Diphenylthiocarbazon (dithizone) (200 mg powder).
2. Sterile dimethyl sulfoxide (DMSO) (20 ml).
3. HBSS (80 ml).
4. 10 ml pipettes × 2.
5. Drummond pipette aid.
6. Weigh scale.
7. Weighing paper.
8. Weighing spatula.
9. 250 ml beaker.
10. 12 cc syringes × 10.
11. Luer Lock syringe caps × 10.
12. 0.45 μm nylon Acrodisc syringe filter.

13. –20 °C freezer.
14. Microscope.
15. 0.45 μm filter.
16. Grid eyepiece reticle for the microscope.
17. Glass slides.
18. 10–100 μl Eppendorf pipetter.
19. Tip extender with a large orifice tip.
20. Lab counter.

2.1.6 Density Gradient Separation and Harvesting of Human Islets

1. Ficoll density gradient solutions.
   (a) Ficoll separating solution, isotonic.
   (b) HEPES 1 M.
   (c) Sodium hydroxide (NaOH) (0.1 N).
   (d) 10× HBSS.
   (e) 70 % ethanol.
   (f) Sterile ice.
   (g) 10 ml pipette.
   (h) 5 ml pipette.
   (i) 2 ml pipette.
   (j) Sterile 250 ml cylinder.
   (k) Sterile 250 ml beaker.
   (l) Sterile aluminum foil.
   (m) Ice bucket.
2. Wash solution (1,050 ml).
3. Final wash solution (MEM) (250 ml).
4. Dithizone.
5. M199, no phenol red (500 ml).
6. HEPES 1 M (10 ml).
7. Human albumin 25 % (100 ml).
8. 10 ml pipette.
9. 5 ml pipette.
10. Sterile mosquitoes.
11. Sterile ice.
12. COBE setup.
13. 250 ml conical tubes × 8.
14. 250 ml conical tube rack.
15. 4 °C cold room or refrigerated COBE.
16. 15 ml conical tube.

17. 15 ml conical tube rack.
18. Ice bucket.
19. 6-well plate.
20. Microscope.
21. Sterile pipette extender and tip with a large orifice tip × 2.
22. 10–100 μl Eppendorf pipetter.
23. LVC guardian 12 l jug.
24. Sterile tubing.
25. Sterile glass rod.
26. Refrigerated centrifuge.
27. 50 ml conical tubes.
28. 50 ml conical tube rack.
29. Transplant media.
    (a) M199, no phenol red (500 ml).
    (b) HEPES 1 M (10 ml).
    (c) Human albumin 25 % (100 ml).

3 Methods

3.1 Human Pancreas Decontamination Setup

1. Clean the Biosafety cabinet (see Note 1).
2. Aseptically drape the workbench with the back table cover and aseptically place the sterile beakers and cylinder on the workbench. Inject 20 ml of HBSS into each of the Ancef and Amphotericin B vials. Work HBSS into powder with syringe until mixed. Draw out solutions and inject into the sterile 600 ml beaker. Aseptically fill this beaker to the 150 ml mark with HBSS (see Note 2).
3. Aseptically place 150 ml of betadine into a sterile 600 ml beaker.
4. Use the sterile 100 ml cylinder to aseptically measure 500 ml of HBSS and add it to the sterile 1,000 ml beaker.

3.2 Formulating Liberase Enzyme (See Note 2)

1. Place ice inside the ice bucket and keep the HBSS and HEPES on ice.
2. Remove the Liberase enzyme from −20 °C freezer and place on ice. Expose to room temperature for approximately 2–5 min and in biological safety cabinet reconstitute with cold sterile HBSS (30 ml). Gently swirl the bottle frequently for 30–45 min and visually inspect solution to be sure that the entire enzyme is dissolved (see Note 3).
3. Aseptically use the sterile 100 ml cylinder to add 270 ml of cold HBSS to the sterile 600 ml beaker. Add 0.16 g of $CaCl_2$ and

8.5 ml of 1 M HEPES to the beaker and swirl to mix. Place this solution at 37 °C until the remaining Liberase is dissolved in the bottle. Place the bottle of HBSS on ice for future use.

4. After 30 min, when the enzyme is dissolved, carefully add this to the 270 ml of HBSS in the water bath.
5. Calibrate the pH meter and pH the Liberase solution to 7.35–7.4 using endotoxin-free NaOH (0.1 N) and endotoxin-free HCl (0.1 N).
6. Pre-wet the bottle top filter on the sterile 500 ml bottle with 50 ml of HBSS and pour the Liberase solution onto the pre-wetted filter.
7. Keep the Liberase solution on ice until required (see Note 4).

3.3 Trim and Decontaminate Human Cadaveric Pancreas

1. Ensure that the biological safety cabinet has been properly cleaned.
2. Non-sterile personnel will open the sterile items listed above and sterilely hand them to the sterile personnel. The ice bag will be placed in the sterile towel and the ice broken apart with the hammer, without compromising the sterility of the ice, by the non-sterile personnel. Once the ice is broken apart, the non-sterile personnel will remove the outer bag, and aseptically give the inner bag of sterile ice to the sterile personnel. This ice will be placed into the large sterile tray. The small sterile tray will be placed on top of the ice.
3. Place surgical supplies aseptically onto the sterile drape.
4. Place the decontamination solutions on the left side of the table on the sterile drape.
5. The non-sterile personnel will sterilely open the outer bag containing the pancreas and allow the sterile personnel to aseptically obtain the pancreas and place it in the small sterile tray on the ice. Aseptically add 30 ml of either Eurocollins or HBSS to this tray to keep the pancreas moist (see Note 5).
6. Sterile personnel will then obtain 5 ml of fluid from the pancreas transport bag and aseptically place it in the 15 ml conical tube. The non-sterile personnel will place parafilm around the lid of the conical tube, label the tube as "transport fluid, isolation ID, date, time, and initials," and place the tube on ice. Record the sample ID number. At the end of the isolation procedure, this tube will be placed in the refrigerator for gram stain analysis.
7. The sterile personnel will remove the duodenum and any other organs attached to the pancreas. Record the time.
8. The pancreas will be placed into the first beaker containing the Ancef and amphotericin, it will then be removed and dipped into the second beaker of betadine, and then finally rinsed in the last beaker of HBSS.

9. The non-sterile personnel will tare the weigh scale with the Saf-T Basin, leaving the basin in the sterile packaging. Once this has been accomplished, the sterile package will be opened under the hood and the basin left in the package.
10. The sterile personnel will aseptically place the pancreas in the basin, and the non-sterile personnel will place the top of the package back to the original position, covering the pancreas.
11. The non-sterile personnel will weigh the pancreas and record the weight. The pancreas will remain in the basin until the next setup has been completed.
12. The non-sterile personnel will quickly remove all decontamination glassware and supplies from the hood.
13. The non-sterile personnel will aspirate the decontamination solutions into the LVC jug and place the glassware, instruments, and trays in the Prezyme solution. Place all the disposables in the biohazard bin.
14. The non-sterile personnel will quickly set up the hood for the cannulation and distension of the pancreas (see Note 6).

3.4 Cannulation and Distension of the Human Cadaveric Pancreas

1. The non-sterile personnel will place a new back table cover onto the table in the biological safety cabinet. The sterile personnel will then open the cover, creating a sterile environment.
2. Non-sterile personnel will open the sterile items listed above under the cannulation step and aseptically hand them to the sterile personnel. The ice bag will be placed in the sterile towel and broken apart with the hammer, without compromising the sterility of the ice, by the non-sterile personnel. Once the ice is broken apart, the non-sterile person will remove the outer bag and aseptically hand the inner bag of the sterile ice to the sterile personnel. The ice will then be placed into the large sterile stainless steel tray by the sterile personnel and the small stainless steel sterile tray will be placed on top of the ice.
3. Place the surgical supplies aseptically onto the sterile drape.
4. The sterile personnel will aseptically take the pancreas from the basin and place it into the small sterile tray and replace the small tray on the ice in the large sterile tray as soon as possible. Record the time that the pancreas was put back on ice. Aseptically add 30 ml of HBSS or Eurocollins into the tray with the pancreas. Record which solution was used on the batch record.
5. Remove as much of the fat and capsule from the pancreas as can be done in a short amount of time (see Note 7).
6. Cannulate the main pancreatic duct with a 14–22-gauge cannula (depending on the size of the duct) and tie the cannula in the duct (Fig. 4). Go about 1/3 of the way to the end of the pancreas and cut down to the pancreatic duct using the scalpel. Pulse some

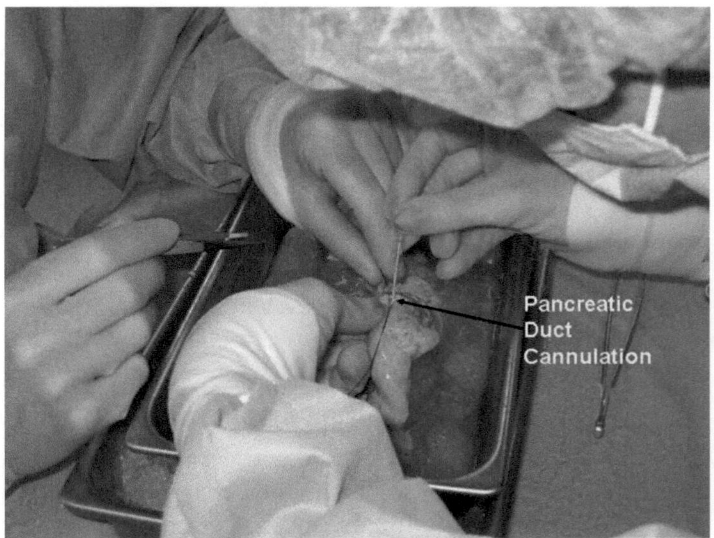

Fig. 4 Cannulation of the pancreatic duct for the introduction of Liberase enzyme

HBSS solution through the cannula at the top of the pancreas to help locate the duct further down the pancreas if needed.

7. Loosely tie two ties around both ends of the isolated duct (isolate about 1 cm of the duct).
8. Using the iris scissors, carefully make a knick in the low end of the isolated pancreatic duct. Butterfly this knick open and cannulate using a 14–22-gauge cannula with the needle removed (the size of cannula is dependent on the size of the duct). Record the size of the cannula used. Tie the cannula into position with the needle partially inserted to avoid tying the duct off completely. Be mindful not to cut completely through the duct when making the knick and butterflying (see Note 8).
9. Follow this same procedure on the other end of the duct.
10. Remove the cannula from the top of the pancreas and tie the main pancreatic duct off with the 4-0 silk.
11. Remove 30 ml of solution from the tray and fill the two 35 cc syringes with the Liberase solution. Quickly distend the pancreas and record the start time (Fig. 5). Use the mosquitoes to clamp off any visible leaks in the pancreas and the 4-0 silk to tie them off.
12. Once the pancreas is fully distended, record the time. Remove any obvious pieces of fat while the digestion chamber is being emptied.
13. Weigh the discarded tissue and fat and record. Calculate the actual weight of the trimmed and cleaned pancreas and record.
14. Cut the pancreas into seven to nine equal pieces and record the number of pieces.

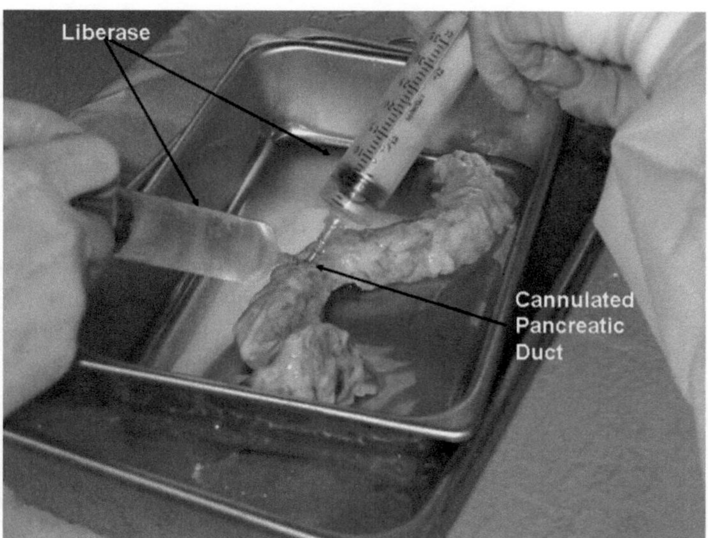

Fig. 5 Adding Liberase through the distension of the pancreatic duct

15. Clear the biological safety hood and place the disposable sharps in the sharps container, the glassware instruments, and trays in the Prezyme and the disposable materials in the biohazard bin.
16. Clean the table twice with isopropyl alcohol.
17. The personnel who cannulated the ducts will don new gowns and gloves.

3.5 Digesting the Human Pancreas

1. Remove the cannulas from the cut pancreas, and place the pancreas in the digestion chamber in a biological safety cabinet. Record the time that the pancreas was added to the chamber.
2. Add all of the remaining Liberase into the chamber. Replace the screen and tighten the lid to the chamber. If all of the Liberase does not fit into the chamber, pump the remaining Liberase through the recirculation line (Fig. 20.1) to fill up the entire chamber system volume. Record the temperature of the chamber upon the addition of the Liberase (found on the temperature monitor).
3. Turn on the peristaltic pump at a rate of approximately 150 ml/min and add the primer to the system until the chamber and tubing is full, and recirculation can begin. To obtain recirculation, remove the blue clamp from the size 17 tubing on the bottom of the large T and clamp the tubing on the right side of the large T. Also remove the clamp from the size 16 tubing on the bottom of the small T and clamp the tubing on the right side of the small T. Rock the chamber (back and forth) to get the air out of the system (see Note 9).
4. Once recirculation has begun, start the timer; record the chamber temperature. Agitate the pancreas/enzyme preparation by

rotating between a shaking and rocking motion of the digestion chamber. Shake at 30-s vigorous shake intervals. Once the temperature in the chamber has reached 37 °C, record the time from the timer. Maintain the temperature of the recirculation media at no greater than 37.5 °C. If temperature exceeds this level, remove heat-exchanger coil from the water bath until the temperature in the chamber falls below 37.2 °C (see Note 10).

5. Samples (1.0 ml of tissue digest) are taken at timed intervals, from the sample port, beginning 8–10 min after recirculation has begun and continuing for every 2–3 min depending on progression of digestion. These samples are placed in a 100×20 mm petri dish and 2–3 ml of dithizone is added to each dish. The sample is then observed under the microscope. Record the time ($T=x$) that each sample was taken and record all observations. When samples show at least one free healthy looking islet (keeping in mind the integrity of the islet, for example, if they are looking ragged or fragmented, as well as the volume of visible tissue and the time elapsed), begin to dilute the digest with ambient temperature MEM to halt the enzymatic digestion. Remove the heat-exchanger coil from the water bath, increase pump rate to approximately 300 ml/min, and dilute and flush the dissociation chamber with ambient temperature MEM solution for the first 4 l. Record the time that the digestion was stopped ($T=x$) (see Note 11).
6. Collect the tissue digest in the 250 ml digest collection conical tubes. Invert each tube of collected digest and place in a crushed ice slush bath as soon as it is filled for rapid cooling.
7. Sample the islets periodically. If the islets are trapped, place the heat coil back in the water bath in order to increase the temperature of the chamber 37 °C. If the islets look good, continue to keep the heat-exchanger coil at room temperature. If the islets look damaged, bring the chamber to cold temperatures by placing the coil on ice. Record all information.
8. After 4 l of MEM has passed through the chamber, invert the chamber to a 45° angle and add air (approximately 150 ml) and continue to shake gently. Record the time that the air was introduced into the system ($T=x$). Let the air in the chamber out of the system by temporarily moving the chamber to the upright position and shaking vigorously.
9. Add the remaining 4 l of 4 °C MEM and record the time ($T=x$). During this process, continue to shake the chamber at a 45° angle and increase the shaking of the chamber to a more vigorous level. There should be a continuous stream of islet tissue exiting the chamber and being collected in the collection tubes. The chamber will begin to feel lighter due to less tissue present and the sound of the marbles against the chamber will have a "twang" sound due to fewer collisions with the tissue.

10. When there is approximately 500 ml of MEM remaining to be added to the chamber, remove inlet tube from the media beaker and allow the system to empty out. This washes the chamber. Once all the liquid is out of the system reinsert the tube and then rinse the system with the remaining 500 ml of MEM. Record the time ($T=x$) that this process started and ended. Also record the time ($T=x$) that the last of the digest was collected into the 250 ml conical tubes (see Note 12).
11. Centrifuge the conical tubes at 245 ×*g* for 1 min at 4 °C. The top 200 ml of supernatant is aspirated in the biological safety cabinet and the islet pellet is gently re-suspended with cold supplemented Wash solution and combined.
12. Combine three to six pellets into one tube and bring the volume in the tube to 250 ml with Wash solution. Spin the pellet as described above. This process is repeated as many times as necessary until all of the digested tissue is collected into one 250 ml conical tube. Record the actual time that all of the pellets were combined into one.
13. Centrifuge the combined pellets at 245 ×*g* for 1 min at 4 °C.
14. Remove 5 ml of the supernatant and place this into the 15 ml conical tube. Parafilm the lid and label this with the sample ID, date, time, and initials. Record the sample ID. Place this tube on ice until the isolation is complete. The sample will then go into a 4 °C refrigerator until the 14-day microbiology results from the final sample have been reported.
15. Aspirate the remaining supernatant and obtain as dry of a pellet as possible. Record the estimated packed cell volume.
16. Tare the weigh scale with an empty 250 ml conical tube with a lid. Weigh the 250 ml conical tube containing the tissue and record the weight.
17. Use the volume lines on the conical tube to bring the volume up to the 200 ml mark on the conical tube with Wash Solution and re-suspend the pellet.
18. Remove 2–100 μl samples from this re-suspension to obtain a pre-purification islet count (see Subheading 3.6 for counting of islets).
19. Calculate the number of 250 ml conical tubes required to distribute the digested tissue to less than 20 g of tissue per conical tube (maximum volume allowed per COBE run). Record this calculation (see Note 13).
20. Calculate the volume to be transferred into each conical tube. Record this calculation. Transfer this amount into each of the conical tubes.
21. Bring the volume of each tube to the 200 ml mark with Wash Solution and spin the tubes at 245 ×*g* for 1 min at 4 °C.

22. Aspirate down to a "dry" pellet and gently re-suspend each pellet in UW. Bring UW up to the 100 ml mark on the conical tube and gently invert to re-suspend. Record the time that the pellets were re-suspended in UW.
23. Place the tubes at an angle in ice for a minimum of 30 min, swirling occasionally.
24. After 30 min, deliver the digested tissue to the COBE operators in a biological safety cabinet and record both the time that tissue was delivered and the elapsed time in UW.
25. Begin the purification of the human islets.
26. Remove all objects from the table in the biological safety cabinet and clean two times with isopropyl alcohol.
27. Record all information, names, manufacturers, lot numbers, expiry dates, and load information.

3.6 Counting Islets

3.6.1 Making Dithizone

1. In the biological safety cabinet, aseptically add 20 ml DMSO to the 250 ml beaker.
2. Weigh out 200 mg of dithizone and add this to the DMSO in the beaker.
3. Swirl this until the dithizone is dissolved.
4. Aseptically add 80 ml HBSS to the beaker and swirl to mix.
5. Draw 10 ml of solution up into each of the 12 cc syringes and cap, leaving an air pocket for expansion after the dithizone has frozen.
6. Store these syringes in a labeled container in a –20 °C freezer until ready to use for up to 6 months.
7. Record names, manufacturers, lot numbers, expiry dates, and load information.

3.6.2 Staining Islets with Dithizone

1. Remove the dithizone from the freezer and allow it to thaw.
2. Place a 0.45 μm filter on the syringe.
3. Add the dithizone to the islet cell suspension provided. Add 1–2 ml of dithizone to the small petri dishes provided. Add two to three drops of dithizone to the glass tubes.
4. Allow the dithizone and islet mixture to mix.
5. Using the microscope, observe the viable islets that are detected by the ability to take up the dithizone stain and turn red.
6. Discard the dithizone into a biohazard bag after use.

3.6.3 Performing Islet Count for Islet Equivalent Calculation

1. Place a glass side on the microscope, and once the cells are stained, remove 20 μl samples and drag them across the microscope so that the sample forms a line half the length of the slide. Repeat this three times per slide until the sample is gone.

2. Label each section of the lab counter with a range for the islet sizes. The ranges should include 50–100 μm, 101–150 μm, 151–200 μm, 201–250 μm, 251–300 μm, 351–400 μm, >401 μm.
3. Under the 10× objective, count the dithizone stained islets. Estimate the size of each counted islet by measuring its average diameter using the eyepiece square grid reticle. Each edge of a small square is 40 μm in length.
4. Calculate the average diameter by estimating the minimum and maximum perpendicular cross-diameters and averaging these two numbers. You may freely rotate and move the reticle to obtain your measurements.
5. Count each islet and categorize it by size on the lab counter. For example, a 120 μm round islet (the length of three squares in each direction) will be categorized into the 101–150 μm section.
6. Count only those islets that are stained red with dithizone and have discernable boundaries. Do not count islets that are embedded in exocrine tissue. Adjust the focus as necessary to be sure to count all islets in each field.
7. Once all the islets have been counted, record the islet numbers of each size category.
8. The islet equivalent (IE) count is used to normalize the total islet count to a value equivalent to islets of 150 μm diameter. Based on the counts in each size category, use the multipliers listed in Table 3 and record the total IE count within each diameter category. Add together the IE values for each category to arrive at the total islet equivalent value.
9. Calculate the total number of islets and the total number of islet equivalents for each set, and then calculate the total of all the sets combined.
10. Calculate the total number of islets and the total number of islet equivalents by taking the total number of all the sets combined and multiplying by a factor that would make the sample volume 1 ml. For example, if the sample volume is 100 μl, then multiply by 10 to get a 1 ml sample volume, if the sample volume is 50 μl then multiply by 20 to get a 1 ml sample volume.
11. Multiply this number by the volume of the solution that was sampled. This will vary for every count.
12. For example, if there is a 100 μl sample from 96.8 ml of solution, and there are 754.8 islet equivalents from all sets, then the calculation would be as follows: $754.8 \times 10 \times 96.8 = 730{,}646$ islet equivalents.
13. Record all data, names, manufacturers, lot numbers, and load information.

Table 3
Islet equivalent calculation

Average diameter category (μm)	IE multiplier
50–100	0.167
101–150	0.648
151–200	1.685
201–250	3.500
251–300	6.315
301–350	10.352
351–400	15.833
>400	22.750

3.7 Density Gradient Separation and Harvesting of Human Islets

3.7.1 Making 1.077 g/ml Ficoll Gradient

1. Using the sterile 250 ml cylinder, aseptically pour 130 ml of 1.077 g/ml density Ficoll into the sterile 250 ml beaker. Using the 10 ml pipette, aseptically add 9.8 ml of 10× HBSS to the 250 ml beaker. Using the 5 ml pipette, aseptically add 3.12 ml of 1 M HEPES to the 250 ml beaker. Using the 2 ml pipette, aseptically add 0.182 ml of endotoxin-free NaOH (0.1 N) to the 250 ml beaker.
2. Be sure to gently swirl the beaker after each addition. Cover the beaker with the sterile aluminum foil, label, and store on ice until the purification process begins.
3. Record the names, manufacturers, lot numbers, expiry dates, and load information.

3.7.2 Making 1.1 g/ml Ficoll Gradient

1. Using the sterile 250 ml cylinder, aseptically pour 290 ml of 1.100 g/ml density Ficoll into the sterile 600 ml beaker. Using the 25 ml pipette, aseptically add 15.08 ml of 10× HBSS to the 600 ml beaker. Using the 10 ml pipette, aseptically add 6.96 ml of 1 M HEPES to the 600 ml beaker. Using the 2 ml pipette, aseptically add 0.4 ml of endotoxin-free NaOH (0.1 N) to the 600 ml beaker.
2. Be sure to gently swirl the beaker after each addition. Cover the beaker with the sterile aluminum foil, label, and store on ice until the purification process begins.
3. Record the names, manufacturers, lot numbers, expiry dates, and load information on the Human Islet Media Preparation Sheets.

3.7.3 Transplant Media

1. Aseptically add the 100 ml of human albumin and the 10 ml of the 1 M HEPES to the Media 199, and gently swirl to mix. Label the media and keep the media at 4 °C for up to 4 weeks. Bring to room temperature 1 h before use.

2. Record the names, manufacturers, lot numbers, expiry dates, and load information.

3.8 Setting up the COBE Prior to Use

1. Take all mobile equipment in the cold room immediately prior to the purification process. The COBE and the surge protector are the only items that should be in the cold room at all times (see Note 14).
2. Turn on the surge protector and the COBE.
3. Purge COBE system before opening the COBE and inserting the COBE bag. The COBE should be stored with the foam insert and plastic cover in place over the diaphragm bag. Make sure these are in place prior to purging the system.
4. Set the dials on the COBE: Speed to 3,000, Super-out to 450 and Super-out volume to 600.
5. Press start. When revolutions are at 3,000, depress button for Super-out. Alarm should sound within 10 s and then press Stop/Reset. If alarm does not sound, repeat procedure again.
6. If alarm still does not sound within 10 s, repeat this process until it does. If there is air in the line, it may require five or six cycles to purge the air out.
7. Once the system is purged, open the centrifuge latch (located on the near end of the bar on the right side of the sliding covers), by rotating the latch knob fully clockwise.
8. Lift the seal weight and open the sliding covers on the COBE.
9. Slide the centrifuge cover into the holder at the base of the control panel. Remove the two white alignment blocks from the centrifuge bowl area.
10. Roll the COBE bag around the hexagonal seal and pass it through the center hole of the centrifuge cover. To make this process easier, hold the cover in your hand.
11. Install the COBE bag in the centrifuge bowl by positioning the four holes in the bag over the four studs on the centrifuge. Make sure that the bag lies flat over the top of the centrifuge. Press the outer edge of the COBE bag completely into the bowl, eliminating as many creases as possible. The spike port should be on top and at a slight angle to allow the bowl cover to close.
12. Position the two white alignment blocks around the center stem of the COBE bag.
13. Place the bowl cover over the four studs.
14. Rotate the bowl cover clockwise until the locking plunger falls into place.
15. Close the rear sliding cover, and while continuing to hold the seal weight up, close the front sliding cover.

16. Lower the seal weight and close the centrifuge latch by rotating the latch knob fully counterclockwise.
17. Place the five-colored tubings loosely into the corresponding metal holders, they are color coordinated. Do not clamp these tubings behind the metal rods.
18. Leave the center tubing free; do not place this in the metal clamp.
19. Using the blue clamps from the COBE pack, clamp each tubing line, except for the green line (top right line), close to the intersection of the main line leading to the bag.
20. From the COBE pack, arrange the gradient maker onto the gradient maker holder and place on top of the stir plate. Make sure that the first beaker with the stir bar is positioned where the stir bar will spin freely.
21. Clamp the line between the two beakers of the gradient maker. Thread the gradient line through the peristaltic pump head.
22. Attach the unclamped green line to the line leading from the gradient beaker.
23. The COBE is now set up and ready for the purification process (Fig. 6).
24. Record the names, manufacturers, lot numbers, expiry dates, and load information.

3.9 Using the COBE for Isolating Islets on a Density Gradient

1. Set up the COBE in the cold room or turn on refrigeration apparatus of the COBE (see Note 15).
2. Bring in the human pancreatic tissue, Ficoll gradient density 1.077 g/ml, and Ficoll gradient density 1.100 g/ml on ice.
3. Set the timer for 5 min and set the COBE speed for 1,500 rpm, Super-out speed at zero and Super-out volume at max.
4. Using one of the 250 ml cylinders, pour 140 ml of the 1.100 g/ml density Ficoll into the first beaker. Set pump at maximum (50 ml/min) and load onto COBE quickly. DO NOT start spinning the COBE.
5. When all of the Ficoll has been loaded, turn off the pump and release the line from the peristaltic pump head. Press the Start/Spin button and then press the Super-out button. Slowly turn up the Super-out speed to 100 until the Ficoll returns up the line into the first beaker pushing air out of the system.
6. Once all the air is out of the system, press the Stop/Reset button, use one of the blue clamps to clamp the line at the front of the gradient maker, tighten the pump head, and then remove the clamp from the front of the gradient maker. Reset the Super-out volume to zero and set the COBE speed to 2,400.
7. Using the same 250 ml cylinder, pour 130 ml of the 1.100 g/ml density Ficoll into the front beaker. Unclamp the clamps

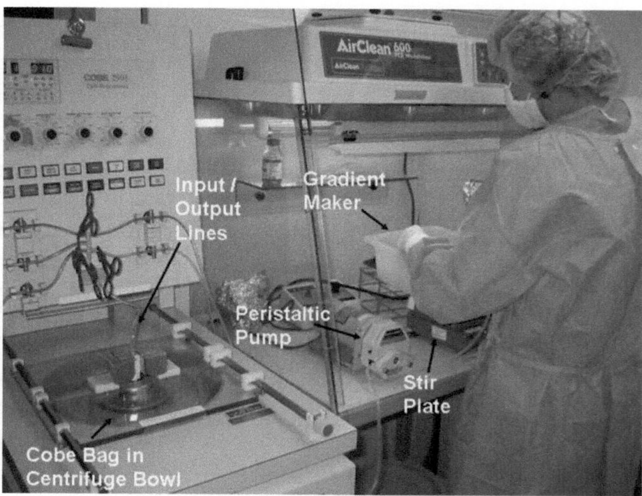

Fig. 6 Setup of the Cobe 2991 cell processor for the gradient separation of islets using Ficoll. The islets and the Ficoll gradient are added to a gradient maker and then pumped into the Cobe bag for eventual separation of the islets from the acinar tissue

just enough between the two beakers to let the Ficoll seep into the connecting line and then re-clamp the line. Using the other 250 ml cylinder, pour 140 ml of the 1.077 g/ml density Ficoll into the second beaker and release the clamp between the two beakers completely.

8. Turn on the magnetic stirrer and check that the Ficoll solutions are mixing (see Note 16). Turn on the peristaltic pump at a rate of 25 ml/min and press the Start/Spin button on the COBE.
9. As the volumes in the beakers get lower, tilt the platform up and place the wedge under the platform, raising the second beaker upward.
10. Once the 1.077 g/ml density Ficoll is out of the second beaker, clamp the line between the beakers.
11. When the last of the Ficoll is ready to load, turn off the stir bar and add small aliquots of the UW/islet prep being careful not to introduce air. Once most of the tissue has gone through the line, use 25 ml of the Wash solution to rinse the UW/islet preparation. Once most of this has gone through the line, rinse the line with the remaining 25 ml of Wash solution.
12. When all of the rinse solution has passed through the lines and passed the interface into the COBE bag, wait 5 s and shut off the pump, clamp the main line to the COBE bag, open the peristaltic pump head, and hit the Super-out button. Slowly release the clamp from the main line. Some liquid will rise up into the line.
13. Start the timer for 5 min.

14. During the 5-min spin, loosen the caps of the 250 ml conical tubes labeled 1–8. Tube 1 will contain 100 ml of the Wash solution, while tubes 2 through 8 will contain 200 ml of the Wash solution.
15. Detach the green line from the gradient beaker line and re-cap the end of the green line. Remove the clamp from the yellow line, the collection line, and clamp the green line close to the main line into the COBE bag.
16. At 5 min slowly turn the Super-out dial up to 100 ml/min. Collect the purified Ficoll/islet prep fractions into each of the tubes. The first tube will contain mostly the rinse solution used to rinse the lines after the tissue was added. When the islets can be visually seen in the main line from the COBE bag, switch to tube number 2. Keep collecting the Ficoll/islet prep until large chunks of unusable tissue can be seen in the main line.
17. Press the Stop/Reset button on the COBE and remove the COBE bag from the machine and place it in a biohazard bag.
18. Tighten the caps on the conical tubes and gently invert tubes to mix.
19. Place the conical tubes on ice until the foam insert and plastic cover have been repositioned over the diaphragm bag, the covers closed and locked, and the COBE machine and surge protector turned off. Remove everything from the cold room except for the COBE and the surge protector.
20. Transport the conical tubes on ice back to the human islet isolation laboratory. If another COBE run is necessary, return the conical tubes to the human islet isolation laboratory and bring the new setup into the cold room. Repeat the COBE setup procedure, without re-priming the system.
21. Clean the Biosafety cabinet and spray all surfaces with 70 % ethanol.
22. Aseptically remove 100 μl from each of the tubes using a sterile pipette extender and an Eppendorf pipetter. To ensure an equal distribution of the tissue throughout the conical tube, gently invert the tubes several times and quickly take the sample.
23. Place the sample into a labeled 6-well plate and add 2–3 ml of dithizone per well.
24. Assess the purity of the islets by gently inverting each tube and quickly removing a 100 μl sample from each tube and place it in the corresponding well.
25. Purity is assessed by observing the amount of exocrine tissue compared to the amount of stained islet tissue.
26. If the well contains greater than 30 % islet tissue, then the corresponding tube will be used for the final cell count.

27. Select tubes that have greater than 30 % pure islets for further processing steps.
28. Centrifuge the designated conical tubes at 550 × *g* for 1 min.
29. Aseptically aspirate the supernatant from the conical tubes and combine the tissue pellets into one 250 ml conical tube. Add Wash solution to the 250 ml mark and gently invert.
30. Spin the 250 ml conical tube at 245 × *g* for 1 min and record the tissue volume.
31. Aseptically aspirate the supernatant from the conical tube and re-suspend the tissue in 250 ml of final wash solution/culture medium.
32. Using a sterile pipette extender and an Eppendorf pipetter, remove 2 samples of 100 μl of tissue for a pre-culture islet count.
33. Add dithizone to the tissue and begin counting the islets. Record the islet count information.
34. Remove 800 islet equivalents and place in a 15 conical tube for glucose stimulation assay.
35. Remove 100 μl of the cell suspension and place in a 15 ml conical tube for mycoplasma testing.
36. Remove another 50 μl of the cell suspension and place this in a 1.5 ml microcentrifuge tube for use in a Live/Dead Assay.
37. Spin the 250 ml conical tube at 245 × *g* for 1 min.
38. Remove 5 ml of the supernatant and place in a 15 ml conical tube for microbiology testing.
39. Remove 4 ml of the supernatant and place in an endotoxin-free tube for the endotoxin assay.
40. Once all of the supernatant is removed, leaving the pellet that cannot be over 10 ml, using transplant media, re-suspend, and aliquot this pellet into a volume that will allow roughly 2.5 ml of tissue into 50 ml conical tubes.
41. Record all names, manufacturers, lot numbers, expiry dates, and load information.

3.10 Loading Islets into Bag for Transplantation

1. Once the islets are in the final pellet remove all of the supernatant, leaving a dry pellet, which cannot be over 10 ml of tissue. Record the estimated volume of this pellet. Re-suspend the tissue using transplant media (2 ml of media per 1 ml of tissue) and aliquot this pellet into a volume that will allow 3.5 ml of tissue or less into 50 ml conical tubes. Record the information.
2. Add 35 ml of transplant media to the tube(s).

3. Aseptically remove the tissue culture bags from the packaging and place in a biological safety cabinet. Perform all procedures involving the loading of the tissue culture bags using aseptic techniques.
4. Clamp the hoses on all of the bags. To do this, rotate the blue roller on the hoses upward toward the bag until it reaches the top of the plastic housing. Once it is snug, the hose is closed.
5. Label the tissue culture bag(s) that will contain the islet cells (one bag per conical tube) with the following information (Do not write on the bags, use a pre-made label—see example below). A tissue culture bag will be used to "flush" the bag containing the islets after transplantation. This bag will not be labeled, as it will be attached to the labeled tissue culture bag.

   Caution: New drug-limited by Federal Law to Investigational Use

   Islet isolation ID number

   Expiration date: Expiration time:

   Bag ____ of _____ Initials:
6. Aseptically fill the flush bag with 100 ml of transplant media. To do this, aseptically remove the blue luer lock cap from the connection port on the hose of the "flush" bag. The connection port is located adjacent to the spike port. Retain the cap in order to sterilely re-cap the connection port after the bag has been filled.
7. Aseptically remove the plunger from the 60 cc syringe and attach the syringe to the connection port of the flush bag.
8. Aseptically add 50 ml of transplant media to a sterile 50 ml conical tube. Use this to pour the media into the 60 cc syringe attached to the flush bag.
9. Open the connection port hose by rotating the blue roller downward away from the bag. Once the roller moves freely, the line is open.
10. Once all of the media is loaded into the bag, close the connection port hosing.
11. Add an additional 50 ml of transplant media to the flush bag by repeating the steps listed above.
12. Once all of the transplant media has been added to the flush bag aseptically, re-cap the luer lock on the connection port.
13. Once the flush bag has been loaded and re-capped, begin loading the cells into the transplant bag. This will be done in the same manner as the loading of the flush bag.
14. Remove the luer lock cap from the connection port of the transplant bag.

15. Aseptically remove the plunger from the 60 cc syringe and attach the syringe to the connection port of the transplant bag.
16. Gently re-suspend the islet cells in the conical tube and aseptically pour the cells into the 60 cc syringe. Open the connection port hose by rotating the blue roller downward away from the bag. Once the roller moves freely, the line is open.
17. Once all of the cells and media have been loaded into the transplant bag, close the connection port by repeating the steps.
18. Rinse the conical tube and flush the connection port by adding 50 ml of transplant media to the conical tube that contained the cells and gently shake to retrieve any islets that may still be in the tube. Add this to the 60 cc syringe and open the connection port to load this into the transplant bag. Repeat this step two more times. (200 ml of islets and transplant media should be in each transplant bag, with a flush bag of 100 ml of transplant media attached.)
19. Aseptically re-cap the connection port and place the bag on the conical tube rack.
20. At intervals of 5–15 min (or as often as is necessary to prevent the cells from clumping) gently rotate the transplant bag. Be mindful not to disturb the spike port of the connecting bag. Record the times of rotation (see Note 17).
21. For local transport to the clinical transplant site, place the bags into a Styrofoam box.

4 Notes

1. Prior to starting this procedure, don the proper attire for a sterile procedure.
2. Do not begin until pancreas arrives in lab and supervisor visually inspects organ.
3. Be careful not to promote foaming.
4. Use Liberase within 2 h of reconstitution.
5. Record time and solutions used (either Eurocollins or HBSS are acceptable).
6. Record all names, manufacturers, lot numbers, expiry dates, and load information.
7. It is better to have the pancreas as clean as possible for the digestion process.
8. It is sometimes challenging to find the duct—be patient.
9. Be cognizant that the clamps are opened before turning on the pump.

10. It can take a while for the temperature to go back down to 37 °C after removal of the coil. It is suggested to have a second person be involved in the digestion, due to the primary person who is shaking the chamber is unable to monitor and manipulate all aspects of the digestion system.
11. Liberase works at around 37 °C. By changing the temperature to RT, this will slow down the digestion rate and the islets that have been liberated from the tissue will be circulating within the perfusion system.
12. After completion of collection of the cells, open the chamber to observe the remaining tissue. If there is a thin strand of tissue left, then the digestion went well. However, if there is still a noticeable tissue structure, then the digestion was not performed to completion. The "twang" in the chamber at the end of the procedure is generally a good indication on how the digestion went.
13. A second COBE run is necessary if there is more than 20 g of tissue to be processed. If there is greater than 20 g of cells to be processed for the COBE, it is suggested to evenly divide the weight into two to perform two COBE runs.
14. Some COBE systems have been modified to maintain the bowl close to 4 °C. This is a preferred method versus operating the system in a cold room. If using the cold room, a portable laminar flow hood is necessary for the loading of the islets into the COBE.
15. The bowl of the COBE needs to be refrigerated during this process. The islets are temperature sensitive and the long-term operation of the gradient can damage the islets if the system is not cooled.
16. It is recommended for the small stir bar to be off-center a little to mix the contents of the two gradients. Mix at a slow rate and in order to be careful not to introduce gas bubbles into the gradients during the mixing.
17. It is very important to keep the islets in suspension and prevent aggregation at this high density used for transplantation. Designate a person to oversee this shaking during the time that the assays are being performed up until transport.

References

1. Gores PF, Najarian JS, Stephanian E, Lloveras JJ, Kelley SL, Sutherland DE (1993) Insulin independence in type I diabetes after transplantation of unpurified islets from single donor with 15-deoxyspergualin. Lancet 341:19–21
2. Shapiro AM, Lakey JR, Ryan EA, Korbutt GS, Toth E, Warnock GL, Kneteman NM, Rajotte RV (2000) Islet transplantation in seven patients with type 1 diabetes mellitus using a glucocorticoid-free immunosuppressive regimen. N Engl J Med 343:230–238
3. Shapiro AM, Ricordi C, Hering BJ, Auchincloss H, Lindblad R, Robertson RP, Secchi A, Brendel MD, Berney T, Brennan DC, Cagliero E, Alejandro R, Ryan EA, DiMercurio B, Morel P, Polonsky KS, Reems JA, Bretzel RG, Bertuzzi F, Froud T,

Kandaswamy R, Sutherland DE, Eisenbarth G, Segal M, Preiksaitis J, Korbutt GS, Barton FB, Viviano L, Seyfert-Margolis V, Bluestone J, Lakey JR (2006) International trial of the Edmonton protocol for islet transplantation. N Engl J Med 355:1318–1330

4. Shapiro AM, Lakey JR, Paty BW, Senior PA, Bigam DL, Ryan EA (2005) Strategic opportunities in clinical islet transplantation. Transplantation 79:1304–1307
5. Ryan EA, Paty BW, Senior PA, Bigam D, Alfadhli E, Kneteman NM, Lakey JR, Shapiro AM (2005) Five-year follow-up after clinical islet transplantation. Diabetes 54: 2060–2069
6. The CITR Research Group (2007) 2007 Update on allogeneic islet transplantation from the Collaborative Islet Transplant Registry (CITR). Cell Transplant 18(7):753–767
7. Alejandro R, Barton FB, Hering BJ, Wease S, Collaborative Islet Transplant Registry Investigators (2008) 2008 Update from the Collaborative Islet Transplant Registry. Transplantation 86:1783–1788

Chapter 21

Microencapsulation of Pancreatic Islets for Use in a Bioartificial Pancreas

Emmanuel C. Opara, John P. McQuilling, and Alan C. Farney

Abstract

Islet transplantation is the most exciting treatment option for individuals afflicted with Type 1 diabetes. However, the severe shortage of human pancreas and the need to use risky immunosuppressive drugs to prevent transplant rejection remain two major obstacles for the routine use of islet transplantation in diabetic patients. Successful development of a bioartificial pancreas using the approach of microencapsulation with perm-selective coating of islets with biopolymers for graft immunoisolation holds tremendous promise for diabetic patients because it has great potential to overcome these two barriers. In this chapter, we provide a detailed description of the microencapsulation process.

Key words Islets, Alginate, Microencapsulation, Immunoisolation, Diabetes, Transplantation

1 Introduction

The pancreas is a dual-function organ featuring both endocrine and exocrine tissue. Approximately one million cell clusters called the islets of Langerhans produce a variety of metabolic hormones including glucagon, pancreatic polypeptide, somatostatin, and insulin. Islets consist of four main cell types, namely; α cells: secrete glucagon (induces hepatic release of glucose); β cells: secrete insulin (promotes glucose uptake); δ cells: secrete somatostatin (regulates α and β cells); and PP cells: secrete pancreatic polypeptide. Thus, the islet plays a diverse and complex role in glucose metabolism and blood glucose homeostasis.

In the pancreas, insulin is released in proportional response to actual blood glucose levels. The insulin is released into the portal vein, where it predominately flows toward the liver, which is the major organ to store glycogen and about 50% of secreted insulin gets used in the liver. In addition, the insulin release is pulsatile which helps to maintain the insulin sensitivity of the hepatic tissue. Owing to severe shortage of human pancreas and the shortcomings

Joydeep Basu and John W. Ludlow (eds.), *Organ Regeneration: Methods and Protocols*, Methods in Molecular Biology, vol. 1001, DOI 10.1007/978-1-62703-363-3_21, © Springer Science+Business Media New York 2013

of insulin therapy, a lot of effort has been made to develop an artificial pancreas.

The artificial pancreas is a technological development to enable Type 1 diabetic patients to automatically control their blood glucose, acting in essence like a healthy pancreas. The goals of the artificial pancreas are: (a) to improve presently popular but inefficient insulin therapy to attain a better glycemic control, thus avoiding the complication due to blood glucose fluctuations, and (b) to mimic normal stimulation of the liver by the pancreas and to normalize carbohydrate and lipid metabolism.

The bioengineering approach for designing a bioartificial pancreas has generally involved the development of either microcapsules, or macrocapsules, or other devices such as biocompatible sheet of encapsulated islets. When implanted, these constructs would substitute for the defective native endocrine pancreas (1). This chapter will focus on the microencapsulated islet construct, as it has advanced into the stage of clinical trials (2–5) and has significant promise to be a good alternative to pancreas transplantation.

Alginate is attractive as a biomaterial for microencapsulation of cells because of its relative ease of gelling under mild conditions such as the presence of divalent cations as well as its biocompatibility (6), hence this chapter will focus on the use of alginate for microencapsulation. Using alginate as the encapsulation polymer, the concept of islet immunoisolation is illustrated in Fig. 1, which essentially incorporates a semipermeable membrane into the process because alginate does not have any appreciable permselectivity towards immune cells and other immunological factors such as antibodies that can potentially destroy the encapsulated cells.

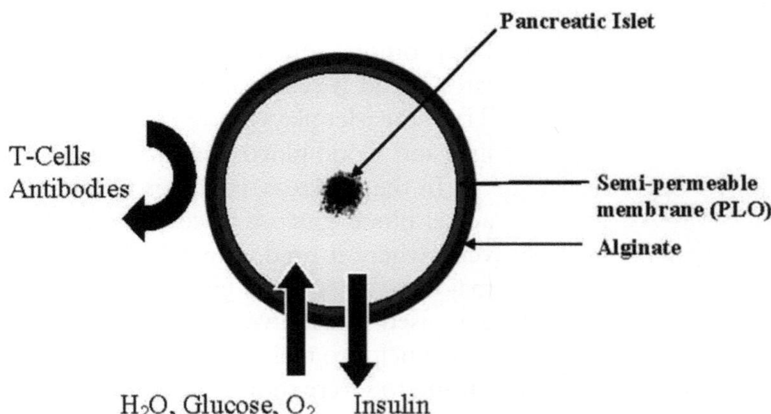

Fig. 1 Illustration of the principle of immunoisolation by microencapsulation

2 Materials

2.1 Chemicals

1. Alginate (Pronova UP LVM and UP LVG, Novamatrix, Sandvika, Norway).
2. Poly-L-Lysine (PLL) (P4957, Sigma-Aldrich, St. Louis, MO, USA).
3. Poly-L-Ornithine (PLO) (P5061, Sigma-Aldrich).
4. 100 mM $CaCl_2$ solution (C614-10, Fischer Scientific, Waltham, MA, USA).
5. 55 mM sodium citrate solution (S467-3, Fischer Scientific).
6. 0.9% sodium chloride solution (normal saline) (71376-5KG, Sigma-Aldrich).
7. 10 mM HEPES solution (H3375-2KG, Sigma-Aldrich).

2.2 Equipment

1. Air-syringe pump droplet generator (see Note 1).
2. Electrostatic generator (see Note 2).
3. Voltage generator (CZE1000R, Spellman High Voltage Electronic Corporation, Hauppauge, NY).
   (a) Syringe pump.
   (b) Stirring hotplate.
   (c) 18 G Blunt tip needles.
4. Microfluidic devices (North Carolina State University, Raleigh, NC).

3 Methods

Our laboratory adopts a four-step process in islet microencapsulation, as illustrated in Fig. 2.

1. Islets are suspended in a solution of sodium alginate (usually from 1.2 to 1.8% w/v made up in normal saline with 5,000 islets suspended in 1 mL of alginate).
2. Microspheres of alginate containing one or two islets (depending upon the alginate–islet ratio in the suspension) are generated and allowed to gel into microbeads in a bath of 100 mM $CaCl_2$ dissolved in 10 mM HEPES solution at 4°C, pH 7.4 (see Note 3).
3. Following two washes with normal saline, the microbeads are perm-selectively coated with variable concentrations of 0.1% PLL or 0.1% PLO for variable duration of time depending on the desired pore-size exclusion limit (12) (see Note 4).
4. Liquefaction of the alginate core of the microcapsules is achieved by a brief (2 min) incubation in 55 mM sodium citrate solution at 4°C.

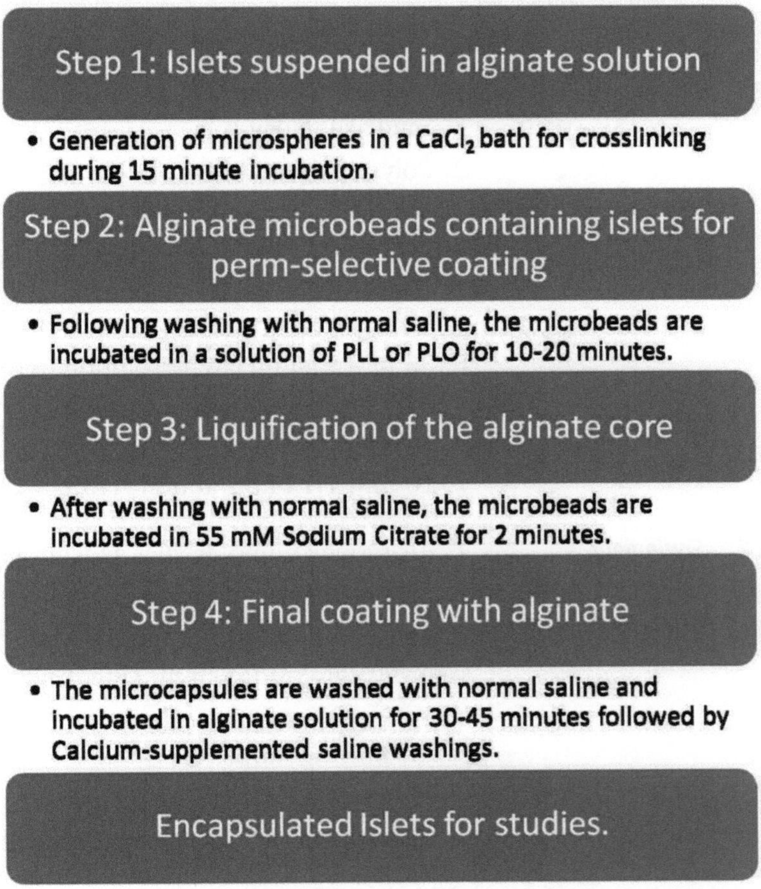

Step 1: Islets suspended in alginate solution

- Generation of microspheres in a $CaCl_2$ bath for crosslinking during 15 minute incubation.

Step 2: Alginate microbeads containing islets for perm-selective coating

- Following washing with normal saline, the microbeads are incubated in a solution of PLL or PLO for 10-20 minutes.

Step 3: Liquification of the alginate core

- After washing with normal saline, the microbeads are incubated in 55 mM Sodium Citrate for 2 minutes.

Step 4: Final coating with alginate

- The microcapsules are washed with normal saline and incubated in alginate solution for 30-45 minutes followed by Calcium-supplemented saline washings.

Encapsulated Islets for studies.

Fig. 2 Schematic representation of the microencapsulation process

5. Wash three times with normal saline, this is accomplished by allowing the capsules to settle on the bottom of a 100 mL beaker over the course of approximately 2 min.
6. External coating with a lower concentration (routinely about 10% of the concentration used in generating the initial microspheres), but our new procedure utilizes only a slightly lower concentration than the initial alginate concentration for microspheres (13, 14). This is accomplished by incubating the capsules with alginate for 5 min or 45 min if angiogenic protein is incorporated into the outer layer at 4°C (see Note 5).
7. To cross-link the external alginate coat, a solution of normal saline with 22 mM calcium chloride has been recently described for the final washings of microcapsules (14), albeit, normal saline is routinely used. Figure 3 shows some islets encapsulated in alginate–PLO–alginate (APA) microcapsules in our laboratory using the four-step process.
8. A critical component of the microencapsulation process is the device used to generate the initial alginate microbeads

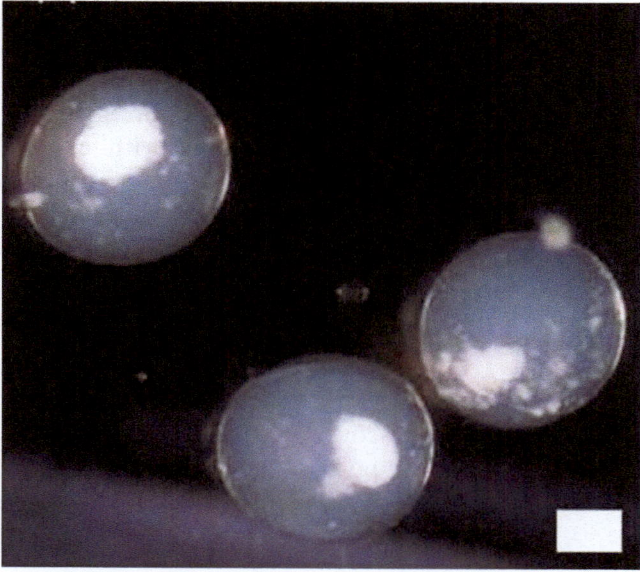

Fig. 3 Encapsulated islets in an alginate microcapsule. Scale = 100 μm

containing the islets. Within our lab we have used an eight-nozzle microfluidic device for the encapsulation of large numbers of islets (15), using this device we are able to encapsulate islets using flow rates of 0.5–1.5 mL/min with air pressures of 4.0–5.0 psi. For smaller number of islets (>10,000) we have utilized an electro-spraying technique.

4 Notes

1. The air-syringe droplet generator here is constructed with in-house materials as previously described (7, 8). This two-nozzle device uses standard syringe needles (gauges 20–27) and generates capsules at air jacket pressures of 10 psi and alginate jacket pressures of 15 psi. An air-syringe droplet generator (CF-01) may also be purchased from Biorep technologies (Miami, FL, USA).
2. The electro-spraying device used here is similar to that previously described (9). Briefly a high voltage source was attached to an 18 G blunt tipped needle, which was positioned above an aqueous $CaCl_2$ solution. Alginate was pumped through the 18-G needle using a syringe pump and droplets were allowed to fall into the $CaCl_2$ solution.
3. One major advantage of using Ca^{2+} as the crosslinking cation is that an inner alginate core encapsulating islets can be liquefied in order to enhance the diffusion of permissible molecules to

and from the microcapsules (10). The process of liquefaction is pretty delicate and has to be performed with utmost caution in order to avoid capsule breakage caused by high internal colloid-osmotic pressure after the "degelling" (11). The perm-selective coating with either PLL or PLO is achieved by incubating the alginate microbeads containing islets in 0.1% solution of the polymer for 20 min in order to obtain microcapsules with pore-size exclusion <100 kDa.

4. The preferred molecular weight range for both PLL and PLO for the purpose of perm-selective coating of alginate microbeads is 15–30 kDa.
5. Both PLL and PLO are polycationic polymers that require covering of their surface with a coat of the more biocompatible polyanionic alginate in order to prevent electrostatic interactions with cells and proteins after in vivo implantation.

References

1. Pareta RA, McQuilling JP, Farney A, Opara EC (2012) Bioartificial pancreas: evaluation of crucial barriers to clinical application. In: Organ donation, chapter 14. INTECH Publishers, pp 241–266. ISBN 979-953-307-081-9
2. Soon-Shiong P, Feldman E, Nelson R et al (1992) Successful reversal of spontaneous diabetes in dogs by intraperitoneal microencapsulated islets. Transplantation 54:769–774
3. Calafiore R, Calabrese G, Basta G et al (2006) Microencapsulated pancreatic islet allograft into non-immunosuppressed patients with Type 1 diabetes. Diabetes Care 29(1):137–138
4. Elliott RB, Escobar L, Tan PLJ et al (2007) Live encapsulated porcine islets from type 1 diabetic patient 9.5 yr after xenotransplantation. Xenotransplantation 14:157–161
5. http://www.newscientist.com/article/dn18730-pig-sushi-diabetes-trial-brings-xenotransplant-hope.html
6. Smidsrod O, Skjak-Braek G (1990) Alginate as immobilization matrix for cells. Trends Biotechnol 8:71–78
7. Moya ML, Morley M, Khanna O et al (2012) Stability of alginate microbead properties in vitro. J Mater Sci Mater Med 23:903–912
8. Wolters GH, Fritschy WM, Gerrits D et al (1991) A versatile alginate droplet generator applicable for microencapsulation of pancreatic islets. J Appl Biomater 3:281–286
9. Hongkwan P, Kim PH, Hwang T et al (2012) Fabrication of cross-linked alginate beads using electrospraying for adenovirus delivery. Int J Pharm 427:417–425
10. Garfinkel MR, Harland RC, Opara EC (1998) Optimization of the microencapsulated islet for transplantation. J Surg Res 76:7–10
11. Zimmermann U, Mimietz S, Zimmermann H et al (2000) Hydrogel-based non-autologous cell and tissue therapy. Biotechniques 29:564–581
12. Darrabie MD, Kendall WF, Opara EC (2005) Characteristics of poly-L-ornithine-coated alginate microcapsules. Biomaterials 26(34):6846–6852
13. Opara EC, Mirmalek-Sani S-H, Khanna O et al (2010) Design of a bioartificial pancreas. J Investig Med 58(7):831–837
14. Khanna O, Moya ML, Opara EC et al (2010) Synthesis of multi-layered alginate microcapsules for the sustained release of fibroblast growth factor-1. J Biomed Mater Res A 95(2):632–640
15. Tendulkar S, Mirmalek-Sani S-H, Childers C et al (2012) A three-dimensional microfluidic approach to scaling up microencapsulation of cells. Biomed Microdevices 14(3):461–469

Chapter 22

Bioengineered Skin Substitutes

Pedro Lei, Hui You, and Stelios T. Andreadis

Abstract

Bioengineered skin has great potential for use in regenerative medicine for treatment of severe wounds such as burns or chronic ulcers. Genetically modified skin substitutes have also been used as cell-based devices or "live bioreactors" to deliver therapeutics locally or systemically. Finally, these tissue constructs are used as realistic models of human skin for toxicological testing, to speed drug development and replace traditional animal-based tests in a variety of industries. Here we describe a method of generating bioengineered skin based on a natural scaffold, namely, decellularized human dermis and epidermal stem cells.

Key words Epidermal stem cells, Three-dimensional tissue constructs, Keratinocytes, Differentiation

1 Introduction

1.1 Skin Structure and Function

The skin is the largest organ outlining the human body. It is composed of three layers, namely, (1) epidermis, (2) dermis, and (3) hypodermis. The epidermis is primarily composed of keratinocytes at various differentiation stages: stratum basale, stratum spinosum, stratum granulosum, stratum lucidum, and stratum corneum. Other cell types that are present within the epidermis in small numbers include pigment-producing melanocytes, Merkel cells, and Langerhans cells. Basal epidermal cells lie on a basement membrane (BM), which separates the epidermis from the dermis and contains laminin, collagen type IV, and other extracellular matrix proteins. The dermis is made of collagen, elastin, and reticular fibers. Cells found in this layer are fibroblasts, marcophages, mast cells, and scattered white blood cells. Additionally the dermis is enriched with nerve fibers and blood vessels supplying nutrients to the skin tissue. Below the dermis is the hypodermis that is mainly made up of fat and connective tissues. Functionally, the skin acts as a protective layer against external insults. It also helps to maintain homeostasis by preventing water loss and to regulate body temperature via capillary networks and sweat glands.

Joydeep Basu and John W. Ludlow (eds.), *Organ Regeneration: Methods and Protocols*, Methods in Molecular Biology, vol. 1001, DOI 10.1007/978-1-62703-363-3_22, © Springer Science+Business Media New York 2013

1.2 Skin Problems and Diseases

Cutaneous problems include burns, chronic wounds, and genetic diseases. Large area of skin burns is painful and can be life-threatening due to excessive water loss, bacteria invasion, or other burn-related complications (1). Chronic wounds such as diabetic ulcers or pressure sores can be very dangerous and in many cases result in limb amputations. Other skin diseases include genetic disorders such as epidermolysis bullosa and lamellar ichthyosis. Epidermolysis bullosa (EB) is characterized by blistering skin, which is caused by defects in multiple proteins (e.g., keratin 5/14, laminin 5, or collagen VII) that are responsible for epidermis–basement membrane interactions (2–4). Ichtyosis is manifested as dried, thickened, and scaly skin and may be due to abnormal keratin (e.g., Kl, K10) expression, transglutaminase-I mutations, or steroid sulfatase deficiency (4, 5). Finally, skin tumors are very frequent possibly due to exposure to the sun or other environmental conditions. The major skin tumors include melanoma, basal cell carcinoma, and squamous cell carcinoma, which arise from various genetic mutations and environmental factors (e.g., UV radiation) (6).

1.3 Different Types of Tissue-Engineered Skin Models

Various tissue-engineered skin substitutes have been proposed to facilitate wound or burn coverage. Additionally, these skin grafts can be employed to evaluate the effect of novel treatments on cutaneous diseases or to assess the feasibility of using skin to produce therapeutic proteins. In general, these engineered tissues are prepared using either polymeric or natural biomaterials as dermal support and seeded with autologous or allogeneic primary keratinocytes in either suspension or as a sheet to generate the epidermis. Examples of dermal supports include fibrin (7), fibroblast-containing nylon (Transcyte™) (8), fibroblast-containing biodegradable polyglactin matrix (9, 10), hyaluronic acid membrane (VivoDerm), collagen (E-Z Derm™) (11, 12), collagen infiltrated with fibroblast (Apligraf™) (13), mixture of pepsinized insoluble collagen atop a collagen sponge (OrCel™) (14), collagen/chondroitin-6-sulfate composite (Integra) (14–17), small intestine submucosa (Oasis™) (18), and decellularized dermis (Alloderm™/Xenoderm™) (19, 20). Detailed description of these engineered skin products can be found in published reviews (21, 22).

1.4 Tissue-Engineered Skin Substitutes as Models for Toxicological Testing

As we discussed above, bioengineered skin substitutes closely resemble the native epidermis in terms of tissue architecture as well as function—they develop barrier function (23)—suggesting that 3-D tissue constructs may be superior than traditional 2-D cell culture systems for use as toxicological test platforms (24). Generally, toxicology tests are conducted in two different settings: (a) direct topical application and (b) in vitro patch test (25) and the effects of compounds are evaluated by measurements of cell viability and interleukin-1αα levels. Using this model, many compounds have been tested including cosmetic ingredients (26); skin irritants, e.g., surfactants (27); environmental irritants, e.g., ozone (28); anti-inflammatory compounds (29); and chemicals with phototoxic potential upon UVA exposure (30). Taken together, these studies suggest that bioengineered skin substitutes can reduce significantly the use of animals in toxicological tests of various chemical and pharmacological compounds that may affect skin function.

1.5 Gene-Modified Skin as Bioreactor for Therapeutic Protein Delivery

As the skin is the outermost layer of the body, it is easily accessible for direct on-site genetic modification. In addition, epidermal stem cells can be readily isolated from the patient, genetically modified, and expanded to large numbers in culture before returning to the patient. Therefore, the skin is a rich source of autologous, highly proliferative cells for cell therapy and regenerative medicine. Notably, in case of any adverse effect the modified cells can be rapidly removed from the patient without undergoing any major surgical procedures.

For these reasons, the skin has been proposed as a bioreactor to deliver therapeutic proteins for treatment of a variety of cutaneous diseases as well as other systemic disorders (5). Promising examples for correction of skin diseases include restitution of collagen type VII for dystrophic epidermolysis bullosa (31) and expression of transglutamase I (32) or steroid sulfatase (33) for laminar or X-linked ichthoysis, respectively. Other encouraging results include ectopic expression of factor VIII (34) and factor IX (35) to overcome hemophilia A and hemaphila B deficiency, respectively; leptin-expressing skin grafts for reversing of obesity (36); regulatable insulin secretion for treatment of diabetes (37, 38); and atrial natriuretic peptide (ANP)-production for normalization of blood pressure in hypertensive animal model (39).

1.6 Description of Our Engineered Skin Model and Its Advantages

Most commercially available skin substitutes are composed of a polymeric support and lacks structural features of natural skin such as rete-ridges, intact elastin, and vascular capillary network (40, 41). In contrast, the skin model presented here employs human cadaveric *de*cellularized human *d*ermis (DED) and primary human keratinocytes isolated from donors. DED maintains the BM components as well as structural skin features such as rete-ridges in the dermo-epidernal junction that may improve transport of nutrients to the epidermis as well as transport of proteins from epidermal cells to the dermis or the system circulation. The dermal compartment of DED maintains the structure of collagen, including both papillary and reticular networks of collagen fibers as well as elastin, both critical ECM components that contribute to robust mechanical properties such as pliability and strength. Finally, a network of pre-existing vascular conduits is maintained in the dermis and may help rapid vascularization of engineered skin tissue after implantation.

To generate the epidermis, the BM side of DED is seeded with primary keratinocytes and the tissue is cultured at the air–liquid

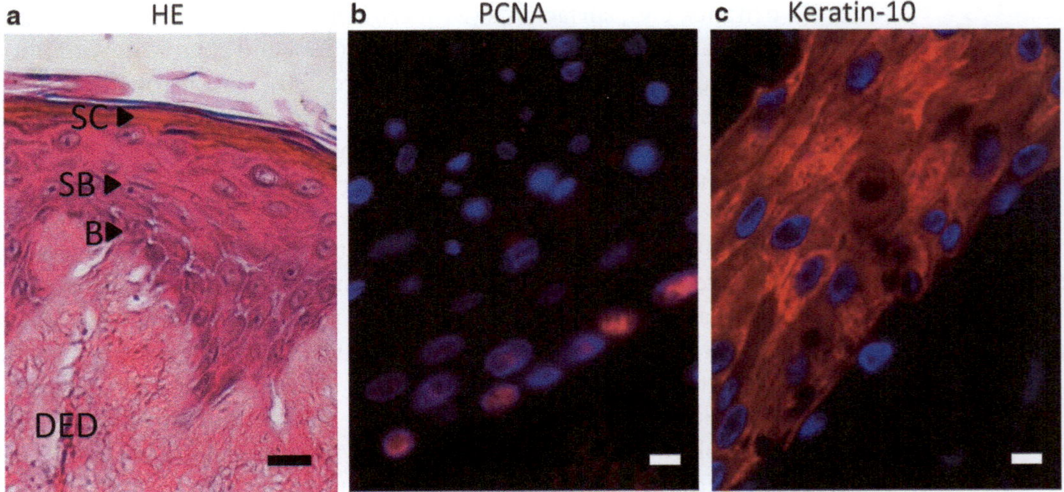

Fig. 1 Histochemistry and immunohistochemistry of bioengineered skin. (**a**) H&E staining reveals the presence of a fully stratified epidermis with basal (B), suprabasal (SB), and stratum corneum (SC). (**b**) The proliferation marker PCNA (*red*) is localized in the basal layer similar to human epidermis. (**c**) Immunostaining for keratin-10 (*red*) shows expression in the suprabasal layers similar to human epidermis. (**b**, **c**) The nuclei were counterstained with Hoechst (*blue*). Scale bar = 10 μm

interface to enable stratification. Under these conditions, keratinocytes proliferate and stratify to form all epidermal layers including basal, suprabasal, granular, and stratum corneum (23, 42, 43) (Fig. 1a). Similar to native epidermis, PCNA staining showed that only basal cells proliferate (Fig. 1b) and K10 immunostaining showed that suprabasal cells differentiate as expected (Fig. 1c). These tissue constructs have been implanted in various animal models where they showed excellent graft take, followed by blood vessel and fibroblast infiltration, resulting in excellent wound healing and long-term wound remodeling (42–44).

Notably, as shown recently this strategy can be adapted into a two-step approach that may be more suitable for wound coverage in vivo (45). In the first step, the wound bed is first covered with angiogenic factor-decorated DED to promote angiogenesis and to serve as a barrier that prevents fluid loss or infection. During this period, autologous keratinocytes can be isolated and expanded in culture before delivering as a suspension in fibrin hydrogel to the vascularized dermis for re-epithelialization. This strategy resulted in excellent graft take while reducing the number of surgical procedures, as compared to the split-thickness autograft approach. Incorporation of angiogenic factors within the DED enhanced vascularization, which in turn correlated with enhanced proliferation and stratification of the neoepidermis originating from the fibrin–keratinocyte cell suspension. This strategy may be useful for treatment of burns inflicting large area of the body where rapid skin coverage is necessary and the donor skin sites are rare.

2 Materials

2.1 Acellular Dermis

1. Human skin tissue (can be obtained from tissue bank).
2. Heat water bath set at 37°C.
3. Liquid nitrogen.
4. Antibiotic cocktail: 100 μg/mL Gentamicin (Sigma-Aldrich, St Louis, MO, USA), 0.001% v/v Ciprofloxacin (Bayer HealthCare Pharmaceuticals Inc., Wayne, NJ, USA), 1% v/v Antibiotic–Antimycotic (gibco® by life technologies, Grand Island, NY, USA).
5. Phosphate-buffered saline (PBS).
6. Sterile plastic Ziploc bags (various sizes).
7. Sterile forceps.

2.2 Human Primary Keratinocyte Isolation and Maintenance

1. 3T3/J2 cells (ATCC, Manassas, VA, USA).
2. Dulbecco's modified Eagle's medium (DMEM; gibco®).
3. Antibiotic–Antimycotic mentioned in Subheading 2.1, Step 4.
4. Bovine serum (BS; gibco®).
5. Tissue culture flasks (75-cm^2; Greiner Bio-One, Monroe, NC, USA).
6. Mitomycin-C (Sigma-Aldrich).
7. PBS.
8. *Keratinocyte culture medium (KCM)*: 3:1 DMEM: Ham's F-12 (gibco®), 10% v/v fetal bovine serum (FBS; gibco®), 1% v/v antibiotic–antimycotic, 24 mg/L adenine (Sigma-Aldrich), 10^{-10} M cholera toxin (Vibrio cholerae, Type Inaba 569 B; Calbiochem, La Jolla, CA, USA), 0.4 μg/mL hydrocortisone (Calbiochem), 5 μg/mL insulin (Sigma-Aldrich), 5 μg/mL transferrin, (Boehringer Mannheim, Indianapolis, IN, USA), 2×10^{-9} M triiodo-L-thyronine (Sigma-Aldrich) (see Note 1).
9. Fresh human newborn foreskin samples from hospital (see Note 2).
10. Tissue culture dishes (100-mm tissue culture dishes; Greiner).
11. Sterile sharp rounded scissors.
12. Conical tubes (15 and 50 mL).
13. 0.25% Trypsin containing 0.5 mM EDTA.
14. Serological pipettes (5 mL).
15. Versene: 0.48 mM EDTA in PBS.
16. Tissue culture roller drum (TC-8, Brunswick Scientific Co. Inc., Edison, NJ, USA).
17. Epidermal growth factor (EGF; BD Biosciences, Mountain View, CA, USA).

2.3 Epidermal Skin Equivalents

1. Polydimethylsiloxane (PDMS; Dow Corning Corporation, Midland, MI, USA).
2. Human acellular dermis (from Subheading 2.1).
3. PBS.
4. Sterile surgical blade.
5. Sterile forceps.
6. Non-tissue culture treated 6-well plates (BD Falcon, Franklin Lakes, NJ, USA).
7. Human primary keratinocytes (from Subheading 2.2).
8. *Seeding medium*: 3:1 DMEM: Ham's F-12 mentioned in Subheading 2.2, Step 8. 1% v/v FBS, 1% v/v antibiotic–antimycotic redundancy, 10^{-10} M cholera toxin mentioned in Subheading 2.2, Step 8. 200 ng/mL hydrocortisone mentioned in Subheading 2.2, Step 8. 5 μg/mL insulin mentioned in Subheading 2.2, Step 8. 50 μg/mL ascorbic acid (Sigma-Aldrich) (see Notes 1 and 3).
9. *Priming medium*: Seeding medium supplemented with 24 μM bovine serum albumin (Calbiochem), 1% v/v FBS, 25 μM oleic acid (Sigma), 15 μM linoleic acid (Sigma-Aldrich), 7 μM, arachidonic acid (Sigma-Aldrich), 25 μM palmitic acid (Sigma-Aldrich), 10 μM L-carnitine (Sigma-Aldrich), 1 mM L-serine (Sigma-Aldrich) (see Notes 1 and 3).
10. Metal support: Bend 2–3 mm of two opposing sides of a ~2 cm × 2 cm metal screen mesh to 90°.
11. *Air/liquid interface medium*: 3:1 DMEM: Ham's F-12, 1% v/v antibiotic–antimycotic, 24 μM bovine serum albumin, 10^{-10} M cholera toxin, 200 ng/mL hydrocortisone mentioned in Subheading 2.2, Step 8. 5 μg/mL insulin mentioned in Subheading 2.2, Step 8. 50 μg/mL ascorbic acid mentioned in Subheading 2.3, Step 8. 25 μM oleic acid mentioned in Subheading 2.3, Step 9. 15 μM linoleic acid mentioned in Subheading 2.3, Step 9. 7 μM arachidonic acid mentioned in Subheading 2.3, Step 9. 25 μM palmitic acid mentioned in Subheading 2.3, Step 9. 10 μM L-carnitine mentioned in Subheading 2.3, Step 9. 1 mM L-serine mentioned in Subheading 2.3, Step 9. 1 ng/mL EGF (see Notes 1 and 3).

3 Methods

3.1 Acellular Dermis

1. Remove skin tissue from −80°C and seal the tissue in sterile plastic Ziploc bags (double or even triple bagged) (see Note 4).
2. Thaw the skin tissue in a water bath at 37°C for 30 min.
3. Freeze and thaw the tissue three times: 10-min freezing in liquid nitrogen; 30-min thawing at 37°C in a water bath.

4. Wash the skin three times in PBS to remove glycerol (see Note 5).
5. Incubate the tissue in 500 mL of antibiotic cocktail at 37°C for 1 week.
6. De-epidermalize the skin by peeling off all the epidermis using sterile forceps (see Note 6).
7. Transfer the decellularized dermis to fresh antibiotic cocktail and incubate for 4 more weeks at 4°C before use.

3.2 Human Primary Keratinocyte Isolation and Maintenance

1. The day before isolation, seed 3T3/J2 feeder cells in DMEM + 1% v/v antibiotic-antimycotic + 10% v/v BS at $1.2–1.8 \times 10^6$ cells per 75-cm^2 tissue culture flask (see Note 7).
2. On the day of isolation, treat 3T3/J2 cells with 15 mL mitomycin-C (15 μg/mL in DMEM + 1% v/v antibiotic-antimycotic + 10% v/v BS) for 2–3 h (see Note 8).
3. Wash the cells extensively with 5 mL PBS three times (see Note 9).
4. Add 15 mL KCM to the flask and incubate for at least 1 h before use.
5. Add 1–3 mL PBS to a 100-mm tissue culture dish.
6. Place a foreskin sample, with the epidermis side facing down, on the tissue culture dish (see Note 10).
7. Trim the tissue to remove fat and connective tissue with a pair of sterile sharp rounded scissors (see Note 11).
8. Transfer the tissue into a 50-mL conical tube.
9. Rinse the tissue with 30 mL PBS eight times (see Note 12).
10. Add 3 mL trypsin to a new 100-mm tissue culture dish.
11. Transfer the tissue to the tissue culture dish.
12. Mince the tissue with a pair of sterile sharp rounded scissors for 10–15 min (see Note 13).
13. Add 3 mL trypsin to a 15-mL conical tube.
14. Pre-wet a 5-mL serological pipette with trypsin.
15. Use the pre-wet pipette to transfer all the minced tissue into the 15-mL conical tube.
16. Wash the tissue culture dish with 3 mL versene.
17. Transfer the versene into the conical tube.
18. Repeat steps 16 and 17. The conical tube should now contain 12 mL solution.
19. Leave the conical tube on a cell culture roller drum (1.0–1.5 rpm) at 37°C.
20. After 30 min, remove the conical tube from the roller drum and allow the tissue to settle down by gravity.
21. Carefully transfer the supernatant into a clean 15-mL tube.

22. Centrifuge the supernatant at 140 × *g* for 5 min.
23. Resuspend the cells in 1 mL conditioned KCM (from the T-75 flask in step 4) and add all cells to the mitomycin-C-treated feeder layer in the T-75 flask.
24. Add 6 mL trypsin and 6 mL versene to the minced tissue and return the conical tube on the rotator at 37°C.
25. Repeat steps 20–24 three more times.
26. Three days after isolation, replace cells with KCM supplemented with 10 ng/mL EGF.
27. Check cells daily.
28. Before cell colonies start to merge, subculture cells at 1:10 dilution either on mitomycin-C-treated 3T3/J2 feeder layers (see steps 1–4) or in keratinocyte-SFM (see Note 14).

3.3 Epidermal Skin Equivalents

1. Before making skin equivalents, prepare some PDMS disks that are ~30 mm diameter and 3 mm thick with a 1 cm × 1 cm opening in the center.
2. Remove the dermis from antibiotic cocktail and wash thoroughly with PBS eight times.
3. Cut the dermis into 1 cm × 1 cm pieces with a surgical blade.
4. Place the PDMS disk in a 6-well non-tissue culture plate and secure the disk by pressing the disk against the bottom of the well.
5. Scratch the open area with a surgical blade (see Note 15).
6. Using a pair of forceps, place the dermis with papillary side up in the open area.
7. Incubate the dermis at 37°C for 1 h (see Note 16).
8. Trypsinize keratinocytes and resuspend cells in seeding medium at 5×10^6 cells/mL.
9. Add 100 μL of cells onto the dermis and leave the sample in the incubator for 1.5 h (see Note 17).
10. Add 3 mL seeding medium to each well.
11. On the next day, remove seeding medium and add 3 mL priming medium to each well.
12. In 2 days, raise the dermis to the air–liquid interface by putting the dermis on the metal support.
13. Add air–liquid interface medium to the well so that it is just enough to come in contact with the bottom of the dermis.
14. Change medium every 3 days (see Note 18).
15. After 7 days at the air–liquid interface examine the tissues with histology and immunostaining using standard protocols.

4 Notes

1. All culture media should be stored at 4°C and used within 1 month.
2. Immediately after surgery, human foreskin tissues are collected in 50 mL conical tubes containing 30 mL of KCM and transported to the laboratory on ice.
3. Serum or bovine serum albumin must be added to DMEM/Ham's F-12 before addition of other medium components. Serum or bovine serum albumin will coat the container wall, thereby preventing the loss of medium supplements through nonspecific binding to the container wall during medium preparation.
4. Minimize air trapped inside the Ziploc bags because the air will expand during thawing, resulting in explosion of the bags. Consequently the tissue inside the bags will be contaminated.
5. Skin tissues are typically stored in cryopreserved medium containing 15% glycerol.
6. Usually after a 1-week incubation in antibiotic cocktail, the epidermis can easily be peeled off. If it is too difficult to detach the epidermis, the tissue can be left in antibiotic cocktail for an extra week.
7. When cells are plated early in the morning, 1.2×10^6 cells/T-75 flask is sufficient. On the other hand, if cells are plated late in the day, 1.8×10^6 cells/flask is required to ensure more than 95% confluence by the next day.
8. Mitomycin-C treatment will inhibit 3T3/J2 cell growth, which would otherwise outgrow keratinocytes. Nonetheless, mitomycin-C-treated fibroblasts can still produce soluble factors that promote keratinocyte proliferation and clonogenic potential.
9. It is important to wash away all the mitomycin-C to avoid inhibition of keratinocyte proliferation.
10. Although it is best to isolate cells on the same day, we have successfully isolated cells from tissues that were stored at 4°C for up to 2 days post harvest. To avoid disease transmission from human tissue wear gloves at all times. After use surgical instruments such as scissors should be treated with bleach, washed thoroughly with soap and water, and autoclaved. Disposable items such as cell culture dishes should be discarded in proper biohazard bags.
11. It is important to remove ALL connective tissue to avoid fibroblast contamination and obtain a more homogenous keratinocyte culture.

12. This washing step is required to remove contaminants from the tissue.
13. It is important to cut the tissue into as small pieces as possible to increase the yield of keratinocyte isolation.
14. When the colonies merge with each other keratinocytes will differentiate, resulting in loss of proliferative potential. Also, do not culture cells beyond passage 5 because they undergo terminal differentiation.
15. Scratching the surface of the well will help to secure the dermis in position.
16. This incubation step helps to evaporate PBS that remains on the dermis; and hence the dermis sticks tighter to the surface.
17. Be careful while transferring cells to the incubator so that the cells remain on top of the dermis.
18. Epidermal skin equivalent forms the best skin structure after culturing at the air–liquid interface for 7 days.

References

1. Medalie DA, Eming SA, Collins ME et al (1997) Differences in dermal analogs influence subsequent pigmentation, epidermal differentiation, basement membrane, and rete ridge formation of transplanted composite skin grafts. Transplantation 64:454–465
2. Hilal L, Rochat A, Duquesnoy P et al (1993) A homozygous insertion-deletion in the type VII collagen gene (COL7A1) in Hallopeau-Siemens dystrophic epidermolysis bullosa. Nat Genet 5:287–293
3. Jarvikallio A, Pulkkinen L, Uitto J (1997) Molecular basis of dystrophic epidermolysis bullosa: mutations in the type VII collagen gene (COL7A1). Hum Mutat 10:338–347
4. Chamcheu JC, Siddiqui IA, Syed DN et al (2011) Keratin gene mutations in disorders of human skin and its appendages. Arch Biochem Biophys 508:123–137
5. Khavari PA, Rollman O, Vahlquist A (2002) Cutaneous gene transfer for skin and systemic diseases. J Intern Med 252:1–10
6. Woodman SE, Lazar AJ, Aldape KD et al (2012) New strategies in melanoma: molecular testing in advanced disease. Clin Cancer Res 18(5):1195–1200
7. Meana A, Iglesias J, Del Rio M et al (1998) Large surface of cultured human epithelium obtained on a dermal matrix based on live fibroblast-containing fibrin gels. Burns 24: 621–630
8. Mansbridge J (2002) Tissue-engineered skin substitutes. Expert Opin Biol Ther 2:25–34
9. Liu K, Yang Y, Mansbridge J (2000) Comparison of the stress response to cryopreservation in monolayer and three-dimensional human fibroblast cultures: stress proteins, MAP kinases, and growth factor gene expression. Tissue Eng 6:539–554
10. Pinney E, Liu K, Sheeman B et al (2000) Human three-dimensional fibroblast cultures express angiogenic activity. J Cell Physiol 183:74–82
11. Healy CM, Boorman JG (1989) Comparison of E-Z Derm and Jelonet dressings for partial skin thickness burns. Burns Incl Therm Inj 15:52–54
12. Vanstraelen P (1992) Comparison of calcium sodium alginate (KALTOSTAT) and porcine xenograft (E-Z DERM) in the healing of split-thickness skin graft donor sites. Burns 18:145–148
13. Falanga V, Sabolinski M (1999) A bilayered living skin construct (APLIGRAF) accelerates complete closure of hard-to-heal venous ulcers. Wound Repair Regen 7:201–207
14. Windsor ML, Eisenberg M, Gordon-Thomson C et al (2009) A novel model of wound healing in the SCID mouse using a cultured human skin substitute. Australas J Dermatol 50:29–35
15. Burke JF, Yannas IV, Quinby WC Jr et al (1981) Successful use of a physiologically

acceptable artificial skin in the treatment of extensive burn injury. Ann Surg 194:413–428
16. Yannas IV, Burke JF (1980) Design of an artificial skin. I. Basic design principles. J Biomed Mater Res 14:65–81
17. Yannas IV, Burke JF, Gordon PL et al (1980) Design of an artificial skin. II. Control of chemical composition. J Biomed Mater Res 14:107–132
18. Mostow EN, Haraway GD, Dalsing M et al (2005) Effectiveness of an extracellular matrix graft (OASIS Wound Matrix) in the treatment of chronic leg ulcers: a randomized clinical trial. J Vasc Surg 41:837–843
19. Livesey SA, Herndon DN, Hollyoak MA et al (1995) Transplanted acellular allograft dermal matrix. Potential as a template for the reconstruction of viable dermis. Transplantation 60:1–9
20. Cuono C, Langdon R, McGuire J (1986) Use of cultured epidermal autografts and dermal allografts as skin replacement after burn injury. Lancet 1:1123–1124
21. Ramos-e-Silva M, Ribeiro de Castro MC (2002) New dressings, including tissue-engineered living skin. Clin Dermatol 20: 715–723
22. Andreadis ST (2004) Gene transfer to epidermal stem cells: implications for tissue engineering. Expert Opin Biol Ther 4:783–800
23. Andreadis ST, Hamoen KE, Yarmush ML et al (2001) Keratinocyte growth factor induces hyperproliferation and delays differentiation in a skin equivalent model system. FASEB J 15:898–906
24. Fentem JH, Botham PA (2002) ECVAM's activities in validating alternative tests for skin corrosion and irritation. Altern Lab Anim 30(Suppl 2):61–67
25. Tornier C, Rosdy M, Maibach HI (2006) In vitro skin irritation testing on reconstituted human epidermis: reproducibility for 50 chemicals tested with two protocols. Toxicol In Vitro 20:401–416
26. Hagino S, Okazaki Y, Itagaki H (2008) An in vitro tier evaluation for the identification of cosmetic ingredients which are not ocular irritants. Altern Lab Anim 36:641–652
27. Roguet R, Cohen C, Dossou KG et al (1994) Episkin, a reconstituted human epidermis for assessing in vitro the irritancy of topically applied compounds. Toxicol In Vitro 8:283–291
28. Cotovio J, Onno L, Justine P et al (2001) Generation of oxidative stress in human cutaneous models following in vitro ozone exposure. Toxicol In Vitro 15:357–362
29. Kim KH, Chung CB, Kim YH et al (2005) Cosmeceutical properties of levan produced by Zymomonas mobilis. J Cosmet Sci 56:395–406
30. Lelievre D, Justine P, Christiaens F et al (2007) The EpiSkin phototoxicity assay (EPA): development of an in vitro tiered strategy using 17 reference chemicals to predict phototoxic potency. Toxicol In Vitro 21:977–995
31. Chen M, Kasahara N, Keene DR et al (2002) Restoration of type VII collagen expression and function in dystrophic epidermolysis bullosa. Nat Genet 32:670–675
32. Choate KA, Medalie DA, Morgan JR et al (1996) Corrective gene transfer in the human skin disorder lamellar ichthyosis. Nat Med 2:1263–1267
33. Freiberg RA, Choate KA, Deng H et al (1997) A model of corrective gene transfer in X-linked ichthyosis. Hum Mol Genet 6:927–933
34. Fakharzadeh SS, Zhang Y, Sarkar R et al (2000) Correction of the coagulation defect in hemophilia A mice through factor VIII expression in skin. Blood 95:2799–2805
35. Gerrard AJ, Hudson DL, Brownlee GG et al (1993) Towards gene therapy for haemophilia B using primary human keratinocytes. Nat Genet 3:180–183
36. Larcher F, Del Rio M, Serrano F et al (2001) A cutaneous gene therapy approach to human leptin deficiencies: correction of the murine ob/ob phenotype using leptin-targeted keratinocyte grafts. FASEB J 15:1529–1538
37. Lei P, Ogunade A, Kirkwood KL et al (2007) Efficient production of bioactive insulin from human epidermal keratinocytes and tissue-engineered skin substitutes: implications for treatment of diabetes. Tissue Eng 13:2119–2131
38. Tian J, Lei P, Laychock SG et al (2008) Regulated insulin delivery from human epidermal cells reverses hyperglycemia. Mol Ther 16:1146–1153
39. Therrien JP, Kim SM, Terunuma A et al (2010) A gene therapy approach for long-term normalization of blood pressure in hypertensive mice by ANP-secreting human skin grafts. Proc Natl Acad Sci U S A 107:1178–1183
40. Andreadis ST (2006) Experimental models and high-throughput diagnostics for tissue regeneration. Expert Opin Biol Ther 6:1071–1086
41. Andreadis ST (2007) Gene-modified tissue-engineered skin: the next generation of skin substitutes. Adv Biochem Eng Biotechnol 103:241–274
42. Geer DJ, Swartz DD, Andreadis ST (2004) In vivo model of wound healing based on trans-

planted tissue-engineered skin. Tissue Eng 10:1006–1017
43. Geer DJ, Swartz DD, Andreadis ST (2005) Biomimetic delivery of keratinocyte growth factor upon cellular demand for accelerated wound healing in vitro and in vivo. Am J Pathol 167:1575–1586
44. Medalie DA, Eming SA, Tompkins RG et al (1996) Evaluation of human skin reconstituted from composite grafts of cultured keratinocytes and human acellular dermis transplanted to athymic mice. J Invest Dermatol 107: 121–127
45. Lugo LM, Lei P, Andreadis ST (2011) Vascularization of the dermal support enhances wound re-epithelialization by in situ delivery of epidermal keratinocytes. Tissue Eng Part A 17:665–675

Chapter 23

Formulation of Selected Renal Cells for Implantation into a Kidney

Craig Halberstadt, Neil Robbins, Darell W. McCoy, Kelly I. Guthrie, Andrew T. Bruce, Toyin A. Knight, and Richard G. Payne

Abstract

Delivery of cells to organs has primarily relied on formulating the cells in a nonviscous liquid carrier. We have developed a methodology to isolate selected renal cells (SRC) that have provided functional stability to damaged kidneys in preclinical models (Kelley et al. Poster presentation at 71st scientific sessions of American diabetes association, 2011; Kelley et al. Oral presentation given at Tissue Engineering and Regenerative Medicine International Society (TERMIS)—North America annual conference, 2010; Presnell et al. Tissue Eng Part C Methods 17:261–273, 2011; Kelley et al. Am J Physiol Renal Physiol 299:F1026–F1039, 2010). In order to facilitate SRC injection into the kidney of patients who have chronic kidney disease, we have developed a strategy to immobilize the cells in a hydrogel matrix. This hydrogel (gelatin) supports cells by maintaining them in a three-dimensional state during storage and shipment (both at cold temperatures) while facilitating the delivery of cells by liquefying when engrafting into the kidney. This chapter will define a method for the formulation of the kidney epithelial cells within a hydrogel.

Key words Kidney disease, Renal epithelial cells, Gelatin, Hydrogel

1 Introduction

There are several primary diseases that impact the function of the kidney. As these diseases progress, the kidney goes through several stages of tissue damage (stages I–V). For later stages of kidney damage, the treatment modalities are limited to either dialysis or eventually, kidney transplant. Based on data generated in animal models (1–4), the transplantation of autologous, homologous selected renal cells (SRC) into the kidney at an early stage of the degeneration of the kidney (stage III or early stage IV) should prolong kidney health. Hence, the transplantation of these cells has a great potential in delaying the eventual use of dialysis by offsetting the loss of function of the kidneys.

Cell transplantation has been used for the amelioration or treatment of many diseases or tissue injury. Some of the

Joydeep Basu and John W. Ludlow (eds.), *Organ Regeneration: Methods and Protocols*, Methods in Molecular Biology, vol. 1001, DOI 10.1007/978-1-62703-363-3_23, © Springer Science+Business Media New York 2013

applications include the implantation of chondrocytes for cartilage repair (5), delivery of non-encapsulated pancreatic islets for the treatment of Type I diabetes (6), and the injection of mesenchymal stem cells for the repair of heart muscle tissue (7). All of these therapies have relied on delivering the cells in a nonviscous solution (such as saline or cell culture media). Some of the challenges with this approach of cell delivery have included the potential for cell settling during the injection phase (which can reduce the even distribution of the cells upon implant) and short-term shelf life due to product settling and cell aggregation.

Gelatin is a non-cross-linked collagen that has been used for many different medical applications. The biocompatible nature of gelatin has been extensively reviewed and this material has been used for in vivo applications (8). One can control the gelation properties of the gelatin (formation of a hydrogel) depending on the concentration and bloom (a measure of force (weight) required to depress a standard plunger of 12.5 mm diameter into the surface of the gel at a distance of 4 mm (AOAC standard)) of the gelatin as well as its temperature. Previously (9–12) we have shown that gelatin is biocompatible with SRC both in vitro and in vivo.

This chapter will describe a method to suspend the cells in gelatin mixed in a phosphate buffered saline solution for storage and shipping at a cold temperature. This formulation has the unique property of remaining in a gelatinous state at a cold temperature and upon warming up to room temperature begins to liquefy (17–21°C). Hence, the product can be shipped and stored cold keeping the cells in a suspension. When required, the product can be removed from the cold and warmed to room temperature where it begins to liquefy.

2 Materials

2.1 Formulation of Rat Renal Cells

2.2 Cell Culture Media and Reagents

1. Dulbecco's Modified Eagle Medium, high glucose (DMEM-HG), containing sodium pyruvate.
2. Keratinocyte Serum-Free Medium (KSFM) containing L-glutamine and supplied with prequalified human recombinant epidermal growth factor 1-53 (EGF 1-53) and bovine pituitary extract (BPE) in separate packaging (Invitrogen) (see Note 1).
3. Fetal bovine serum (FBS).
4. Insulin–Transferrin–Selenium G solution (ITS) 100× (10 mL per 1 L Media).

5. Antibiotic/Antimycotic, 100× (Invitrogen) (10 mL per 1 L Media).
6. Renal Cell Culture Media—50:50 Mix of DMEM and KSFM with 5% FBS, with EGF and BPE, and ITS.
7. 0.25% Trypsin–EDTA.

2.3 Tissue Digestion

1. Collagenase Type IV—300 U in 600 μL.
2. Dispase, 5 mg/mL concentration, in Hank's buffered saline solution.
3. Neutralizing Media (Renal Cell Culture Media).

2.4 Cell Separation

1. Iodixanol (Optiprep™ 60% wt/vol) in KSFM (see Notes 2 and 3).
2. Dulbecco's phosphate buffered saline (DPBS) (For washing of cells after gradient separation).
3. Steriflip (Millipore).

2.5 Cell Formulation

1. Porcine Skin Gelatin (Type A) from Gelita USA (Sergeant Bluff, IA, USA).
2. Mix Gelatin with DPBS from 0.625 to 1.5% (see Note 4).

2.6 Creating a Gel

1. Syringe (1–10 cc syringes).
2. Sterile luer lock cap.

2.7 Equipment

1. Class 100 inverted laminar flow biological safety cabinet (BSC).
2. Tube rotator (Customized with an RPM setting of 2–25 RPM) (see Note 5).
3. Refrigerator with power outlet.
4. Humidified 37°C 5% CO_2 incubator.
5. Humidified 37°C 2% O_2 incubator.
6. Water bath (37°C).
7. Dry bath (28°C).
8. Swinging bucket centrifuge system (Sorvall, see Note 6).
9. Hot stir plate (needs 50°C).

2.8 Plastic Supplies and Cultureware

1. T-75, T-500 flasks.
2. 150 mm plastic petri dishes (pre-weighed).
3. Microcentrifuge tubes.
4. 15 mL centrifuge tubes.
5. Pipettes (1, 5, 10, 25 mL).
6. Syringes (1, 5, 10 mL).

7. Pipetters (Rainin or other brand; P10, P100, P1000).
8. Pipette tips for pipetters (see Note 7).
9. Glass beaker (100, 500 mL).
10. Magnetic stir bar.
11. 0.22 μm sterile filter units (Corning).

2.9 Stainless Steel Instruments and Surgical Material

1. Scissors.
2. Scalpel.
3. Forceps.
4. Hemostats.
5. Sterile gauze pads, 12-ply, 4×4 in.

3 Method

3.1 Rat Renal Cell Isolation Method

1. Rat kidneys are harvested after euthanasia from an appropriately certified vendor following all institutional guidelines for safe handling of animals.
2. Upon kidney arrival to the laboratory, aspirate the shipping medium and pour kidneys into a 150 mm dish.
3. Remove and discard connective tissue, calyx, and capsule (if applicable) around the kidney using forceps and scalpel.
4. Wash the kidneys one time in HBSS or PBS to remove any debris.
5. Manually mince together remaining kidneys using scalpel blade, making finely chopped slurry of tissue.
6. Add 1.0 g (±0.1 g) minced kidney tissue to each pre-weighed 50 mL conical tube.
7. Add 20 mL of pre-made digestion buffer to each 50 mL tube.
8. Perform two sequential digestions of the kidney tissue.
9. *Digestion 1:* Incubate digestion tubes at 37°C on a rocker for *20 min.*
10. Remove tubes from incubator/rocker and place in the BSC.
11. Allow undigested tissue to settle to the bottom of the tube via gravity.
12. Aspirate supernatant from each tube and discard.
13. Add 20 mL of warm digestion buffer (warmed to 37°C) to each 50 mL tube.
14. *Digestion 2:* Incubate digestion tubes at 37°C on a rocker for *30 min.*

15. Remove tubes from rocker/incubator, combine two digestion tubes into one, and run the contents of each tube through a 100 μm SteriFlip.
16. Transfer the cell suspension into a sterile bottle.
17. Neutralize the enzyme with an equal amount of neutralization buffer.
18. Centrifuge in 50 mL tubes at 300×*g* for 5 min (or in 225 mL tubes at 300×*g* for 8 min). Aspirate supernatant and discard.
19. Resuspend the cell pellet in desired amount of KSFM (break up clumps).
20. Count cells using a hemocytometer.
21. To clean up the tissue prior to plating the cells, the digested tissue is placed onto a 15% Mixing Gradient (OptiPrep).
22. Based on total remaining cell number, calculate how many 15 mL gradient tubes are needed to load 75×10^6 cells/tube (or in 50 mL tubes at 225×10^6 cells per tube).
23. Aliquot 75×10^6 cells into each 15 mL tube and bring each cell suspension up to 5 mL with KSFM or aliquot 225×10^6 cells into each designated 50 mL tube and bring each cell suspension up to 22.5 mL with KSFM.
24. Add an equal volume of 30% Optiprep to each tube in order to achieve a 15% OptiPrep solution.
25. Mix tubes by inversion six times.
26. Carefully layer 1 mL of PBS on top of Optiprep/cell mixture (or 5 mL for a 50 mL tube).
27. Centrifuge at 800×*g* for 15 min with NO brake.
28. Carefully remove the tube(s) and place back into the BSC.
29. Collect and combine all cell bands via a sterile transfer pipette into a clean 50 mL tube.
30. Aspirate and discard the remaining supernatant (leaving the cell pellet(s) in the tube(s)).
31. Resuspend all pellets with KSFM and combine with the bands collected above.
32. Add enough KSFM to reach a 4:1 ratio of KSFM to cell suspension (Split sample into multiple 50 mL tubes if necessary).
33. Mix tubes by inversion six times (see Note 8).
34. Centrifuge at 300×*g* for 5 min.
35. Aspirate supernatant leaving pellet in the tube.
36. Resuspend cell pellet with KSFM to desired volume. This is the 15% Band+Pellet sample.
37. Count cells using a hemocytometer.

38. The Band + Pellet cells are now ready to freeze, plate, analyze, or use for experimentation.
39. Plate 25,000 cells/cm^2 in a desired TC-treated vessel using complete Renal Cell Culture Media at a volume that is recommended by vessel manufacturer. (For example: T500 Nunc flask, 12.5×10^6 cells/flask, 100 mL of Renal Cell Culture Media.) The culture vessels containing cells (passage 0) are placed in a 5% CO_2 incubator at 37°C in a humidified environment (see Note 9). After 48 h, a complete media exchange occurs.

3.2 Harvesting the Cells for Formulation

1. After culturing the cells for 2–3 days, the cells are placed in a hypoxic (2% oxygen) environment overnight (see Note 10).
2. Prepare four gradient solutions (16, 13, 11, and 7%) of Iodixanol (Optiprep™ 60% wt/vol) in KSFM.
3. Each density step gradient requires a cell number ranging from 60 to 75 million cells per tube. Once the amount of cells has been determined, calculate the number of gradient tubes to be generated.
4. Make a density step gradient(s) by first pipetting 2 mL of 16% density gradient medium into a 15 mL conical centrifuge tube(s).
5. Carefully layer 2 mL of the 13% density gradient medium onto the 16% density gradient medium by tilting the tube at a 45° angle and letting the medium slowly run down the side of the tube. This will minimize mixing at the interface between the two different densities.
6. Once the 13% density gradient medium has been layered, continue with the 11% density gradient medium using the same layering method and finish with the 7% density gradient medium.
7. Once the gradients are formed, take care in moving as not to disturb the boundary interfaces.
8. Carefully pipette 2 mL of cell suspension containing between 60 and 75 million cells in KSFM medium on top of the step gradient. Continue until all of the gradients have been loaded with cell suspension.
9. Once the cells have been loaded, carefully place tubes into the centrifuge and spin at $800 \times g$ for 20 min without brake.
10. After centrifugation, collect tubes and visually inspect gradient bands to verify banding pattern.
11. Collect gradient bands by aspirating each band using either sterile bulb or 5 mL pipette.
12. Combine bands 2, 3, and 4 (B2, B3, and B4).
13. Wash the cells (B2, B3, and B4) three times using DPBS by centrifugation.

3.3 Formulation in Gelatin Hydrogel

1. Batches of gelatin are pre-made in DPBS at concentrations ranging from 0.625 to 1.5% v/v. Aliquot the gelatin into smaller vials to be used for the final formulation steps.
2. Briefly, gelatin is measured and dissolved in DPBS. The gelatin solution is dissolved at 50°C while being mixed and then sterile filtered using a 0.22 μm filter into a sterile container. The sterile gelatin solution is then aseptically added to smaller vials for storage at 4°C until use.
3. After washing, the pelleted SRC are re-suspended and counted to determine cell concentration to be formulated in the gelatin solution. At this time, heat up the pre-made gelatin solution aliquot(s) using a dry bath set at 25–28°C for >1 h.

3.3.1 After Counting the Cells a Final Wash Should Be Performed Using Gelatin

1. Following the final centrifugation, the gelatin solution supernatant is removed and sufficient volume of 0.625–1.5% gelatin is added to targeted volume/cell concentration with a maximum cell to gelatin ratio of 50:50. For example, an estimated packed volume for rat SRC is 300 million cells for a volume of 1.0 mL. If the total cell number required per injection is 20 million cells, then the minimum volume of product that could be injected into a kidney is 0.13 mL (0.067 mL of packed cells + 0.067 mL gelatin). Bring up the total volume to 0.13 mL by adding gelatin (see Note 10).
2. The final SRC product is used as a gel. To create the gel, add the gelatin/cell solution, use a small pipette tip that fits inside the luer lock of a syringe (see Note 11).
3. Carefully place a sterile luer lock cap onto the syringe. Make sure the cap is tight.
4. Place the container on a rotator at a minimum of 2.0 RPM in a 2–5°C environment overnight. The rotation maintains the cells in suspension during the gelation period.
5. Once gelation has occurred, the SRC/gelatin product can be removed from the rotator and stored cold until use (see Note 12).

4 Notes

1. KSFM comes with a bullet kit that is kept frozen until use. It is recommended to use the media within a couple of weeks after mixing the components.
2. Make sure Optiprep is made in an osmotically correct buffer such as KSFM.
3. Make solutions up prior to use.

4. The higher the concentration of gelatin, the stiffer the gel. At the higher concentration, the gelatin will take longer to liquefy at room temperature.
5. We used a modified electric drill with a polycarbonate disk attached to it.
6. Swinging bucket is preferable for establishing the different bands of the density gradient.
7. Small tips are used for loading the syringes with the cell/gelatin mixture.
8. If too little KSFM is added and/or if tubes are not mixed well, the Optiprep mixture will re-band.
9. Isolated rat renal cells do not passage (re-plate onto plastic). Hence, cell harvesting takes place at the end of the initial plating process (4 days after seeding onto the plastic).
10. The cells should not be greater than 90% confluent when placed in the low oxygen environment.
11. The user should take into account the dead volume of a syringe and the needle. For example, a 1 cc syringe with a 27 G needle has approximately 80 μL of dead volume. Hence, an additional 80 μL of product should be made to take this into account.
12. Ensure gelatin is chilled at or below 4°C for proper gelation. Do not exceed below 0°C during the process. Ensure gelatin + cells remain below 8°C for proper storage and shipping. Ensure gelatin + cells warm up to room temperature prior to injection. This will support the delivery of the cells into the kidney parenchyma.

References

1. Kelley R, Spencer T, Werdin E, Bruce A, Ilagan R, Wallace S, Watts B, Choudhury S, Cox B, Guthrie K, Jayo M, Bertram T, Presnell S (2011) Intra-renal transplantation of bioactive renal cells preserves renal functions and extends survival in the ZSF1 model of progressive diabetic nephropathy. Poster presentation at 71st scientific sessions of American diabetes association. www.tengion.com/news/documents/ADA%202011%20rwk_Tengion%20FINAL.pdf
2. Kelley RW, Werdin ES, Bruce AT, Wallace SM, Jayo MJ, Bertram TA, Presnell SC (2010) Bioactive renal cells augment renal function in the ZSF1 model of diabetic nephropathy. Oral presentation given at Tissue Engineering and Regenerative Medicine International Society (TERMIS)—North America annual conference. www.tengion.com/pdfs/Kelley-2010-TERMIS-FINAL.pdf
3. Presnell SC, Bruce AT, Wallace SM, Choudhury S, Genheimer C, Cox B, Guthrie KI, Werdin ES, Tatsumi-Ficht P, Ilagan R, Kelley RW, Rivera EA, Ludlow JW, Wagner BJ, Jayo MJ, Bertram T (2011) Isolation, characterization, and expansion (ICE) methods for defined primary renal cell populations from rodent, canine, and human normal and diseased kidneys. Tissue Eng Part C Methods 17:261–273
4. Kelley R, Werdin ES, Bruce AT, Choudhury S, Wallace SM, Ilagan RM, Cox BR, Tatsumi-Ficht P, Rivera EA, Spencer T, Rapoport HS, Wagner BJ, Guthrie K, Jayo MJ, Bertram TA, Presnell SC (2010) A tubular cell-enriched subpopulation of primary renal cells improves survival and augments kidney function in a rodent model of chronic kidney disease. Am J Physiol Renal Physiol 299:F1026–F1039
5. Zaslav K, Cole B, Brewster R, DeBerardino TM, Farr J, Fowler P, Nissen C (2009) A prospective

study of autologous chondrocyte implantation in patients with failed prior treatment for articular cartilage defect of the knee: results of the Study of the Treatment of Articular Repair (STAR) clinical trial. Am J Sports Med 37: 42–55

6. Shapiro J, Lakey J, Ryan E, Korbutt G, Toth E, Warnock G, Kneteman N, Rajotte R (2000) Islet transplantation in seven patients with type 1 diabetes mellitus using a glucocorticoid-free immunosuppressive regimen. NEJM 343:230–238
7. Williams AR, Hare JM (2011) Mesenchymal stem cells: biology, pathophysiology, translational findings, and therapeutic implications for cardiac disease. Circ Res 109:923–940
8. Van Vlierberghe S, Dubruel P, Schacht E (2011) Biopolymer-based hydrogels as scaffolds for tissue engineering applications: a review. Biomacromolecules 12:1387–1408
9. Basu J, Genheimer CW, Rivera EA, Payne R, Mihalko K, Guthrie K, Bruce AT, Robbins N, McCoy D, Sangha N, Ilagan R, Knight T, Spencer T, Wagner BJ, Jayo MJ, Jain D, Ludlow JW, Halberstadt C (2011) Functional evaluation of primary renal cell/biomaterial Neo-Kidney Augment prototypes for renal tissue engineering. Cell Transplant 20:1771–1790
10. Payne R, Knight T, Basu J, Rivera E, Robbins N, McCoy D, Jayo M, Halberstadt C, Jain D (2011) Bioresponse of mammalian kidney to implantation of polymeric materials. Society for biomaterials annual meeting & exposition. www.tengion.com/pdfs/Bioresponse-of-Mammalian-Kidney-to-Implantation-of-Polymeric-Materials.pdf
11. Basu J, Payne R, Rivera E, Robbins N, McCoy D, Knight T, Jayo MJ, Jain D, Halberstadt CR (2010) In vivo evaluation of biomaterials in mammalian kidney. Tissue Engineering and Regenerative Medicine International Society (TERMIS)—North America annual conference. www.tengion.com/pdfs/Basu-2010-TERMIS-Biomaterials-poster-FINAL.pdf
12. Basu J, Rivera E, Genheimer CW, Robbins N, Payne R, McCoy D, Guthrie KI, Sangha N, Knight T, Jayo M, Ludlow JW, Jain D, Halberstadt CR (2010) Mammalian kidney nephrogenic response to primary renal cell/biomaterial-based Neo-Kidney Augment™ Prototype. Tissue Engineering and Regenerative Medicine International Society (TERMIS)—North America annual conference. www.tengion.com/pdfs/Basu-2010-cell-biomaterial-NKA-poster-FINAL.pdf

Chapter 24

Human Liver Bioengineering Using a Whole Liver Decellularized Bioscaffold

Pedro M. Baptista, Dipen Vyas, Emma Moran, Zhan Wang, and Shay Soker

Abstract

As a result of significant progress made in the last years in developing methods of whole organ decellularization techniques, organ bioengineering may now look more feasible than ever before. In this chapter, we describe in detail the necessary steps in human liver bioengineering. These include ferret liver decellularization by detergent perfusion, human liver progenitor and endothelial cell isolation, and finally, liver bioscaffold recellularization in a perfusion bioreactor.

Key words Human liver progenitors, Decellularization, Scaffold, Bioengineering, Endothelial cells, Perfusion bioreactor, Whole organ, Recellularization

1 Introduction

Tissue decellularization is a successful method to prepare matrices and scaffolds for research and regenerative medicine applications (1). Over the past few years, some of the techniques used have been optimized to a point where the decellularization of whole organs is now possible, generating scaffolds for organ bioengineering (2–4). The introduction of perfusion decellularization method, where a detergent solution is actively "pushed" into the vasculature of an organ or a tissue with a pump, allowed for decellularization of thick tissues and organs, which was previously unattainable with passive diffusion of detergents. The organ scaffolds prepared in this fashion can then be readily recellularized with primary human cells.

For liver bioengineering, liver progenitor and endothelial cells are isolated from human livers and full-term placentas, respectively.

*Pedro M. Baptista and Dipen Vyas have contributed equally to this work.

Joydeep Basu and John W. Ludlow (eds.), *Organ Regeneration: Methods and Protocols*, Methods in Molecular Biology, vol. 1001, DOI 10.1007/978-1-62703-363-3_24, © Springer Science+Business Media New York 2013

Collagenase digestion of the tissues can provide a highly functional cell source with great potential for growth and differentiation. However, their delivery into the liver scaffolds requires development and use of a perfusion seeding bioreactor for effective recellularization of such scaffolds.

Once cell seeding is completed, a 7–14-day maintenance period, with continuous culture media perfusion of the bioengineered livers, is necessary to effectively differentiate the hepatic progenitors to mature cell populations (hepatocytes and biliary epithelial cells). The differentiated cells will then exhibit typical hepatic functions like albumin and urea secretion, as well as diverse phenotypic markers of biliary cholangiocytes and hepatocytes (4).

2 Materials

2.1 Bioreactor System

1. Bioreactor vessel (Glass Ball Spinner, 250 mL, Bellco Biotechnology Inc., Vineland, NJ, USA).
2. Peristaltic pump with 14 G or 16 G tubing (Masterflex L/S with Masterflex L/S easy load pump head, Cole Parmer, Vernon Hills, IL, USA).
3. Pulse dampener (Cole Parmer).
4. Silicone tubing (Silicone Tubing Size 14 G, Cole Parmer) for connections.
5. 3-Way valves (Cole Parmer) to direct flow to the reservoir for media changes.
6. Smart Site Connection (Cole Parmer).
7. Luer lock syringe (30 mL).
8. Incubator.

2.2 Decellularization

1. Sterile distilled water.
2. Detergent solution: 1% Triton X-100, 0.1% ammonium hydroxide, in distilled water.
3. Gamma radiation source.
4. Peristaltic Pump (Same description as 2.1 -2.)
5. Silicone tubing (Silicone tubing size 14 and 16 G, Cole Palmer).

2.3 Human Tissue Processing and Cell Isolation

1. Human fetal livers (Advanced Bioscience Resources, CA, USA).
2. Enzyme solution: 0.8 U/mL dispase in Advanced RPMI (Life Technologies, Inc., Alameda, CA, USA).
3. Collagenase type IV/deoxyribonuclease digestion solution: 6 mg/mL Collagenase type IV (Worthington Biochemical Corp., Lakewood, NJ, USA) and 2,000 U/total deoxyribonuclease

(Roche Diagnostics, Inc., Indianapolis, IN, USA) solubilized in Advanced RPMI media (Life Technologies, Inc., Grand Island, NY, USA).

4. Neutralization buffer: DMEM/F12 containing 10% FBS.
5. Hepatocyte Wash Medium (Life Technologies).
6. Histopaque-1077 (Sigma-Aldrich, St. Louis, MO, USA).
7. 40 and 100 μm cell strainer (BD Biosciences, San Diego, CA, USA).
8. Human placenta.
9. Endothelial Growth Medium 2 (EGM2; Lonza, Switzerland).
10. Fibronectin (Millipore, Billerica, MA, USA).
11. Anti-human CD31 antibody (BD Biosciences).
12. Liver progenitor cell seeding medium: Advanced RPMI containing 5% FBS, 1% antibiotics/antimicotic (Invitrogen, Corp., Carlsbad, CA, USA), 0.04 mg/L dexamethasone, 2.45 mg/L cAMP, 10 IU/L hProlactin, 1 mg/L hGlucagon, 10 mM niacinamide, 0.105 mg/L alpha lipoic acid, 67 ng/L triiodothyronine (Sigma-Aldrich), 40 ng/mL hEGF (R&D Systems, Inc., Minneapolis, MN, USA), 10 mg/L hHDL (Cell Sciences, Canton, MA, USA), 20 ng/mL hHGF, 3.33 ng/mL hGH (eBiosciences, San Diego, CA, USA), and 1.2 mg/50 mL of matrigel™ (BD Biosciences).
13. Hepatic progenitor cell maintenance medium: Same formulation of liver progenitor cell seeding medium without FBS and matrigel (BD Biosciences, CA, USA).
14. Vascular cannulas 16 and 20 G (Terumo Medical Corp., Somerset, NJ, USA).
15. 50 mL conical centrifuge tubes.
16. Syringe with 18 G needle.
17. Phosphate-buffered saline (PBS), no calcium or magnesium.
18. 0.05% Trypsin/EDTA.
19. Scalpel.
20. Scissors.
21. Forceps.
22. Silk suture 4-0.
23. Collagen-IV (5 μg/cm^2) and Laminin (1 μg/cm^2) coated 15-cm culture plates.
24. Fibronectin-coated 6-well plates (5 μg/cm^2).
25. Centrifuge.
26. Incubator shaker.

3 Methods

3.1 Harvesting Cadaveric Animal Livers

All procedures are performed under aseptic conditions.

1. A longitudinal abdominal incision is made to visualize the liver, lower abdominal cavity, and rib cage.
2. The supra hepatic vena cava is transected as close to the atrium as possible, along with falciform and cardiac ligaments.
3. The diaphragm is carefully dissected around the esophagus in order to separate it from the liver and diaphragm.
4. The common bile duct is dissected and transected as close to the duodenum as possible.
5. The adipose tissue layers surrounding the portal vein are dissected carefully in order to visualize the vein and its branches. The lateral branches are ligated with silk suture 4-0 and cut closer to the intestines (distal end regarding the liver).
6. The portal vein is transected at about 1.5–2 cm away from the liver.
7. The infra hepatic vena cava located under the right lobe of the liver is carefully dissected and transected without damaging the liver lobe.
8. Before removing the intact liver, confirm that no additional attachments to the liver are present. Carefully dissect any remaining attachment and gently remove the liver holding it by the diaphragm.
9. The portal vein is cannulated with a 16–20 G cannula for ferret and rat livers depending on the diameter of the vein.

3.2 Liver Decellularization

The decellularization steps are carried out at 4°C and the decellularized liver scaffold is stored sterile in deionized water at 4°C until use.

1. The cannulae in the portal veins are attached to a peristaltic pump by using 14 G or 16 G tubing.
2. 2 L of distilled water is perfused through the portal vein at the rate of 6 mL/min (rat and ferret livers).
3. 4 L of detergent solution is perfused following the initial wash with water.
4. 8 L of distilled water is perfused through the liver to remove all of the decellularization detergent present (see Fig. 1).

3.3 Isolation of Human Fetal Liver Progenitor Cells

1. Human fetal livers are cut into small fragments with a scissor in a petri dish containing 25 mL of collagenase type IV/deoxyribonuclease digestion solution.

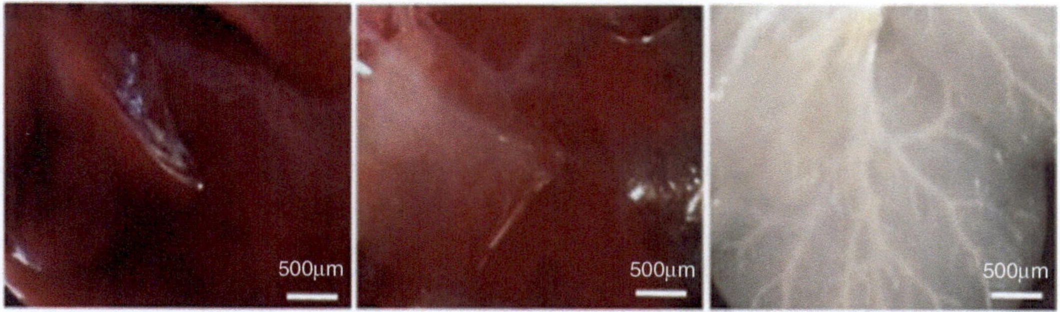

Fig. 1 Appearance of ferret liver immediately after isolation (*left panel*), following 20 min of decellularization (*middle panel*), and post-detergent perfusion (*right panel*)

2. The suspension of small tissue chunks in this solution is placed in a 50 mL conical tube in an incubator at 37°C on a shaker with agitation at 200 rpm for 45 min.
3. The heterogenous suspension is gently triturated with a pipet to break the remaining tissue pieces and 25 mL of neutralization buffer is added to followed by centrifugation at 300 × *g* for 10 min.
4. The supernatant is aspirated and the pellet is resuspended/washed with 25 mL of Hepatocyte Wash Medium. The suspension is then centrifuged at 300 × *g* for 10 min.
5. The supernatant is aspirated and the pellet is resuspended again in 25 mL of Hepatocyte Wash Medium.
6. The cell suspension is then passed through a 100 μm cell strainer and subsequently through a 40 μm cell strainer. The resulting suspension is then centrifuged at 300 × *g* for 10 min.
7. The resulting cell pellet is resuspended in 25 mL of Hepatocyte Wash Medium.
8. 12.5 mL is layered over 25 mL of Histopaque-1077 in two separate tubes to form a uniform layer over the histopaque solution (the number of tubes with histopaque for separation depends on the size/volume of the pellet obtained) (see Note 1).
9. The tubes are centrifuged for 30 min at 400 × *g* at 21°C.
10. These layers are separated into different tubes and washed with Hepatocyte Wash Medium and centrifuged at 400 × *g* for 10 min.
11. The lower fraction cell pellet is resuspended in Liver Seeding Medium and plated on Collagen-IV-coated and Laminin 15-cm culture plates and incubated at 37°C. These cells are assigned passage 0.

3.4 hUVEC Isolation and Expansion

1. Full-term placenta is obtained under institutional review board guidelines and permission, and the cord is separated.
2. Locate cord vein and wash it thoroughly with 20 mL PBS from syringe via 18 G needle.
3. Insert an 18 G needle into the vein with the syringe containing 20 mL pre-warmed enzyme solution.
4. Allow enzyme solution to flush the cord vein very slowly for approximately 7 min. Collect the enzyme solution throughout.
5. Neutralize the enzyme by adding 10 mL FBS into the enzyme solution and centrifuge at 400 × *g* for 5 min.
6. Resuspend the cells in 10 mL EGM-2 with 20% FBS.
7. Transfer the cells to a fibronectin-coated 6-well dish and place it into the cell incubator at 37°C.
8. Allow the cells to grow for 3–4 days and change the medium at day 4.
9. After colony appears, use EGM-2 medium with 10% FBS and change the medium every 3 days.
10. Perform regular subculture procedure with trypsin when the cells grow to more than 80% confluence (see Note 2).
11. If there is fibroblast-like cell contamination, then perform CD31 antibody-based cell sorting.

3.5 Scaffold Preparation and Sterilization for Bioreactor

1. After decellularization, the left, quadrates and caudate lobe removal is required to decrease the size of the liver scaffold and hence, the total number of cells necessary for reseeding.
2. The decellularized remnants of the diaphragm are cut and removed from around the supra hepatic vena cava.
3. A silk suture line 4-0 is then passed around the vascular structures of the left lobe to ligate it. These vessels are then cut and the left lobe removed.
4. The vascular structures of the caudate and quadrate lobe are then ligated in the same fashion and the lobes are removed.
5. The right lobe of the decellularized liver is then put in a 50 mL conical tube with its cannula in deionized water.
6. The acellular scaffold is then sterilized with 1.5 Mrad of gamma radiation with a cobalt 60 gamma irradiator (see Note 3).

3.6 Bioreactor Assembly and Recellularization

The perfusion bioreactor illustrated below provides the in vitro environment that is necessary for appropriate liver tissue bioengineering. This process consists of two phases, the first is perfusion seeding of cells and the second is maintenance of the construct to allow for tissue formation and maturation.

1. The components of the bioreactor listed in Subheading 2 are sterilized by steam at 121°C (see Notes 4 and 5).

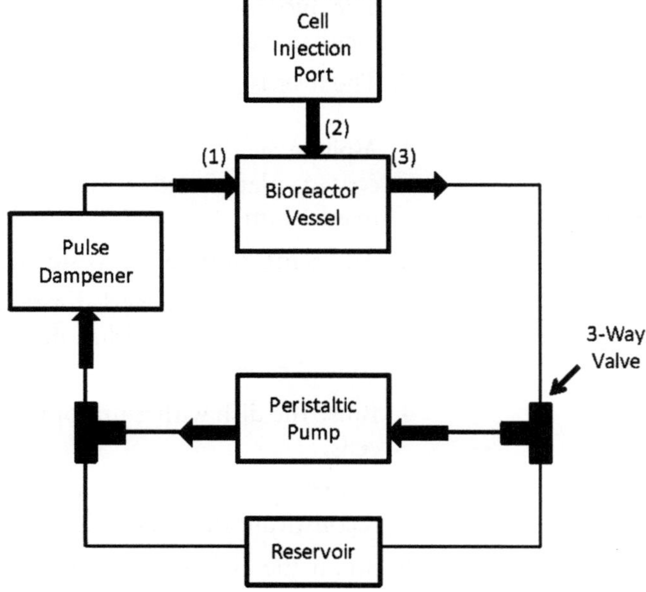

Fig. 2 Schematic diagram of the perfusion bioreactor used in the recellularization of acellular liver scaffolds. The bioreactor vessel consists of three ports: (1) the entry of fluid into the bioscaffold via portal vein cannulation, (2) cell injection port for seeding of cells into the suspension media via smart connector, and (3) outflow of media to recirculate throughout the bioreactor. Media is normally circulated through the top loop in the diagram (pump → pulse dampener → bioreactor vessel → pump). During media changes, the 3-way valve is changed so that the media flows directly from the vessel to the reservoir

2. Bioreactor assembly is performed in a tissue culture biosafety cabinet using sterile gloves to reduce the likelihood of contamination.
3. All components of the bioreactor assembly, except for the peristaltic pump, are configured according to Fig. 2.
4. The tubing is connected to the 3-way valves using male and female luer locks (see Note 6).
5. The pulse dampener and reservoir container are then linked to the tubing.
6. The tubing is then connected to all three ports of the bioreactor vessel. The cell seeding port contains a short piece of tubing connected to a "smart site" connector, which remains closed unless cell seeding is taking place.
7. The tubing is primed with culture medium with a 30 mL luer lock syringe until all the air is removed from the system.
8. The liver is attached to the inlet port tubing via a cannula in the portal vein and then is suspended in seeding medium within the bioreactor vessel.

9. The bioreactor assembly is transferred to a 37°C/5% CO_2 incubator and attached to the peristaltic pump.
10. The liver is perfused at 3 mL/min overnight prior to seeding.

3.7 hUVECs Preparation for Bioreactor Seeding

1. Aspirate the EGM-2 culture medium from the 15-cm fibronectin-coated plates and wash once with PBS (no calcium or magnesium).
2. Add 5 mL of 0.05% Trypsin/EDTA and incubate at 37°C.
3. Carefully pipet up and down and wash any unattached cells remaining on the culture dish. Transfer the cells to a 50 mL conical tube.
4. Rinse the dish with 5 mL of DMEM + 10% FBS for any leftover cells.
5. Centrifuge at 300 × *g* for 5 min. Aspirate supernatant and reconstitute the pellet with 10 mL of Liver Seeding Media.
6. Strain the cells through a 40 μm cell strainer to remove cell aggregates and count the cells (see Note 7).
7. Keep the cells on ice until ready for injection in the bioreactor.

3.8 Human Liver Progenitor Cell Preparation for Bioreactor Seeding

1. Remove the Liver Maintenance Medium from the 15 cm plates coated with collagen IV and wash once with PBS 5 mM EDTA.
2. Aspirate and repeat this washing step two more times. On the third time, allow the cells to incubate at room temperature for 5 min.
3. Aspirate the rinsing buffer and replace it with 5 mL of collagenase IV and dispase solution, respectively.
4. Incubate at 37°C for 30 min.
5. Pipet up and down to ensure detachment of most of the cells and transfer to a 50 mL falcon tube. Neutralize the enzymes with 10 mL DMEM + 10% FBS.
6. Centrifuge at 300 × *g* for 5 min. Aspirate supernatant and reconstitute the pellet with 10 mL of HDM.
7. Centrifuge once again at 300 × *g* for 5 min. Aspirate supernatant and reconstitute the cell pellet in 10 mL of seeding medium.
8. Strain the cells through a 40 μm cell strainer to remove cell aggregates and count the cells (see Note 7).
9. Keep the cells on ice until ready for injection in the bioreactor.

3.9 Bioscaffold Seeding of Human Liver Progenitor Cells and hUVECs

1. Approximately 70×10^6 human liver progenitor cells and 30×10^6 hUVECs (all from the same batch) are co-seeded through the portal vein of ferret acellular scaffolds by perfusion at 6 mL/min (see Note 8). Specify that the cells are

injected into the media through the cell-seeding port which eventually recirculates into the portal vein.

2. Cells are co-infused through the portal vein of the acellular scaffold every 4 h, with a total of four repetitions (16 h).
3. Following the seeding period, the flow rate is reduced to 0.5 mL/min and the liver is perfused with the Liver Maintenance Medium (supplemented with 1% FBS).
4. The media is changed every 48 h to ensure adequate oxygen and nutrient delivery to cells.
5. After 1 week, a small piece of the seeded scaffold is collected for DNA extraction and the remaining bioscaffold is fixed in 4% paraformaldehyde and processed for paraffin embedding.

4 Notes

1. After histopaque 1077 gradient separation, the resulting bilayered solution is made up of lower fraction primarily enriched with hepatoblasts and hepatic progenitor cells along with red blood cells while the upper fraction is predominantly made up of mononuclear cells (stromal, endothelial, etc.).
2. After hUVECs isolation and plating, colonies usually appear between 5 and 21 days.
3. Scaffold sterilization with gamma irradiation is sometimes not available. However, we strongly recommend this method. The use of chemical disinfectants sometimes has the undesirable effect of changing the biomaterial mechanical properties (fixation, etc.).
4. The 3-way valves and smart connectors, drawn in the bioreactor components, cannot be autoclaved, and we sterilize them in the gamma irradiator while we are sterilizing the liver scaffolds.
5. The reservoir container consists of a sterile media bottle (empty) with three holes drilled in the top: two of which are for tubing leading to the 3-way valves and one small piece of tubing connected to a filter.
6. We recommend sterilizing an extra set of fittings and luer locks in case parts are damaged during the sterilization process.
7. Straining the cells prior injection is essential to remove cell aggregates that might clump the scaffold vasculature.
8. When injecting the cells into media during cell seeding, wipe the smart-fit connector with alcohol before and after cell injection to reduce the likelihood of contamination.

References

1. Gilbert TW, Sellaro TL, Badylak SF (2006) Decellularization of tissues and organs. Biomaterials 27(19):3675–3683
2. Baptista PM, Orlando G, Mirmalek-Sani SH et al (2009) Whole organ decellularization—a tool for bioscaffold fabrication and organ bioengineering. Conf Proc IEEE Eng Med Biol Soc 2009:6526–6529
3. Ott HC, Matthiesen TS, Goh SK et al (2008) Perfusion-decellularized matrix: using nature's platform to engineer a bioartificial heart. Nat Med 14(2):213–221
4. Baptista PM, Siddiqui MM, Lozier G et al (2011) The use of whole organ decellularization for the generation of a vascularized liver organoid. Hepatology 53(2):604–617

Chapter 25

A "Living Bioreactor" for the Production of Tissue-Engineered Small Intestine

Daniel E. Levin, Frederic G. Sala, Erik R. Barthel, Allison L. Speer, Xiaogang Hou, Yasuhiro Torashima, and Tracy C. Grikscheit

Abstract

Here, we describe the use of a mouse model as a living bioreactor for the generation of tissue-engineered small intestine. Small intestine is harvested from donor mice with subsequent isolation of organoid units (a cluster of mesenchymal and epithelial cells). Some of these organoid units contain pluripotent stem cells with a preserved relationship with the mesenchymal stem cell niche. A preparation of organoid units is seeded onto a biodegradable scaffold and implanted intraperitoneally within the omentum of the host animal. The cells are nourished initially via imbibition until neovascularization occurs. This technique allows the growth of fully differentiated epithelium (composed of Paneth cells, goblet cells, enterocytes and enteroendocrine cells), muscle, nerve, and blood vessels of donor origin. Variations of this technique have been used to generate tissue-engineered stomach, large intestine, and esophagus. The variations include harvest technique, length of digestion, and harvest times.

Key words Tissue engineering, Small intestine, Organoid unit, Short bowel syndrome, Living bioreactor

1 Introduction

The management of organ failure is complex and varies greatly by organ system and disease severity. Treatments may involve the use of medications such as insulin and digestive enzymes for pancreatic failure. Extracorporeal devices such as hemodialysis or mechanical ventilation may be used to support the failing kidney or lung. Implantable synthetic devices are well established for the bypass of atherosclerotic blood vessels or replacement of diseased joints. Perhaps one of the greatest achievements in medicine was the successful transplant of human organs from donor to recipient. In the 1950s, Dr. Joseph Murray was the first to perform a successful kidney transplant, ushering in the field of transplant medicine (1, 2).

All of the methods summarized above provide great benefit to patients; however, they are also fraught with a number of

Joydeep Basu and John W. Ludlow (eds.), *Organ Regeneration: Methods and Protocols*, Methods in Molecular Biology, vol. 1001, DOI 10.1007/978-1-62703-363-3_25, © Springer Science+Business Media New York 2013

disadvantages. Medications are associated with side effects and issues of patient compliance. External devices are inconvenient, costly, and often a temporary bridge to more definitive therapy. Synthetic, implantable material and devices suffer mechanical failure and are prone to infection. Lastly, transplantation requires lifelong immunosuppression, and carries a high cost and the potential for allograft rejection. Donor supplies remain a critical issue.

Intestinal failure is particularly challenging and treatment options are limited. Short bowel syndrome (SBS) is a leading cause of intestinal failure characterized by the malabsorption of nutrients and water that frequently occurs following the loss of 70%–75% of normal intestinal length (3, 4). Supplementation with total parenteral nutrition is associated with hepatic failure, sepsis, and death (5). Intestinal transplant is associated with the aforementioned risks and poor survival. A 2006 review of 141 intestinal and multivisceral transplants in 123 children revealed a 1 year patient survival between 44 and 83% and a 3 year survival between 32 and 60% (6). Although these survival rates are improving, both patients and providers would welcome improvements.

Ideally, the intestinal replacement would be derived from the recipient's autologous cells and therefore be of the same genetic and immunologic background. Tissue-engineered intestine would function as native intestine, grow with the recipient, and eliminate the need for immunosuppression or graft management. In an attempt to improve the treatment of short bowel syndrome, efforts to grow tissue-engineered small intestine have been pursued since the 1980s. Dr. Joseph Vacanti's lab adapted an organoid unit isolation protocol, originally described by Evans et al., and began seeding biodegradable scaffolds for implantation into living hosts (7, 8).

Since these initial experiments, additional techniques to grow tissue-engineered intestine have involved a variety of in vitro methods. Current strategies in tissue engineering have been performed using in vitro cultures, rotating mechanical bioreactors and living animal models (9–11). Multiple species, including rat, mice, Yorkshire swine, and beagle have been used as a host for the in vivo growth of tissue-engineered small intestine (11–13). The technique described here, however, takes advantage of the body as a living bioreactor. In contrast to the majority of in vitro methods, this in vivo method allows the growth of a fully differentiated intestinal epithelium in addition to nerve and mesenchymal components such as muscularis, and some blood vessels.

Small intestine is harvested from neonatal mice and enzymatically digested, triturated, and centrifuged to produce organoid units. The organoid units, which contain all cell types of full thickness intestine, are loaded onto a biodegradable scaffold composed of polyglycolic acid coated with poly-L-lactic acid and type 1 collagen. This scaffold is designed to be porous enough to permit the imbibition of nutrients by donor cells, yet rigid enough to allow adequate surface tension at the interface of donor cells with host tissue. The omentum provides a well-vascularized surface for implantation and subsequent neovascularization of the growing tissue-engineered intestine. It also has the advantages of being located within the peritoneum. This in vivo location will allow for a future gastrointestinal anastomosis with an intact vascular pedicle.

The tissue-engineered small intestine may be harvested for histologic evaluation, immunohistochemistry, or additional RNA and protein analysis. Although the murine TESI is generated from either syngeneic or allogeneic mice, the isolation of organoid units from a living donor and subsequent implantation of autologous units at a single operation has been demonstrated in a Yorkshire swine model (13). In addition, this technique has been used to successfully grow tissue-engineered esophagus, stomach, and colon in addition to small intestine (13–16).

2 Materials

Reagents are to be kept on ice unless otherwise noted. All waste materials should be disposed of according to accepted disposal regulations. These steps should be carried out using sterile technique.

Animal protocols must be approved by the Institutional Animal Care and Use Committee (IACUC) and all participants must be approved for the practice of animal survival surgery. The animal care facility must be equipped for survival surgery and operations are to be performed in a clean, disinfected hood. Donor and host may be of identical genetic background (syngeneic) or different genetic background (allogeneic). When using allogeneic mice, we recommend an immunodeficient host: Nonobese Diabetic/Severe Combined Immunodeficiency gamma chain deficient mice (Jackson Laboratory) that have received 350 CGy of full-body irradiation prior to implantation.

Surgical instruments should be sterilized initially via autoclave followed by glass bead sterilizer in between each new animal. An anesthetic machine capable of delivering 2.5–5 vol% isoflurane and 1 LPM O_2 is needed. The operating surface should be warmed to prevent hypothermia with careful use of a heat lamp or warm water operating surface. Additional materials are listed in detail below.

When needed for sacrifice, animals should be euthanized in accordance with the standard euthanasia guidelines as mandated by the IACUC: for mice, the inhalation of CO_2 from a pressurized tank in an uncrowded cage with confirmation of death prior to procurement of both donor intestine or eventual harvest of tissue-engineered intestine from host mice.

2.1 Experimental Animals

1. Donor mice: 2–3-week-old donor mice of either gender (see Note 1).
2. Host mice: Syngeneic mice or adult Nonobese Diabetic/ Severe Combined Immunodeficiency gamma mice (NOD/ SCID gamma, Jackson Laboratory, Bar Harbor, ME, USA,) for allogeneic implantations.

2.2 Organoid Units

1. Hanks balanced salt solution (HBSS, Invitrogen, Life Technologies Corp., Grand Island, NY, USA), 500 mL, stored on ice.
2. Antibiotic/antimycotic (A/A), penicillin G (base) 10,000 U, streptomycin sulfate (base) 10,000 μg, and amphotericin B 25μg in 0.85% saline (Invitrogen). 5 mL aliquots can be stored at −20°C.
3. HBSS with antibiotic/antimycotic: 5 mL of A/A is added to 500 mL HBSS stored on ice.
4. 60 mm × 15 mm sterile tissue culture dishes.
5. 15 mL centrifugation tubes.
6. Refrigerant brick (EverCold, Therapak Corp., Buford, GA, USA).

2.3 Digestion Enzyme

1. Collagenase type 4 (Worthington, Biochemical Corp, Lakewood, NJ, USA), 225 u/mg.
2. Dispase (Invitrogen), 1.77 u/mg.
3. Enzyme solution: Dissolve 142 mg of 225 u/mg collagenase type 4 (Worthington) and 5 mg of 1.77 u/mg dispase in 40 mL HBSS, to create a collagenase (800 u/mL) and dispase (0.22 u/mL) solution stored at 4°C.
4. Enzymatic arresting solution: 500 mL of Dulbecco's modified Eagle's medium (DMEM, Invitrogen) with 10% Fetal Bovine serum (FBS, Invitrogen) and 1× A/A (500 mL DMEM with 50 mL FBS and 5 mL A/A).

2.4 Operating Instruments and Medications

1. Autoclaved forceps ×2 and 4.5 in. straight operating scissors.
2. Hair clipper (Wahl Clipper Corp, Sterling, IL, USA).
3. Clear adhesive tape.
4. Number 4-0 polyglactin 910 (Vicryl) suture (VCP214H Ethicon inc. Somerville, NJ, USA).
5. Number 5-0 Poliglecaprone 25 (Monocryl) suture (Y213H, Ethicon).
6. Petrolatum ophthalmic ointment sterile ocular lubricant (17033-211-38, Puralube Vet Ointment, Dechra Veterinary Products, Overland Park, KS, USA).
7. Clear plastic drape, 8 in. × 8 in. with oval adhesive fenestration (88VCSTF, Gepco, Philadelphia, PA, USA).

8. Chlorhexidine scrub and alcohol pads.
9. Ketoprofen (0856-4396-01, Ketophen, Fort Dodge, Overland Park, KS,USA), 100 mg/mL, diluted 1:50 with distilled, sterile water loaded in a 100 U insulin syringe for subcutaneous injection.
10. Trimethoprim/sulfamethoxazole suspension (13310-146, Septra, Pfizer, New York, NY, USA,) diluted in animal drinking bottle, 1 mL Septra per 100 mL drinking water.

3 Methods

At all times during the preparation of the intestine, once it has been removed from the body, it should be kept in HBSS with A/A on ice unless otherwise noted. Maintaining the cells at 4°C enhances survival. For implantation, we favor the use of a biodegradable scaffold such as polyglycolic acid coated with poly-L-lactic acid and type 1 collagen that have been made into 4 mm×4 mm tubes, as previously described in the literature (3).

3.1 Recovery of Small Intestine from Donor Mice

1. Select four to five donor mice of interest and sacrifice by CO_2 inhalation in accordance with IACUC protocol. 70% ethanol spray is applied to ventral surface of the abdomen. Lower abdominal incision is made with dissecting scissors and ventral skin is pulled rostrally exposing the peritoneal cavity.
2. The peritoneal cavity is entered and the first portion of the duodenum is exposed and divided. The small intestine is eviscerated in its entirety with gentle traction to dissect away the mesentery (see Note 2). The terminal ileum is divided and the cecum is not included in the specimen (Fig. 1).

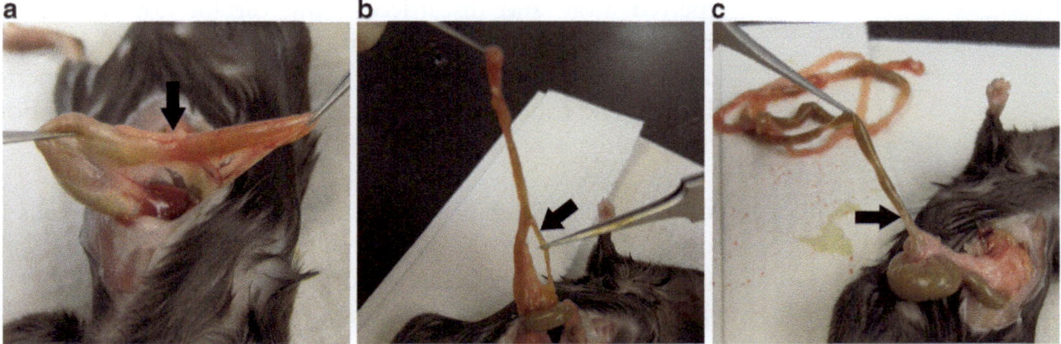

Fig. 1 Recovery of small intestine for the production of organoid units. (**a**) Demonstrates the transition from stomach to duodenum. The intestine is transected (*black arrow*) in the first portion of the duodenum. (**b**) The mesenteric vessels (*black arrow*) may be avulsed and are not included in the specimen. (**c**) Dissection is carried distally and transected at the terminal ileum (*black arrow*)

3. The intestine is placed in a culture dish with HBSS with A/A on ice. This step is repeated until the intestine from all mice is harvested. Intestine from different mice can be placed into the same culture dish.

3.2 Removal of Fecal Material from Intestine

1. Using a dissecting microscope, a segment of small intestine is placed in a separate culture dish with HBSS with A/A and divided longitudinally along its antimesenteric border using forceps and scissors. The divided intestine is shaken within the HBSS culture dish to remove bulk fecal material and lumenal contents (see Note 3).
2. The intestine is then transferred to a 15 mL centrifugation tube with 10 mL of HBSS with A/A and placed on ice. All divided intestine is placed into the same 15 mL centrifugation tube. The tube is shaken vigorously by hand and supernatant discarded (see Note 4). Additional 10 mL of HBSS is added. This washing step is repeated as needed to remove all fecal material and loose villi tips.

3.3 Mincing of Intestine

1. Upon satisfactory washing of the intestine, the intestine is transferred to a small 60 mm × 15 mm culture dish with 5 mL of cold HBSS with A/A. Using dissecting scissors, the large segments of intestine are minced into a fine pulp of intestine (see Note 5). This is then pipetted back into a 15 mL tube (see Note 6) and washed 4 additional times with cold HBSS with A/A and stored on ice.

3.4 Enzyme Digestion

1. Supernatant from the 15 mL tube is poured off and 10 mL of the enzyme solution is added and mixed vigorously. This is incubated at 37.0°C on a rocker for 20 min.
2. The solution is triturated using a 10 mL pipette until the tissue is mostly digested (see Note 7). A layer of foam and mesenchymal debris generally collects at the top of the tube. This may be pipetted away and discarded, being careful not to remove underlying solution. DMEM with 10% FBS, 1% A/A is added to stop the enzymatic reaction and bring the total volume to 15 mL. The tube is inverted a few times to mix the solution and the tube is centrifuged at 500 rpm for 10 min at 4°C.

3.5 Isolation of Organoid Units

1. Following centrifugation, the supernatant with single cells is poured off, leaving the pellet behind. DMEM with A/A and 10% FBS is added to bring the total volume to 15 mL.
2. The pellet is resuspended with vigorous shaking. A low speed vortex may be used, but high speed is avoided to prevent shearing of organoid units. The solution is centrifuged at 800 rpm for 5 min at 4°C.

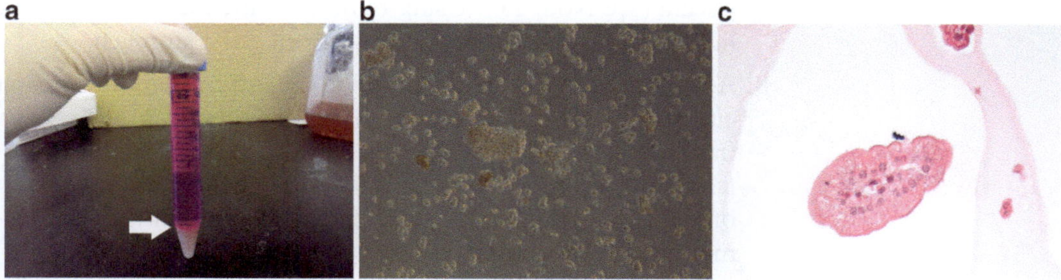

Fig. 2 Organoid unit appearance. (**a**) Demonstrates the pelleted organoid units (*white arrow*). (**b**) Organoid units as they appear under the inverted microscope (10×). (**c**) Organoid unit stained with hematoxylin and eosin (40×)

3. Supernatant is poured off leaving behind the pellet of organoid units (Fig. 2). 0.5 mL of DMEM with A/A and 10% FBS should be intentionally left behind with the pellet to facilitate pipetting.
4. A 1,000 μL pipette, set at 200 μL is used to mix the solution and evenly suspend the organoid units in the remaining DMEM.
5. The scaffolds are transferred to a culture dish placed atop a foam refrigerant brick (see Note 8).
6. Two hundred microliters of organoid units are applied to each scaffold. First, the central lumen of the scaffold is filled and remainder of the 200 μL is applied over and around the entire scaffold. These loaded scaffolds are ready for implantation.

3.6 Implantation of Loaded Scaffold

1. Syngeneic or allogeneic recipient mice are selected for implantation (see Note 9).
2. Following successful induction of anesthesia with 5% isoflurane and 1 LPM of O_2, the host animal is placed supine on a warmed operating platform with the 2.5% isoflurane and 1 LPM of O_2 flowing through the nose cone.
3. Lubricating gel is applied to the eyes to prevent corneal desiccation.
4. Forceps are used to pinch a hind leg to ensure adequate anesthesia as evidenced by lack of movement or reflexive withdrawal of the limb.
5. Clippers are used to remove hair overlying the ventral surface (see Note 10). The abdomen is prepped three times with chlorhexidine and alcohol and a sterile drape is placed.
6. A 2 mg/kg injection of ketophen is administered subcutaneously.
7. A 1 cm upper midline laparotomy incision is made through the skin with dissecting scissors. The peritoneum is entered using scissors as well, being careful not to lacerate the underlying

viscera (see Note 11). Using blunt forceps, the left lobe of the liver is gently retracted rostrally exposing the stomach.

8. A second forcep is used to grasp the omentum along the greater curvature. The mouse omentum is very thin, translucent, and fragile. Care is taken not to avulse the artery running circumferentially around the omentum or tearing the delicate film of omentum. The omentum is fanned out to accommodate placement of the scaffold.
9. The scaffold, loaded with organoid units, is placed atop the omentum. The omentum is folded over the scaffold and then the sides of the omentum are tucked around the sides of the scaffold.
10. Five-zero monocryl suture is used to secure the scaffold to the omentum (see Note 12). The implanted scaffold is positioned in the peritoneal cavity.
11. Abdominal muscles are closed with a 4-0 vicryl running suture. The skin is closed with 4-0 vicryl interrupted suture.
12. Blood is cleaned from the abdomen with an alcohol pad and the animal is placed into its recovery cage. Half of the cage is on a warm heating pad to allow the recovering animal the option of resting on a warm or cool surface.
13. Water should be treated with 1 mg/100 cm^3 of Septra 24 h prior to implantation and for 1 week after implantation. Some food should be placed directly into the cage so that the animal does not have to extend its body to access the food compartment.
14. The animal is monitored in accordance with the animal facility protocols for the institution until it is time to sacrifice and harvest the tissue-engineered small intestine for embedding, sectioning and staining.

3.7 Recovery of TESI

1. Mice with implanted TESI are sacrificed by CO_2 inhalation in accordance with IACUC protocol. 70% ethanol spray is applied to ventral surface of the abdomen. A midline scar is anticipated, and the peritoneal cavity is entered 2 mm left lateral to the scar.
2. The TESI is generally located along the greater curvature of the stomach and identified as a sphere of tissue (Fig. 3). Adhesions are carefully dissected away and the TESI is removed.

4 Notes

1. Mice have been selected as donor and host as they offer a broad variety of transgenic strains to evaluate the impact of varied gene expression on the growth and development of tissue-engineered small intestine.

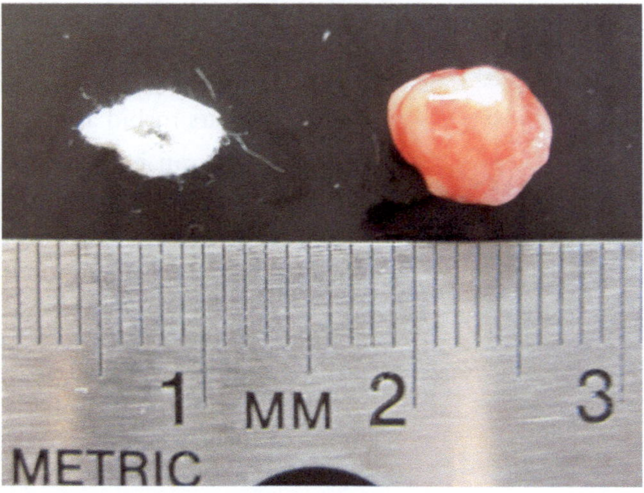

Fig. 3 Tissue-engineered small intestine harvested 4 weeks after implantation (*right*) adjacent to the biodegradable scaffold (*left*)

2. Care should be taken not to include any of the stomach. The most proximal portion of the intestine can be grasped with forceps and gently pulled from the body. As this is done, the mesentery generally just disconnects. If it does not, it can be gently tugged and freed easily. If the intestine is ripped at this point, this is not a problem. Simply put the piece of intestine into cold HBSS with A/A then continue with the evisceration where the intestine was avulsed.
3. It is not critical that the intestine be divided along its antimesenteric border; it is just easier to do it this way. If the HBSS with A/A that the longitudinal dissection is being performed in becomes cloudy with feces and lumenal content, the solution should be discarded and new HBSS with A/A added.
4. When rinsing, the solution will initially be cloudy with lumenal content and loose villi. Allow the sample to sit on ice for a few minutes, giving time for sedimentation of the tissue to occur. Supernatant should be poured off without losing the bulk of the intestine. With each subsequent rinse, the supernatant becomes clearer. If it is difficult to pour off the supernatant without losing the intestine, one can perform a "manual centrifugation" by simply spinning the tube by hand. Mechanical centrifuge should not be used for these rinsing steps.
5. The use of a tissue homogenizer has been attempted, but tends the shear apart the organoid units and is not recommended.
6. If the minced intestine is not easily passed through the tip of the 10 mL serological pipette (Falcon) then the intestine has not been minced adequately.

7. During trituration (repeatedly applying suction into and out of a 10 mL pipette such that turbulence provides some mechanical digestion), the larger particles should be fully digested. The solution appears turbid and tan in color.
8. A paper towel can be placed over the brick and beneath the culture wells to prevent the organoid units from becoming too cold and freezing accidentally.
9. When using allogeneic donor and recipient mice, we favor the use of recipient Nonobese Diabetic/Severe Combined Immunodeficiency (NOD/SCID) gamma chain deficient mice (Jackson Laboratory) that have received a full-body irradiation at 350CGy prior to implantation. Syngeneic donor and recipient mice (such as C57BL/6 mice, Jackson Laboratory) do not require irradiation. Alternatively, the successful implantation of autologous organoid units produced from a Yorkshire swine and implanted at the same operation has been published (5).
10. Adhesive tape may be used to remove loose hair from the operating field since a loose hair in the peritoneal cavity may be a nidus for infection.
11. When entering the peritoneum, an avascular pale white line is observed (linea alba) in the midline. Entering sharply along this line without cutting into muscle will help keep blood loss minimal.
12. While suturing the scaffold to the omentum with monocryl suture, the knot must be placed such that it is not tying off the artery supplying the omentum. The goal is to have the omentum draped over the entire surface of the scaffold with few wrinkles such that they are closely approximated. Any gaps between omentum and scaffold will prevent the process of imbibition from occurring in that area. Secure wrapping will also prevent organoid units from leaking freely into the peritoneum.

Acknowledgments

This work was supported by grants from the California Institute for Regenerative Medicine (#RN2-00946-1) (TG2-01168) and the Saban Research Institute

References

1. Atala A (2009) Engineering organs. Curr Opin Biotechnol 20:575–592
2. Merrill JP, Harrison JH, Murray J et al (1956) Successful homotransplantation of the kidney in an identical twin. Trans Am Clin Climatol Assoc 67:166–173
3. Stollman N, Neustater B, Rogers A (1996) Short-bowel syndrome. Gastroenterologist 2:118–128
4. Lykins T, Stockwell J (1998) Comprehensive modified diet simplifies nutrition management of adults with short-bowel syndrome. J Am Diet Assoc 3:309–315

5. Peyret B, Collardeau S, Touzet S et al (2011) Prevalence of liver complications in children receiving long-term parenteral nutrition. Eur J Clin Nutr 65:743–749
6. Kato T, Tzakis A, Selvaggi G et al (2006) Intestinal and multivisceral transplantation in children. Ann Surg 243:764–766
7. Vacanti J, Morse M, Saltzman W et al (1988) Selective cell transplantation using bioabsorbable artificial polymers as matrices. J Pediatr Surg 23:3–9
8. Evans G, Flint N, Somers A et al (1992) The development of a method for the preparation of rat intestinal epithelial cell primary cultures. J Cell Sci 101:219–231
9. Sato T, Vries R, Snippert H et al (2009) Single Lgr5 stem cells build crypt-villus structures in vitro without a mesenchymal niche. Nature 459:262–265
10. Salerno-Goncalves R, Fasano A, Sztein M (2011) Engineering of a multicellular organotypic model of the human intestinal mucosa. Gastroenterology 141:18–21
11. Sala F, Matthews J, Speer A et al (2011) A multicellular approach forms a significant amount of tissue-engineered small intestine in the mouse. Tissue Eng 17:1841–1849
12. Grikscheit TC, Siddique A, Ochoa ER et al (2004) Tissue-engineered small intestine improves recovery after massive small bowel resection. Ann Surg 240:748–754
13. Sala FG, Kunisaki SM, Ochoa ER et al (2009) Tissue-engineered small intestine and stomach form from autologous tissue in a preclinical large animal model. J Surg Res 156:205–212
14. Agopian V, Chen D, Avansino J et al (2009) Intestinal stem cell organoid transplantation generates neomucosa in dogs. J Gastrointest Surg 13:971–982
15. Speer AL, Sala FG, Matthews JA et al (2011) Murine tissue-engineered stomach demonstrates epithelial differentiation. J Surg Res 156:205–212
16. Grikscheit TC, Ochoa ER, Ramsanahie A et al (2003) Tissue-engineered large intestine resembles native colon with appropriate in vitro physiology and architecture. Ann Surg 238:35–41

Chapter 26

Tissue Engineering of Esophagus and Small Intestine in Rodent Injury Models

Joydeep Basu, Kim L. Mihalko, Elias A. Rivera, Kelly I. Guthrie, Christopher W. Genheimer, Namrata Sangha, and John W. Ludlow

Abstract

Regenerative constructs composed of synthetically sourced, biodegradable biomaterials seeded with smooth muscle-like cells have been leveraged to mediate regeneration of bladder and bladder-like neo-organs. Here, we describe how such constructs may be applied to catalyze regeneration of esophagus and small intestine in preclinical rodent models.

Key words Esophagus, Small intestine, Tissue engineering, Regeneration, Smooth muscle cell, Construct, Biomaterials

1 Introduction

1.1 Esophagus

In the pediatric population, the demand for esophageal tissue replacement is greatest in patients presenting with congenital long-gap esophageal atresia, since direct anastomosis of the open ends to each other is not a treatment option (1). Injuries resulting from acid or alkali ingestion by children are another indication for which esophageal replacement is needed (2). The most common cause for esophageal replacement in adults is surgical resection of the esophagus due to cancer (3). Standard of treatment for both pediatric and adult patients relies on esophageal lengthening techniques, esophageal substitution with intestinal tissue, or transposition of the stomach to overcome long gaps in the esophagus (4). Unfortunately, these treatment options are often followed by postoperative complications which negatively impact the patient's quality of life (5–7).

1.2 Small Intestine

Short bowel syndrome (SBS) is a consequence of massive small bowel resection performed in patients presenting with inflammatory bowel disease, trauma, mesenteric vascular disease, volvulus, congenital

Joydeep Basu and John W. Ludlow (eds.), *Organ Regeneration: Methods and Protocols*, Methods in Molecular Biology, vol. 1001, DOI 10.1007/978-1-62703-363-3_26, © Springer Science+Business Media New York 2013

atresias, and neonatal necrotizing enterocolitis (8). SBS may affect both pediatric and adult patients presenting with less than a third remnant jejunum and ileum (9). Massive intestinal resection typically leads to disruption of normal nutrient and fluid absorption, including deficiencies in calcium, magnesium, zinc, iron, vitamin B_{12}, and fat-soluble vitamins (10). Onset of SBS may be associated with diarrhea, dehydration, malabsorption of nutrients, and concomitant progressive malnutrition. Long-term survival of patients with less than 50 cm of residual small bowel is no more than 45% (11). The clinical severity of SBS may be correlated with the extent of resection, presence of ileo-cecal valve and jejunum, and the condition of the residual small bowel (10). Total parenteral nutrition has improved clinical outcomes within the SBS patient population, but total parenteral nutrition is itself associated with numerous clinical complications including cholestasis, fibrotic liver disease, venous thrombosis, and catheter-related sepsis (9).

Surgical strategies for intervention in patients presenting with SBS include techniques for lengthening of the remnant small bowel, construction of intestinal valves to modulate passage of intestinal material, and tapering methodologies to improve peristalsis (8). However, such approaches are rarely feasible in adults owing to clinical complications including fistula formation and sepsis typically associated with Crohn's disease, the most prevalent underlying pathology within this patient subpopulation (12). Within pediatric SBS patients treated with intestinal lengthening procedures, long-term survival was 45% (13). Transplantation of small bowel offers the potential for definitive functional rescue but may also be associated with numerous technical complications, high rates of graft rejection, and auxiliary complications associated with long-term immunosuppression (14). Clearly, there remains a significant current clinical need for novel approaches to the treatment and management of SBS.

For both esophagus and small intestine, tissue engineering technology is the next logical step towards developing gastrointestinal tissue replacements (15). This technology typically requires a cell source and a biocompatible scaffold to support tissue regeneration (16). Tissue engineering principles have been successfully used in developing implantable cell-seeded matrices for the reconstruction, repair, augmentation, or replacement of laminarly organized luminal organs and tissue structures, such as a bladder or a bladder component, typically composed of urothelial and smooth muscle layers (17–22). Our preclinical canine study demonstrated that a construct comprised of a polyglycolic acid (PGA) and poly(lactic-*co*-glycolic acid) (PLGA)-based biodegradable mesh scaffold seeded with autologous urothelial and smooth muscle cells is capable of regenerating urinary bladder structure and function as early as 6 months post implantation (23). In addition, the neo-bladder formation stimulated by implantation of this construct has long-term durability and adapts to the size of the animal during growth from juvenile to adult.

We have since identified a population of committed, smooth muscle-like cells that can be isolated from adipose (Ad-SMC) in a reliable and reproducible manner (24). These Ad-SMC cell populations have been successfully used to seed tubular, PGA/PLGA scaffolds to create the Neo-Urinary Conduit (NUC™), a novel neo-organ facilitating urinary diversion in a porcine cystectomy model (25). As with neo-bladder, regeneration of NUC™ is characterized by the de novo induction of native-like urinary tissue, presenting a luminal urothelial layer and abluminal layers of smooth muscle cells (25). Phase I clinical trials are currently being conducted to evaluate functional outcomes associated with NUC™ implantation in patients requiring urinary diversion secondary to bladder cancer (26).

In the method below, we describe in detail how this organ regeneration platform technology composed of an autologous smooth muscle cell/biomaterial combination construct may be applied towards the regeneration of esophagus and small intestine, both being hollow organs with a laminar histo-architecture directly comparable to the bladder. We describe in vivo rodent esophagus and small intestine (SI) injury models that may be used to evaluate tissue regeneration in the context of patch and/or tubular Ad-SMC-seeded bio-polymeric constructs.

For esophagus, 10–16 weeks after implantation in this rodent model, a patch-based biodegradable scaffold seeded with Ad-SMC leads to neo-esophagus regeneration, as characterized by the formation of all three esophageal wall layers: mucosa, muscularis, and serosa.

For SI, patch-based constructs mediate complete regeneration of intestinal epithelia and muscular wall organization after 8 weeks post-implantation in this rodent model. Tubular constructs are associated with complete regeneration of intestinal epithelia and partial regeneration of smooth muscle layers by 10 weeks post-implantation.

2 Materials

2.1 Ad-SMC Isolation and Culture

1. Dulbecco's Modified Eagles' Media (DMEM-HG).
2. Fetal bovine serum (FBS).
3. Phosphate-buffered saline (PBS).
4. Gentamicin, 100× solution.
5. Collagenase I (Worthington Biochemical).
6. Prepare collagenase solution by adding 1 g of BSA and 0.1 g of collagenase per 100 ml of PBS. Filter the solution through a 0.2 μm filter unit. Warm to 37°C.

7. Bovine serum albumin (BSA).
8. 50 ml conical tubes.
9. 100 μm Steri-flip vacuum filter (Millipore, Billerca, MA).
10. Desktop cell culture centrifuge (Sorvall, or equivalent).
11. Parafilm.
12. 37°C tissue culture incubator.
13. Biosafety cabinet for isolation and culture of adipose-sourced cells.
14. Standard laboratory rocker.

2.2 Biomaterials

1. Vicryl® mesh (polyglactin 910) (Ethicon, Somerville, NJ).
2. PLGA (poly-lactic-*co*-glycolic acid), 50:50 (DURECT Corp., Cupertino, CA).
3. Methylene chloride (Spectrum Laboratory Products).
4. SI tube: 4 mm inner diameter, custom fabricated braided tube, 100% polyglycolide (Concordia Medical, Warwick, RI).

2.3 Animal Model

1. Healthy adult Lewis rats (approximately 28 days old).
2. Surgical suite appropriate for small animal surgery.
3. Nonabsorbable 7-0 silk sutures (Ethicon).

3 Methods

3.1 Isolation of Ad-SMC

1. Abdominal adipose may be procured subcutaneously from 28-day-old Lewis rats (syngeneic with the putative recipients of any tissue engineered construct). After euthanasia under CO_2, isoflurane or injection with Euthasol (must be performed by licensed veterinarian), use surgical scissors to open each rat along the abdomen from groin to neck.
2. Remove subcutaneous adipose by dissection with surgical scissors and forceps and place in 50 ml conical tube with approximately equal volume of sterile PBS/gentamicin (1×).
3. Adipose samples from multiple animals may be pooled into a single conical tube or tubes. We suggest harvesting adipose from 10 to 20 animals at a time (see Note 1).
4. Adipose samples are now washed at least 3 times with an equal volume of DMEM-HG/gentamicin (1×) in 50 ml conical tubes (see Notes 2–4).

   Waste wash solution may be pipetted off or vacuumed off if your biosafety cabinet is equipped with a vacuum aspirator device.

5. Add an equivalent volume of Collagenase I solution per unit volume adipose to each 50 ml conical tube (i.e., 10 ml of collagenase solution per 10 ml adipose tissue).
6. Wipe the tubes with disinfectant, cap, wrap with parafilm and place in a 37°C incubator on a rocker for 60 min.
7. Centrifuge at 300×*g* at room temperature for 5 min in a bench-top Sorvall centrifuge.
8. Take the conical tubes out of the centrifuge and shake them vigorously for 10 s to thoroughly mix the cells. This is to complete the separation of stromal cells from the primary adipocytes.
9. Centrifuge again at 300×*g* for 5 min. Carefully aspirate off the oil on top, the primary adipocytes (yellow layer of floating cells), and the collagenase solution. Leave behind approximately 10 ml of the brown collagenase solution above the pellet so that the stromal-vascular fraction (dark red cells on bottom) is not disturbed.
10. Resuspend the pellet of cells in PBS with 1% BSA and filter using a 100 μm Steri-flip vacuum sterilization unit.
11. Centrifuge the cells at 300×*g* for 5 min and aspirate the remaining collagenase solution. When aspirating, the tip of the pipette should aspirate from the top so that the oil is removed as thoroughly as possible. The cell pellet should be tightly packed at the bottom.
12. Add 10 ml of DMEM-HG/10% FBS/1× gentamicin (5 μg/ml) to each centrifuge tube and resuspend the cells. Pool the cells to one tube and centrifuge again (see Note 5).
13. Aspirate supernatant. Resuspend the cells in 10 ml of medium.
14. Divide the cells equally and accordingly to the appropriate number of flasks. 24–72 h after plating, aspirate medium from flask. Wash with PBS and aspirate.
15. Add the original volume per flask of fresh medium.
16. Cells will be grown to 80–90% confluence and then either passaged (3,000–4,000 cells/cm^2) or cryo-preserved (see Note 6).

3.2 Preparation of the Biomaterials Scaffolds

For patch constructs:

1. PLGA coating. All steps are carried out under a chemical fume hood. PLGA is dissolved in methylene chloride (Spectrum Laboratory Products, $MeCl_2$) to a final concentration of 4.25%. Immerse a 1 cm×1 cm piece of Vicryl mesh in the PLGA solution (just a quick dip, enough to get the mesh coated) and remove and evaporate off the excess methylene chloride under cool, forced air (any form of cooled air will work—a blow dryer is also acceptable).

2. Repeat the coating process (1–2 coats is generally adequate).
3. Place the coated mesh under vacuum for at least 8 h to remove residual solvent.
4. With a hot knife (MM Newman Corp., Model #HK-60), cut the coated mesh into 4×5 mm strips. Application of a hot knife seals the edges of the biomaterial.

For tubular constructs:

5. Coat the 4 mm inner diameter tight-braided PGA tube with PLGA as described in steps 1–4 above. After at least 8 h in vacuum, cut the tube into 1 cm long segments (suitable for anastomosis to a rodent small intestine) using a hot knife to prevent unraveling of tube fibers.

3.3 Preparation of the Construct

Scaffolds are prepared under a biosafety cabinet, with each scaffold placed in a single well of a sterile, 6-well tissue culture plate.

1. Scaffolds (patch or tubular) are sterilized by incubation in 70% aqueous ethanol solution under a biosafety cabinet for 30 min. This step also serves to pre-wet the scaffold prior to hydration in step 2 below.
2. Scaffold hydration. Using sterile tweezers, transfer the scaffold to a clean well containing SMC growth medium (DMEM-HG/10% FBS/1× gentamicin) and allow the scaffold to hydrate under the biosafety cabinet for 30 min. The scaffold should sink to the bottom of the well if properly hydrated. Failure to do so indicates presence of residual air bubbles within the biomaterials matrix.
3. Scaffold seeding. Each construct (patch or tubular) is seeded with approximately 50,000 SMC. SMC may be freshly isolated as described above or recovered from a frozen aliquot. Resuspend 50,000 SMC in 50 or 200 μl of DMEM-HG/10% FBS/1× gentamicin.
4. Using a P100 or P200 Pipetman, spot the SMC suspension gently upon the hydrated scaffold in a clean well of a 6-well tissue culture plate. Ensure that the cell suspension is evenly distributed throughout the biomaterial as far as possible (see Note 7).
5. Incubate the seeded construct at 37°C in a humidified, 5% CO_2-containing atmosphere for 5 h. This allows the SMC to adhere to the scaffold biomaterial.
6. After the 5 h incubation period, add enough DMEM-HG/10% FBS/1× gentamicin to completely cover the seeded construct. Return the construct to 37°C in a humidified, 5% CO_2-containing atmosphere tissue culture incubator for 5 days to allow for maturation of the construct (see Notes 8 and 9).

3.4 Animal Surgery

This is meant to be a general outline of the surgical procedure. Animal surgery cannot be learned from a protocols book but requires many hours of hands-on experience and authorization from your institution's animal care committee. Consult with your institution's veterinarian or other authorized animal care representative for details regarding animal surgery protocols as practiced by your institution.

Esophagus

1. Using an adult Lewis rat (approximately 28 days old), under isoflurane anesthesia, perform an upper midline abdominal laparotomy, by using a surgical scalpel to cut from the upper abdominal region towards the thoracic area, exposing the esophagus.
2. Using a surgical scalpel, carefully create a full thickness defect measuring approximately 3 mm in width and 5 mm in length in the abdominal esophagus 5 mm proximal to the cardia (see Note 10).
3. The matured, seeded patch construct is implanted over the esophageal defect using interrupted sutures of nonabsorbable 7-0 silk (Ethicon) (see Note 11).

   The nonadsorbable sutures should be introduced in an unambiguous pattern surrounding the regenerative construct. This will make identification of the area of regeneration straightforward at the time of harvest.
4. Cover the construct with omentum. Omentum may be sutured into place using absorbable suture as needed (see Note 12). For the esophagus, it may be necessary to pull up omentum from lower down in the abdominal cavity (see Note 13).
5. Close the abdomen with surgical staples after intraperitoneal injection of gentamicin (0.1 mg).
6. Postoperative care: Rodents are returned to their cage to recover and are allowed unrestricted oral soft food and water intake for 7 days.
7. After 7 days recovery, rodents are allowed unrestricted oral hard rat chow and water intake.

Small Intestine

1. Using an adult Lewis rat (approximately 28 days old), under isoflurane anesthesia, perform a lower midline abdominal laparotomy, by using a surgical scalpel to cut from the upper abdominal region towards the thoracic area, exposing the small intestine.
2. Gently pull out several centimeters of small intestine and lay on a piece of sterile surgical gauze.
3. For patch constructs, create a defect of approximately 4 × 5 mm within the wall of the small intestine. For tubular constructs,

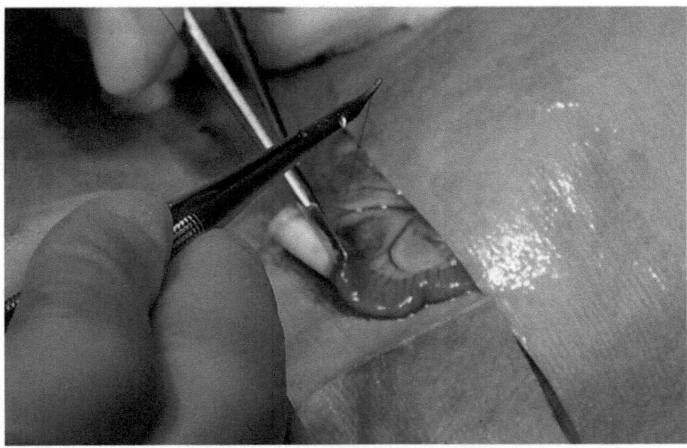

Fig. 1 Anastomosis of SI construct (1 cm) to rodent small intestine

dissect out an approximately 1 cm tubular fragment of small intestine (see Note 14).

4. Anastamose the construct by using interrupted sutures of non-adsorbable 7-0 silk (see example, Fig. 1).

   The nonadsorbable sutures should be introduced in an unambiguous pattern surrounding the regenerative construct. This will make identification of the area of regeneration straightforward at the time of harvest.

5. Cover the construct with omentum. Omentum may be sutured into place using absorbable suture as needed (see Note 12).

   For the small intestine, particularly with younger (<1 month old), female rats, a ready source of nearby omentum may not be forthcoming. Adipose tissue is usually also associated with the vascularized membranes attached to one side of the intestine. This may be leveraged in place of omentum.

6. Close the abdomen with surgical staples after intraperitoneal injection of gentamicin (0.1 mg).
7. Postoperative care: Rodents are returned to their cage to recover and are allowed unrestricted oral soft food and water intake for 7 days.
8. After 7 days recovery, rodents are allowed unrestricted oral hard rat chow and water intake. (see additional information in Notes 15–24).

4 Notes

1. Harvested adipose may be stored overnight at 4°C in PBS/gentamicin. I do not recommend isolating Ad-SMC from adipose specimens older than 1 day.

2. All washes and subsequent transfer steps should be performed under sterile conditions within a biosafety cabinet. Lab coats and protective gloves should be worn at all times.
3. The isolation procedure described above has been reported in the literature; see ref. 24.
4. Rigorous washing of the harvested adipose (step 4) is crucial to avoid contamination of the Ad-SMC with bacteria. Such contamination, evidenced by cloudy, smelly cell culture media in Ad-SMC flasks within 1–2 days after preparation, may usually be traced back to inadequate washing of the original adipose tissue. Harvesting of tissue should be performed as a sterile surgery, despite the animal having been euthanized.
5. Culture medium used is DMEM supplemented with 10% FBS and 5 μg/ml gentamicin; no additional growth factors or differentiation stimulators are added to the culturing medium. As discussed in (24), expansion of adipose-sourced cells at relatively high cell densities in a principally basal medium without addition of growth factors, preselected lots of serum or high concentrations of FBS or other extraneous agents is important for isolation of a principally smooth muscle cell-like population instead of a population composed of mesenchymal stem cells (MSC), endothelial cells, or other adipose-sourced cell population.
6. SMC may be cryo-preserved in a solution of 10% DMSO (Sigma, cell culture grade) in DMEM-HG/10% FBS. For convenience, SMC are cryo-preserved in aliquots of $0.5–1.0 \times 10^6$ cells. Spin cells down in a 15 ml Falcon tube at $300 \times g$ for 5 min, aspirate off media, and resuspend cell pellet in 1 ml freezing media. Transfer to 2 ml cryogenic vial (Corning, catalog # 430488) and place in isopropanol-containing cell-freezing container (e.g., "Mr. Frosty") at −80°C overnight. The next day, transfer cryogenic vial to liquid nitrogen for long-term storage.
7. The volumes suggested above are appropriate for the patch and tubular constructs described here. Should your construct be of different size, modify the seeding volume accordingly to ensure even distribution of cells.
8. During the construct maturation step, SMC proliferate throughout the biomaterial and deposit extracellular matrix (ECM) throughout the scaffold fibers (see Fig. 2). Initial degradation of the biomaterial also occurs at this time. The presence of this ECM is likely to be important for inducing the regenerative response at the time of implantation.

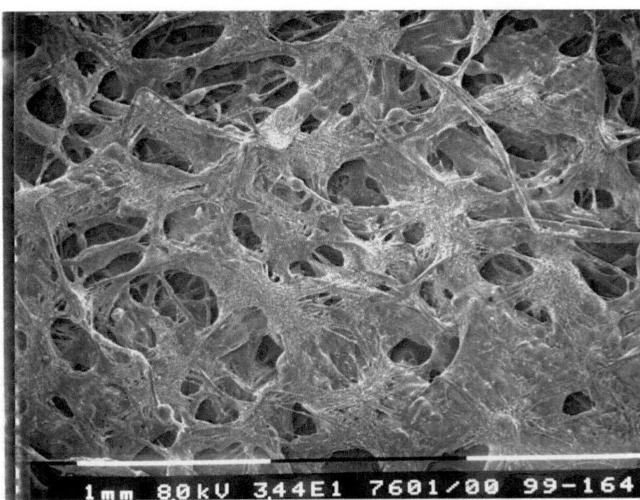

Fig. 2 SEM (34.4×) of PGA scaffold seeded with SMC. Note fibrous structure of scaffold and ECM deposited across scaffold by proliferating SMC

9. Although I recommend a 5 day maturation period, my experience has been that 3 or 4 days is usually adequate should you have time constraints. You can seed with 75,000 Ad-SMC to compensate.
10. Use of magnifying surgical spectacles or in-suite dissecting microscope may be required.
11. Consult with your veterinarian or surgical technician for an explanation of suturing techniques.
12. The presence of omentum is required to facilitate regenerative outcomes. Omentum serves as a source of vascularization, stem and progenitor cell populations, and paracrine bioactivity conducive to the regenerative process (27).
13. I do not like using fibrin glue to secure omentum or to position the regenerative construct owing to difficulties in delivering the glue to a defined location. I have also observed that esophageal cells fail to migrate through fibrin from rodent esophageal tissue explants maintained in a 6-well tissue culture plate. In contrast, robust migration of esophageal cells from such explants not treated by fibrin is observed.
14. Dissection of the small intestine is typically accompanied by extensive bleeding. Press tightly on the cut edges with sterile surgical gauze to ameliorate bleeding.
15. We have evaluated the extent of regeneration of esophagus and small intestine at time periods ranging from 1 to 50 weeks post-implantation. In general, we have observed complete regeneration (i.e., restoration of laminar organization, with luminal epithelial layer surrounded by concentric layers of

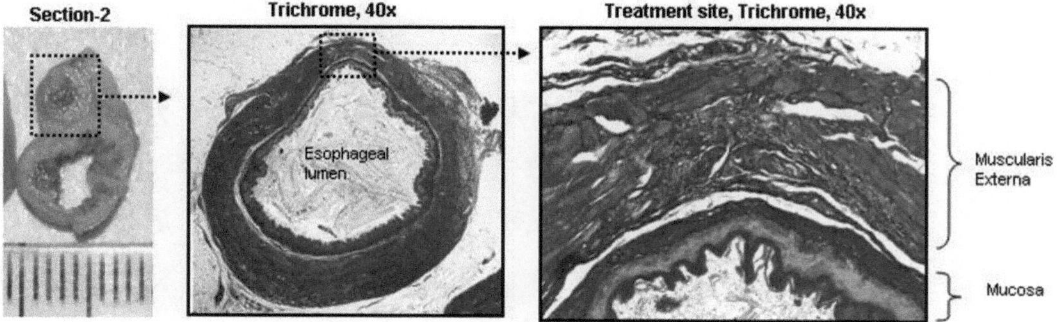

Fig. 3 Transverse section through rodent esophagus highlighting area of regeneration (*right panel*) at 10 weeks post-implantation of patch construct. Note regenerated neo-mucosa and smooth muscle layers

smooth muscle, see example above, Fig. 3) of esophagus at 10–16 weeks post-implantation (see ref. 28) and complete regeneration of small intestine from patch constructs at 8–10 weeks post-implantation (see ref. 29).

16. Despite extensive efforts, anastomosis of a tubular construct to a rodent esophagus was found to be too technically demanding to warrant further exploration.
17. Functional evaluation of regenerative outcomes. Preliminary indications of the success or failure of the implant may be procured from an observation of animal behavior. For the esophagus, does the animal swallow appropriately? Are solid food and water being ingested? Does the animal show obvious signs of discomfort or distress? Failure to swallow or constant "hiccupping" behavior may indicate esophageal blockage, entrapment of food or other matter, or excessive inflammation at the implant site.
18. For the small intestine, the ability to pass fecal material normally is the first indication that the construct may be functional. Failure to pass fecal material will lead to obstruction and death of the animal within days of implantation. As with the esophagus, failure to ingest and swallow food may also indicate potential problems with the construct.
19. Intestinal adhesions are an almost unavoidable complication of intestinal surgery. A number of methodologies have been reported in the literature purporting to alleviate adhesion formation in rodents following intestinal surgery. These include application of COX2 inhibitors (Celebrex), progesterone, omentum, fibrin glue, hyaluronic acid, honey, and paraffin oil (30–35). These reports are generally ambiguous and sometimes contradictory. We have not undertaken a systematic

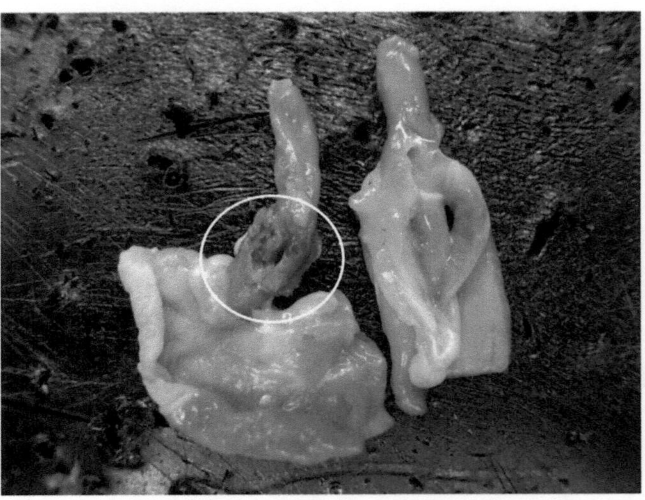

Fig. 4 Rodent esophagus (*dissected open*) showing bedding (*circled*) trapped within the esophagus. Animal displayed obvious signs of distress and was humanely euthanized shortly after surgery

evaluation of these strategies; we have observed that a certain degree of adhesion is tolerable provided it does not lead to intestinal kinkage and obstruction.

20. Role of bedding. We have observed that following esophageal surgery, rodents are inclined to chew and swallow their bedding material. This can lead to entrapment of bedding material at the implant site (see Fig. 4), obstruction, and death of the animal. For this reason, we suggest having the animal recover on wire cages for 7 days post-implantation. Ensure that animal is sufficiently warm during this recovery period.
21. The choice of specific time points to harvest for analysis of regeneration is up to the investigator.
22. For molecular analysis, after euthanasia the regenerated tissue (identified from the nonadsorbable sutures) is removed immediately and flash frozen in liquid nitrogen. Specific regions may then be removed for preparation of RNA, DNA, or protein.
23. For histological analysis, the regenerated tissue (identified from the nonadsorbable sutures) is removed immediately and fixed in 10% buffered formalin overnight. Selected areas of tissue may then be removed, dehydrated in ascending ethanol series, and embedded in paraffin. Sections (5 μm) may then be cut and stained with hematoxylin/eosin and/or Masson's trichrome or other stains as desired by the investigator. See Fig. 3 for examples of regenerated esophageal tissue.

24. Evaluation of other biomaterials (and cell types). It should be self-evident to the investigator that the methods described in this chapter may be applied towards the screening of biomaterials other than PGA/PLGA and cell types other than SMC for their impact on regenerative outcomes within esophagus and small intestine.

References

1. Spitz L (2006) Esophageal atresia: lessons I have learned in a 40-year experience. J Pediatr Surg 41:1635–1640
2. Poley JW, Steyerberg EX, Kuipers EJ, Dees J, Hartmans R, Tilanus HW et al (2004) Ingestion of acid and alkaline agents: outcome and prognostic value of early upper endoscopy. Gastrointest Endosc 60:372–377
3. Hulscher JB, van Sandick JW, de Boer AG, Wijnhoven BP, Tijssen JG, Fockens P et al (2002) Extended transthoracic resection compared with limited transhiatal resection for adenocarcinoma of the esophagus. N Engl J Med 347:1662–1669
4. Spitz L (2006) Esophageal replacement. In: Grosfeld JL, O'Neill JA, Fonkalsrud EW et al (eds) Pediatric surgery. Mosby Elsevier, Philadelphia, PA, pp 1093–1106
5. Deurloo JA, Ekkelkamp S, Hartman EE, Sprangers MA, Aronson DC (2005) Quality of life in adult survivors of correction of oesophageal atresia. Arch Surg 140:976–980
6. Somppi E, Tammela O, Ruuska T, Rahnasto J, Laitinen J, Turjanmaa V et al (1990) Outcome of patients operated on for oesophageal atresia: 30 years' experience. J Pediatr Surg 33:1341–1346
7. Romeo C, Bonanno N, Baldari S, Centorrino A, Scalfari G, Antonuccio P et al (2000) Gastric motility disorders in patients operated on for esophageal atresia and tracheoesophageal fistula: long term evaluation. J Pediatr Surg 35:740–744
8. Warner BW (2004) Tissue engineered small intestine, a viable clinical option? Ann Surg 240:755–756
9. Gabe SM, Day RM, Boccacini A (2004) Tissue engineering of the small intestine. In: Wnek GE, Bowlin GL (eds) Encyclopedia of biomaterials and biomedicine. Informa Healthcare, London
10. Grikscheit TC, Siddique A, Ochoa ER et al (2004) Tissue engineered small intestine improves recovery after massive small bowel resection. Ann Surg 2405:748–754
11. Messing B, Crenn P, Beau P et al (1999) Long term survival and parenteral nutrition dependence in adult patients with the short bowel syndrome. Gastroentology 1175:1043–1050
12. Krupnick AS, Morris JB (2000) The long term results of resection and multiple resections in Crohn's disease. Semin Gastrointest Dis 11: 41–51
13. Bianchi A (1999) Experience with longitudinal intestinal lengthening and tailoring. Eur J Pediatr Surg 9:256–259
14. Warner BW, Vanderhoof JA, Reyes JD (2000) What's new in the management of short gut syndrome in children. J Am Coll Surg 190:725–736
15. Beckstead BL, Pan S, Bhrany AD, Bratt-Leal AM, Ratner BD, Giachelli CM (2005) Esophageal epithelial cell interaction with synthetic and natural scaffolds for tissue engineering. Biomaterials 26:6217–6228
16. Furth ME, Atala A, Van Dyke ME (2007) Smart biomaterials design for tissue engineering and regenerative medicine. Biomaterials 28:5068–5073
17. Atala A (2006) Recent developments in tissue engineering and regenerative medicine. Curr Opin Pediatr 18:167–171
18. Becker C, Jakse G (2007) Stem cells for regeneration of urological structures. Eur Urol 51:1217–1228
19. Frimberger D, Lin HK, Kropp BR (2006) The use of tissue engineering and stem cells in bladder regeneration. Regen Med 1:425–435
20. Roth CC, Kropp BP (2009) Recent advances in urologic tissue engineering. Curr Urol Rep 10:119–125
21. Wood D, Southgate J (2008) Current status of tissue engineering in urology. Curr Opin Urol 18:564–569
22. Basu J, Ludlow JW (2010) Platform technologies for tubular organ regeneration. Trends Biotechnol 28:528–533
23. Jayo MJ, Jain D, Ludlow JW, Payne R, Wagner BJ, McLorie G et al (2008) Long term durability, tissue regeneration and neo-organ growth during skeletal maturation with a neo-bladder augmentation construct. Regen Med 3:671–682
24. Basu J, Genheimer CW, Guthrie KI, Sangha N, Quinlan SF, Bruce AT et al (2011) Expansion of the human adipose derived stromal vascular cell fraction yields a population of smooth muscle-like cells with markedly distinct pheno-

typic and functional properties relative to mesenchymal stem cells. Tissue Eng Part C Methods 17:843–860
25. Basu J, Jayo MJ, Ilagan R, Guthrie KI et al (2012) Regeneration of native-like neo-urinary tissue from non-bladder cell sources. Tissue Eng Part A 18:1025–1034
26. www.clinicaltrials.gov/ct2/show/NCT01087697?term=tengion&rank=3
27. Litbarg NO, Gudehithlu KP, Sethupathi P et al (2007) Activated omentum becomes rich in factors that promote healing and tissue regeneration. Cell Tissue Res 328:487–497
28. Basu J, Mihalko KL, Payne R, Rivera T, Knight T, Genheimer CW, Guthrie KI, Sangha N, Jayo MJ, Jain D, Bertram TA, Ludlow JW (2012) Extension of bladder-based organ regeneration platform for tissue engineering of esophagus. Med Hypotheses 78:231–234
29. Basu J, Mihalko KL, Payne R, Rivera E, Knight T, Genheimer CW, Guthrie KI, Sangha N, Jayo MJ, Jain D, Bertram TA, Ludlow JW (2011) Regeneration of rodent small intestine tissue following implantation of scaffolds seeded with a novel source of smooth muscle cells. Regen Med 6:721–731
30. de Virgilio C, Elbassir M, Hidalgo A, Schaber B et al (1999) Fibrin glue reduces the severity of intra-abdominal adhesions in a rat model. Am J Surg 178:577–580
31. Reijnen M, Jacques F, Postma VA, van Goor H (1999) Prevention of intra-abdominal abscesses and adhesions using a hyaluronic acid solution in a rat peritonitis model. Arch Surg 134: 997–1001
32. Gollu A, Kismet K, Kilicoglu B, Erel S et al (2008) Effect of honey on intestinal morphology, intraabdominal adhesions and anastomotic healing. Phytother Res 22:1243–1247
33. Shafik A, El-Sibai O, Shafik AA (2002) A novel method for preventing postoperative adhesions in the rat model. J Gynecologic Surg 18: 65–68
34. Aydinli B, Ozturk G, Basoglu M et al (2007) Prevention of adhesions by omentoplasty: an incisional hernia model in rats. Turk J Med Sci 37:93–97
35. Greene AK, Alwayn IP, Nose V, Flynn E et al (2005) Prevention of intra-abdominal adhesions using the antiangiogenic COX-2 inhibitor Celecoxib. Ann Surg 242:140–146

Chapter 27

Scanning Electron Microscopy Evaluation of Endothelialized Tissue-Engineered Constructs

Sandra L. Johnson

Abstract

Scanning electron microscopy (SEM) is an important technique for evaluation of the efficiency of endothelialization of tissue-engineered constructs incorporating a surface endothelial cell layer. Here, we describe methodologies for the preparation of such constructs for SEM analysis that are applicable to a broad range of tissue-engineered constructs.

Key words Scanning electron microscopy, Endothelial cell, Endothelium, Tissue engineering, Osmium tetroxide, Glutaraldehyde

1 Introduction

Some tissue-engineered constructs, such as the luminal surface of blood vessels, benefit from endothelialization prior to implant due to reduced or eliminated need for concomitant anticoagulant pharmaceuticals. Such compounds may have side effects for the patient and in addition may complicate analysis and interpretation of graft performance. With tissue-engineered heart valve leaflets, for example, a confluent autologous endothelial layer would provide a non-immunogenic and functionally non-thrombogenic surface which would likely result in improved outcomes (1). In another example, for small diameter vascular grafts, the need for preimplant endothelialization is critical due to the increased occlusion problems in low flow conditions (2).

Preimplant endothelialization requires the initial attachment of endothelial cells (ECs) as well as formation of a biochemically active endothelium with tight junctions and complete coverage of the underlying thrombogenic engineered tissue. This coverage should remain intact under flow. Continuing with the example of vascular grafts, any analysis of in vivo outcomes associated with implantation of endothelialized vascular grafts within small or

Joydeep Basu and John W. Ludlow (eds.), *Organ Regeneration: Methods and Protocols*, Methods in Molecular Biology, vol. 1001, DOI 10.1007/978-1-62703-363-3_27, © Springer Science+Business Media New York 2013

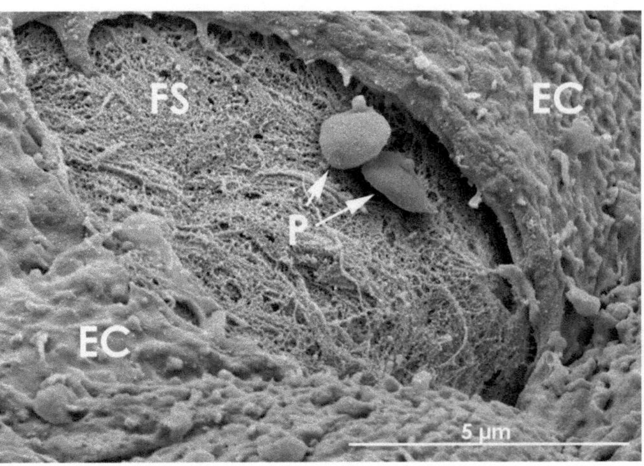

Fig. 1 Fibrin based tissue engineered vascular graft after animal implantation. Specimen preparation was according to the described method. This resulted in good cell membrane preservation, enabling a clear distinction between the ECs and the fibrous substratum (FS). Two platelets (P) have bound to the exposed substratum. Acquired on Hitachi S-4700 at 3 kV with ~10 nm gold–palladium coating

large animal models or within human patients must include a systematic evaluation of the extent and efficiency of endothelialization associated with the construct. In cases where grafts are not endothelialized prior to implant, endothelialization of the graft as mediated by the host needs to be evaluated to determine when anticoagulant therapy can be safely reduced.

To this end, preliminary evaluation of the luminal surface of the construct may be performed with standard immunofluorescence techniques typically leveraging antibodies against EC junction proteins such as VE-cadherin (3). However, an analysis using scanning electron microscopy (SEM) can be beneficial for the following reasons:

1. Antibodies recognizing EC junctions or other relevant antigens may not be available for the species under examination.
2. Developing immuno-staining conditions that result in adequate staining with low background can be problematic.
3. Small imperfections in the endothelium that are sufficient for binding platelets and initiating thrombosis may be undetectable at the fluorescence level (see Fig. 1).

If SEM is to be used for evaluation, preservation of cellular ultrastructure with an adequate fixation protocol is necessary. Poor fixation and/or drying techniques can result in artifacts including loss of ultrastructure, EC shrinkage, and even ECs peeling off from

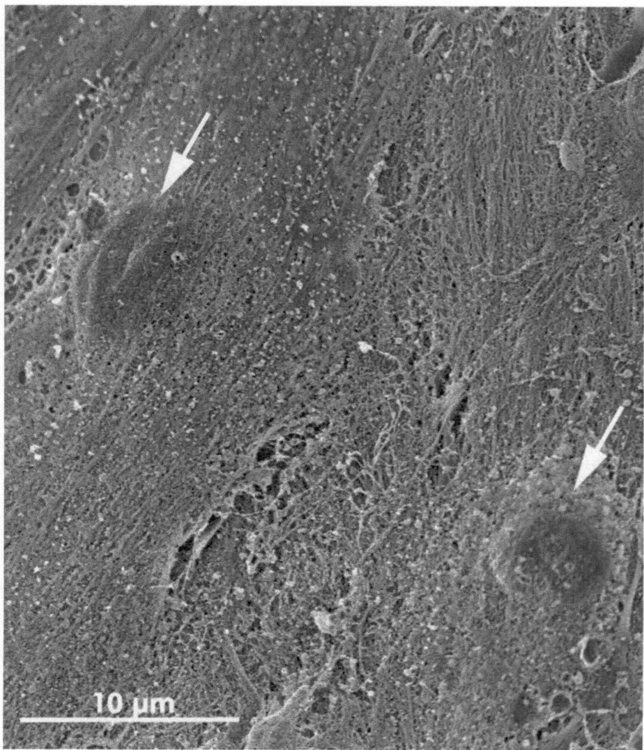

Fig. 2 Blood outgrowth endothelial cells (BOECs) seeded onto a fibrin-based tissue engineered vascular graft and cultured in a flow chamber. Specimen preparation was as in Fig. 1 except post-fixation was at room temperature rather than 4°C. The nuclear bulges are present (*arrows*), but excessive reaction with osmium tetroxide has extracted much of the cellular membrane, making it difficult to discern cellular boundaries from substratum fibers. Acquired on Hitachi S-900 at 3 kV with ~2.5 nm platinum coating

the substratum, thereby preventing proper evaluation of the specimen (see Fig. 2).

This protocol is derived from earlier work (4) with modifications based on the author's experience. This method reliably gives good preservation of EC ultrastructural characteristics (see Figs. 1, 3, 4, and 5).

The SEM protocol below is an immersion fixation methodology designed for constructs with surfaces that can have direct access to fixative, such as valve leaflets or bladder surfaces, or for testing various tissue-engineered specimens from flow chambers. This methodology is also appropriate for vascular grafts that can be cut into "donuts" of proper size to allow for direct contact of lumen with fixative. For grafts where this is not possible, perfusion fixation needs to be performed. Methods of in vivo perfusion fixation need to take into account the position of the graft, the experimental animal's vasculature, as well as proper animal

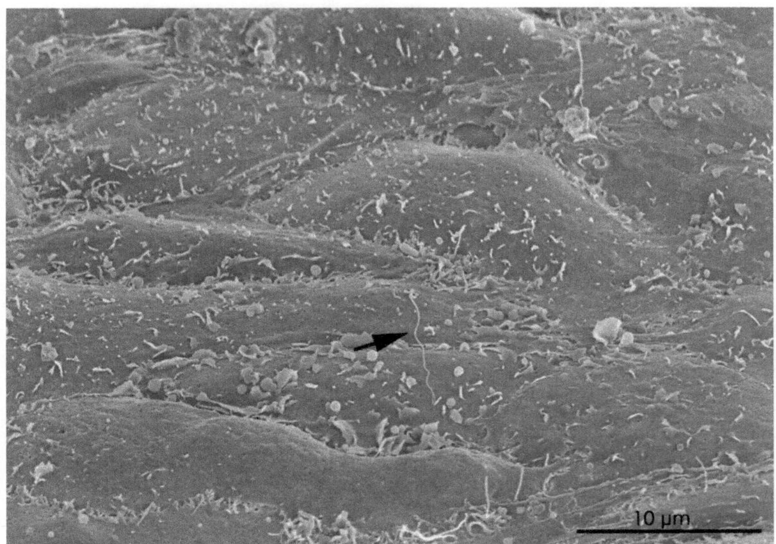

Fig. 3 BOECs seeded as in Fig. 2. Specimen preparation was according to the described method in this chapter. Note the well-preserved microvillous projections and even an intact primary cilium (*arrow*). Osmium tetroxide post-fixation provided good contrast without loss of cellular components. Acquired on Hitachi S-4700 at 3 kV with ~2.5 nm platinum coating

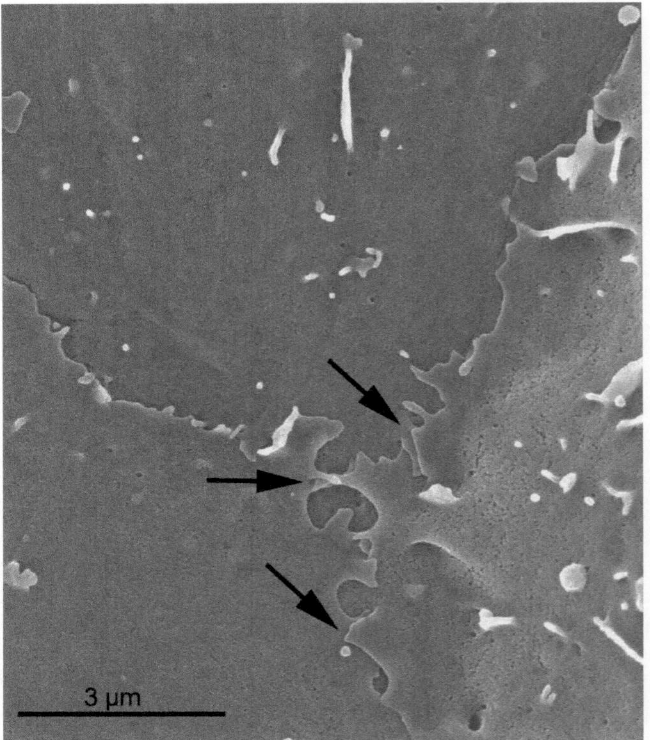

Fig. 4 Rotation of the specimen stage to permit top down viewing enhances imaging of tight junctions (*arrows*). BOECs cultured and prepared as in Fig. 3. Acquired on Hitachi S-900 at 3 kV with ~2.5 nm platinum coating

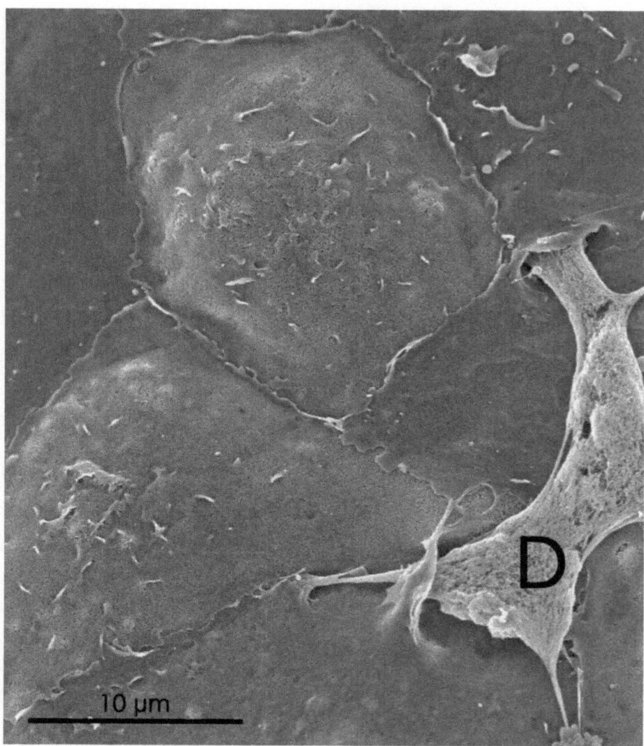

Fig. 5 BOECs on fibrin-based vascular graft with bulging nuclear region and cobblestone morphology typical of endothelial cells cultured in the absence of flow. As has been seen with ECs in vivo (7, 8), neighboring cells are concurrently reestablishing tight junctions (*arrows*) as a dead cell (D) is shed, thereby maintaining an intact endothelium. Acquired on Hitachi S-900 at 3 kV with ~2.5 nm platinum coating

care. The graft can also be removed from the animal and then perfusion fixed with the use of a pump. As with immersion fixation, proper account must be made for adequate pH buffering and osmolarity. In addition, proper viscosity and flow rates are required for good preservation of ultrastructure. Consult the literature for perfusion protocols appropriate for your model system.

2 Materials

1. Hanks' balanced salt solution (HBSS) (37°C) (see Note 1).
2. 0.4 M sodium cacodylate (aqueous), pH 7.3. Cacodylate is arsenic-based buffer; handle with caution (see Notes 2 and 3).
3. 10% glutaraldehyde (aqueous). EM grade glutaraldehyde can be purchased as 10 mL glass ampoules from Electron Microscopy Sciences. Using a freshly opened ampoule is

preferred. Glutaraldehyde is toxic and needs to be handled with care (see Notes 4 and 13).

4. Sucrose.
5. Osmium tetroxide (2 or 4% aqueous solution). Glass ampoules may be procured from Electron Microscopy Sciences. Osmium tetroxide is both volatile and extremely toxic. Its use requires careful handling and waste disposal (see Notes 5–7).
6. 30%, 50%, 70%, 80%, and 95% ethanol in ddH_2O (double distilled water, MilliQ or equivalent) and absolute (100%) ethanol opened on day of use (see Note 8).
7. Critical point drying (CPD) device (e.g., Samdri®).
8. Conductive tape for sample mounting. Check with your EM facility for their preferred material and vendor.
9. Sample holders, compatible with your particular SEM; check with personnel in your EM facility.
10. Dissecting microscope for mounting specimens on sample holders.
11. Antistatic gun if available (e.g., Zerostat®).
12. Sputter coater (consult your EM facility).
13. Vacuum desiccator and silica gel for sample storage.

Other required supplies:

1. Saran wrap.
2. Tissue culture plates or other container, sized to handle approximately 10 volumes of fixative per sample volume.
3. Ice.
4. Ice bucket.
5. Fine-tip forceps.
6. Micro-spatula with the blade bent at 70–90° angle to transfer samples by lifting from bottom. This is especially useful for very small samples that lack extra surface area for forceps handling.
7. Plastic transfer pipettes.
8. Plastic Petri dish.
9. Disposable measuring pipettes and electronic or bulb aspiration/ dispensing device.
10. Surgical or razor blades.
11. Protective face mask for use during specimen mounting (after CPD).
12. Disposable measuring pipettes and pipettor.
13. 50 mL conical tubes.

Prior to initiation of this protocol, prepare the following items:

1. Warm HBSS (37°C).
2. Prepare Primary Fixative: 3% glutaraldehyde in 0.1 M sodium cacodylate pH 7.3 (aqueous, prepared with ddH_2O, MilliQ, or equivalent). Must be cold (4°C or on ice). The volume of this fixative should be at least 10 times the specimen volume. Ideally this is prepared on day of use (see Notes 3, 4, 9, and 13).
3. Sucrose Rinse Solution: 5% sucrose in 0.1 M sodium cacodylate pH 7.3 (aqueous, prepared with ddH_2O, MilliQ, or equivalent). Must be cold (4°C or on ice). Needed for multiple rinses after glutaraldehyde fixation; make 6–10× volume of fixative used, which in turn is dependant on the volume of the specimen and the number of specimens being examined. For example, for a tissue-engineered vascular graft of volume 1 mL, use 10 mL fixative and make 60–100 mL of Rinse Solution.

3 Methods

To preserve EC ultrastructure, flow alignment, and activation state, the construct must be removed from tissue culture (if being matured ex vivo) or from the animal (if being matured in vivo) and fixed as quickly as possible. Avoid physically disturbing the specimen, especially the EC surface, since ECs are readily shed from the substratum. Handle with fine tipped forceps an unimportant surface or with a bent micro-spatula. To avoid drying artifacts, keep the specimen wet during this and all subsequent procedures.

3.1 Primary Fixation

1. Collect the specimen into warm (37°C) HBSS in a beaker, glass vial, or appropriate-size well within a multi-well tissue culture plate for a quick rinse to remove unwanted material. With in vivo samples, this removes blood cells and serum proteins which can obscure visualization of the ECs. For ex vivo samples, the HBSS removes serum and culture debris. This wash should be quick and gentle (less than 5 min), since an excessively long wash may cause ECs to become activated, alter their flow-related morphology, or be shed from the substratum.
2. Carefully and quickly remove the HBSS and replace with Primary Fixative, or transfer the rinsed specimen to the Primary Fixative in another vessel. If using the same vessel, decant the HBSS and gently pipette the fixative over the EC surface. Use of a transfer pipette at this step helps ensure better control of fixative efflux. Application of fixative should be fast enough to prevent drying of ECs, but sufficiently gentle such that ECs are not dislodged from the surface. If fixing a "donut" specimen from a vascular graft, use forceps to lift the specimen up and

down a couple of times in the Primary Fixative to ensure the fixative gets into the lumen.

3. Place vessel containing the specimen undergoing fixation on ice. Fixation continues for 2 h with occasional gentle swirling. Agitation should not be so vigorous as to dislodge ECs prior to complete fixation onto the specimen's substratum (see Notes 4 and 9).
4. At the end of primary fixation, replace Primary Fixative with cold (4°C or on ice) Sucrose Rinse Solution. Dispose of Primary Fixative according to your institution's rules for disposal of hazardous waste (*Important*: Continue to keep the surface of the specimen wet to avoid drying artifacts. These can occur even on fixed tissue).
5. Rinse the specimen with multiple changes of cold (4°C or on ice) Sucrose Rinse Solution to remove unreacted glutaraldehyde. Rinses are performed on ice in a chemical fume hood, with 6–10 changes over a 1 h time period. To facilitate quick liquid changes while keeping the specimen wet, tilt the sample container at an angle such that the specimen slides to the bottom of the angled container and stays in a minimal volume of liquid during the liquid change. If you can, withdraw the spent rinse solution with one hand while adding fresh rinse solution with the other, using the ice to support the container at an angle. Plastic transfer pipettes work well for this (see Note 10).

3.2 Osmium Tetroxide Post-fixation

Post-fixation is with 1% osmium tetroxide in 0.1 M sodium cacodylate pH 7.3 (aqueous), on ice, in a chemical fume hood for 1 h (see Notes 5–7, 11).

1. To prepare the Post-fixation Solution, add the appropriate volume of 0.4 M sodium cacodylate and ddH_2O (MilliQ or equivalent) to a 50 mL conical on ice such that an entire ampoule of osmium tetroxide will be used, with a final concentration of 1% osmium tetroxide in 0.1 M sodium cacodylate, pH 7.3. At time of use, the solution must be cold (4°C or on ice) (see Notes 12 and 13).
2. Just prior to use, carefully and in a chemical fume hood with protective gloves and a lab coat, open the glass ampoule of osmium tetroxide, and using a glass pipette (e.g., Pasteur pipette), transfer the entire contents of the ampoule to the 50 mL conical tube containing the cacodylate/water solution (see Note 13).
3. Remove the last sucrose rinse from the specimen and add the Post-fixation Solution. Do this quickly to avoid specimen drying. It is best to remove the rinse using a transfer pipette in one hand and add the fixative with the other.
4. Wrap the multi-well plate in Saran wrap to trap the osmium vapors and place in ice; cover the ice bucket to maintain a

temperature of 4°C. Keep in the fume hood and incubate for 1 h with occasional mixing. Time and temperature are critical. Warming or increasing incubation time will result in excessive osmium tetroxide reaction with concomitant extraction of cellular components (see Notes 14–18 and Fig. 2).

All remaining steps are at room temperature (RT).

5. Remove the osmium tetroxide solution and replace with ddH_2O (MilliQ or equivalent). Rinse with multiple changes of ddH_2O for approximately 30 min to remove the bulk of unreacted osmium tetroxide and buffer salts. Dispose of fixative and water rinses according to the rules of your institution. If possible, transfer the specimen to a clean container or multi-well plate to facilitate osmium tetroxide removal.

3.3 Ethanol Dehydration

During this step the water in the specimen is gradually replaced with ethanol using a graded series of ethanol solutions. This is to prepare the specimen for CPD (see Note 19).

1. Transfer specimens across the following aqueous ethanol series: 30%, 50%, 70%, 80%, 95% ethanol, then absolute ethanol (100%) from a bottle opened on day of use. The time spent in each ethanol solution is dependant on specimen size, but in general will be approximately 15–20 min, with the ethanol being changed at least three times. The volume is about 10× specimen volume (see Notes 8, 20–23).
2. The final absolute ethanol wash should be done with multiple changes over the course of 45–60 min.
3. During the absolute ethanol wash, trim the specimen to an appropriate size for your SEM sample holders. This is best done in a plastic Petri dish with the specimen immersed in absolute ethanol. Place a sharp scalpel blade (e.g., a size 22 blade on a #4 handle) or razor blade on the desired location and cut with a single blunt cut (starting from the EC surface if the specimen is a flat tissue). Do not use scissors as this will pinch the specimen resulting in compression artifacts (see Note 24).

3.4 Critical Point Drying

In this step, the ethanol in the specimen is replaced by liquid CO_2 which is removed at its critical point using a critical point dryer (e.g., Samdri) (see Note 19). Consult your EM facility for specific apparatus, training, and availability as well as any special specimen preparation methodologies. The following are some general methods of specimen handling for CPD.

1. The specimens will have to be loaded into a CPD holder. Your EM facility will have these. To load the specimens, place the holder in a small beaker or other suitable container tall enough to fully immerse the holder and add freshly opened absolute

ethanol (opened on day of use). Use a bent micro-spatula or fine-tip forceps to quickly transfer specimens from the ethanol solution in the multi-well plate to an appropriate slot in the immersed CPD holder. It is essential to perform this step quickly so the specimens remain moist. Be careful not to jar the CPD holder when loading the samples as the specimen pieces readily float out of the wells (see Note 25).

2. Once transfer of specimen pieces is complete, secure the top on the CPD holder and proceed according to the recommended procedure of your institution's EM facility. When transferring the CPD holder to the CPD, do it quickly so samples do not dry out.

3.5 Sputter Coating

After CPD, the samples are mounted on appropriate specimen holders for your microscope and then sputter coated to put a layer of metal on the viewing surface. The type of metal and thickness should be discussed with your EM facility. The thinner the metal layer, the more sample detail will be visible. However, this needs to be balanced against charging, which is reduced with increased coating. Your EM facility will be able to give you information for appropriate specimen holders and mounting material as well as appropriate training for sputter coater use (see Note 26).

1. Prior to specimen mounting, label the specimen holders with sample ID information on the underside using a standard lab Sharpie or equivalent. If using conductive tape to adhere the specimens to the holders, cut the tape to the correct size with a scissors and attach it to the holder using forceps (see Notes 27–29).
2. Transfer the CPD specimens to a clean Petri dish. If you have equivalent samples in the CPD holder they can be dumped out, otherwise you will need to carefully remove them with fine-tip forceps (see Note 30).
3. To mount specimens, it is best to use a dissecting microscope. Place the Petri dish containing the specimens under the microscope and add the prepared specimen holders, or place these nearby such that the holders can also be viewed under the microscope. Remove protective covering from the conductive tape.
4. Examine the specimens under the dissecting microscope to find the desired surfaces and mount these facing up on the specimen holder. To do this, carefully pick up the specimen with a fine tip forceps on an unimportant edge and position on the adhesive without distorting the specimen. Note that the sample adhesive is very strong and repositioning of a specimen will not be possible. Multiple specimens may be placed on the same specimen holder. Carefully tack down the specimen with

the forceps in a couple of places, being careful not to damage or distort the surface of interest. The specimen holder can be turned upside down to test adequate adhesion of the specimen to the holder (see Notes 31 and 32).

5. Sputter-coated specimens can be stored indefinitely in vacuum desiccators with ample drying agents such as silica gel. Make sure to bring down the vapor pressure in the desiccators as much as possible, e.g., use at least 15–20 min of a standard lab "house" vacuum before closing off the desiccator's valve (see Note 33).
6. You are now ready to observe the specimen under supervision of your EM facility.

4 Notes

1. Must be fresh HBSS, as HBSS is buffered with carbonates that are released as CO_2 with a resulting increase in the pH of the solution.
2. The sodium cacodylate buffer may be replaced with sodium phosphate buffer at the same concentration. However, if no buffering system is present during the primary fixation, the pH will be lowered resulting in numerous artifacts (5).
3. When dried, sodium cacodylate is readily aerolized. This can result in inhalation of toxic arsenic. Rinse all pipettes and glassware with water prior to discarding. Avoid inhalation and use protective gloves.
4. When handling glutaraldehyde solutions, be sure to work in a well ventilated chemical fume hood and use protective gloves. Repeated exposure to glutaraldehyde may cause contact dermatitis and respiratory problems. Avoid exposure to vapors and contact with eyes.
5. Osmium tetroxide is volatile and extremely toxic; consult with your institution's Occupational Health and Safety Office for proper handling of osmium tetroxide and waste. Osmium tetroxide vapors are particularly dangerous to the conjunctiva and respiratory membranes. Always use in properly ventilated chemical hoods. Avoid contact by using protective gloves and a lab coat. Spills can be treated by addition of corn oil and disposed of according to your institution's rules.
6. Opened glass ampoules and anything that comes in contact with osmium tetroxide should be rinsed with water and this rinse disposed of along with the rest of the osmium liquid waste. Some institutions may require rinsing with corn oil, using twice the volume of corn oil as osmium tetroxide, and disposing of this as chemical waste.

7. Your institution may require all materials exposed to osmium tetroxide, even after rinsing, be treated as hazardous. Ziploc freezer bags can be used to dispose of contaminated plastic transfer pipettes and tissue culture plates. Rinsed ampoules of osmium tetroxide and glass pipettes should be stored in a covered container while awaiting waste pick up. An empty liquid laundry detergent bottle works well for this.
8. Do not attempt to dry the ethanol with molecular sieves as this can release small particles that can decorate your sample. The freshly opened bottle of absolute ethanol is a requirement since absolute ethanol readily absorbs humidity, especially in summer months. Cap the bottle when not in use. A newly opened bottle is usable for 1 day. Any remaining ethanol from this bottle should only be used in preparing the diluted aqueous ethanol solution for the dehydration series.
9. The primary fixation should be done in a chemical fume hood if possible, especially if using a multi-well plate. If a hood is not available, e.g., if fixing an explanted graft in a surgical facility, then use a tight-sealing nonreactive vial such as Wheaton glass vials with snap caps (Fisher Scientific 03-335-10).
10. Excess unreacted glutaraldehyde must be removed before osmium tetroxide is added, otherwise it will react with the glutaraldehyde, leaving a black deposit on your specimen. Continual, gentle mixing during washes (such as with an orbital shaker) will improve efficiency of the sucrose washes. After post-fixation, the specimens can be mixed quite vigorously. However surface damage (e.g., from pipettes) can still occur on fixed tissue.
11. The penetration rate of osmium tetroxide into dense tissues will be less than 0.5 mm in 1 h at 4°C (5). You may wish to trim your specimen accordingly to yield complete osmium tetroxide penetration if you desire to preserve and view the EC substratum. If this is the case, trim the specimen with a razor blade, avoiding compression artifacts. If microscopic evaluation of the endothelial surface is all that is needed, then it is best to delay trimming until after osmium tetroxide treatment, since it will further fix the specimen, making it more rigid and easier to cut. Waiting until the final steps of the ethanol dehydration series to do final trimming is advantageous since many little pieces are harder to keep track of and easier to damage with a pipette during the many solvent exchanges.
12. Multi-well tissue culture plates work well since they are available in a variety of sizes to accommodate specimen size and minimal osmium use. Post-fixation is done in less volume than the primary fixation, usually just enough to cover the specimen by a couple of millimeters. Because of the toxicity, volatility,

and reactivity of osmium tetroxide, it is best to plan to use the entire contents of an ampoule once it is opened.

13. Osmium tetroxide will react with the plastic walls of the conical tube so it is added just prior to use. Osmium tetroxide and glutaraldehyde ampoules and all other glass ampoules should be opened in a chemical fume hood using paper towel/gauze between your hand and the vial. One hand holds one side of the narrow neck; the other hand is on the other side of narrow neck; firmly snap the vial open at the narrowest point.
14. Osmium tetroxide works as a secondary fixative primarily by oxidation of double bonds in unsaturated lipids. Saturated lipids can also be preserved indirectly due to the solubility of osmium tetroxide in the lipid bilayers. It is later reduced during ethanol dehydrations. It can also react with double bonds in proteins, and certain amino acids (6). However, excessive exposure to osmium tetroxide can cause extraction of cellular components. This often is manifested as holes in the cellular membranes (see Fig. 2). If the recommended 1 h at 4°C results in too many cells with this artifact, the post-fixation incubation time can be decreased by 5 or 10 min. Less than 45 min of osmium tetroxide will likely decrease the contrast to an unsuitable level.
15. You will see presence of osmium tetroxide vapors very quickly, as the Saran wrap turns yellow, then black as it traps and then reacts with osmium tetroxide vapors. Dispose of contaminated Saran wrap in hazardous waste along with contaminated plastic ware. Use a Styrofoam container for an ice bucket to allow visualization of osmium contamination (osmium contamination will be black).
16. As the osmium reaction proceeds, the specimen will turn from brown to dark black. If it immediately turns black, this indicates insufficient removal of unreacted glutaraldehyde.
17. After treatment with osmium tetroxide, tissues will appear dark brown/black as the heavy metal osmium is deposited. If working with decellularized tissue or an acellular scaffold however, the specimen may appear only slightly darker in color.
18. After the osmium tetroxide fixation, cells appear dark and some surface topography and cellularity can be seen under a dissecting microscope. However, this can only safely be done after making sure the unreacted osmium tetroxide has been reduced to metallic osmium. Changing to a clean multi-well plate during ddH_2O washes facilitates this. Ethanol also reduces osmium tetroxide. Specimen viewing outside of a fume hood should not be done until after storage overnight in 70% ethanol or during the 95% ethanol wash. A piece of filter paper can be soaked with corn oil and then held over the sample; if the filter paper turns brown or black then volatile osmium tetroxide is still present.

19. SEM samples need to be dry to avoid outgassing of the specimen in the high vacuum conditions required for the microscope. To preserve ultrastructure, the specimen needs to be dried in a way that avoids large surface tension forces that can destroy soft biological tissues. This can be done by CPD. The critical point of a fluid is the temperature and pressure combination where the density of the liquid and vapor phases are equal. This allows liquid in the specimen to transition to vapor without excessive surface tension forces. Liquid CO_2 is typically used in CPD since its critical point temperature (31°C) and pressure (1,072 psi or 7,391 kPa) are nondestructive to most biological tissues and both can be readily achieved in a laboratory setting. However, liquid CO_2 is not very miscible with water, but is with ethanol. Therefore specimens are dehydrated in ethanol prior to CPD. For more discussion on CPD, see References (5, 6) and www.emsdiasum.com/microscopy/technical/datasheet/critical_drying.aspx.
20. Use two transfer pipettes and the left-hand/right-hand change of solution technique with plate tilted (supported by its lid or bottle cap) such that the sample stays submerged in liquid as described in step 5, subheading 3.1. Once specimens are in 70% ethanol, surface drying happens readily, so be very careful to always keep specimens immersed in liquid.
21. Ethanol solutions should be discarded as osmium waste, as small amounts of osmium metal will continue to be formed during dehydrations.
22. All the steps up to 70% ethanol need to be sequentially completed without specimen storage. If the specimens cannot be processed through CPD, they can be stored in 70% ethanol overnight at 4°C. Wrap the multi-well plate with parafilm; the next day, let the plate and contents warm to room temperature before removing the parafilm, doing another 70% ethanol wash and continuing the dehydration series.
23. After overnight storage in 70% ethanol, there likely will be black osmium deposits between the wells, and in the ethanol solution. This is especially true if the dehydrations are done in the same multi-well plate as used for the post-fixation. This is osmium tetroxide that has been reduced to metallic osmium.
24. This is the last point the specimens may be trimmed to appropriate size for your microscope holders. Cutting after CPD will yield extensive artifacts.
25. The CPD holders may contain numbered slots, allowing you to keep track of specimens. Alternatively, you can write specimen information on a small piece of white paper, using a regular leaded pencil, and place this in along with the specimens. The paper will not harm your sample.

26. Biological samples are generally good insulators; when subjected to the electron beam they can build up a charge that interferes with the scanning beam and/or signal collection. This results in distortions in the acquired image. Charging is decreased by increasing the conductivity of the sample, as, for example, when metallic coating is applied. There are also microscope settings which can mitigate charging. Consult your EM facility for recommended coating conditions and microscope settings.
27. For handling CPD specimens and mounting stubs, do not use your hands as latent skin oils will contaminate the specimen. Always use forceps. Gloves are not recommended as they tend to increase the likelihood of troublesome static charges.
28. If using a specimen holder with a pin at the bottom (stub), cardboard or plastic stub holder boxes may be purchased for holding and storing mounted specimens. Consult your EM Facility. They may also have a holder for use during specimen mounting.
29. If using a brass or aluminum chip specimen holder, storage can be as follows: take a post-it note, cut away non-sticky portion, and place regular double-sided tape on back of the sticky side of the post-it note. Cut to fit into a Petri dish, place double-stick tape side down in the Petri dish and the sticky side of post-it note will now be available to adhere chips such that they are secure but readily removable with a forceps.
30. After CPD, specimens may be full of static and jump around and adhere to surfaces. Use a static gun if available (e.g., Zerostat). Also, it is easy to distort a specimen with forceps, and it can become airborne when handling. You can use a surgical mask to avoid blowing on the specimens and thereby losing them.
31. After CPD, tissues will be spongy, black or dark gray, and noticeably shrunk, sometimes distorted. Do not try to straighten them out as this will cause tearing and creasing artifacts.
32. You may find it useful to draw a map of the mounted specimens with notes on their appearance or areas of interest as seen under the dissecting microscope to aid in the SEM viewing.
33. CPD specimens can be stored in a vacuum desiccator with silica gel if needed prior to sputter coating. This is not recommended, but can be done if needed: bring the vacuum down as low as possible and have ample silica gel present to keep the specimen dry.

References

1. Lichtenberg A, Cebotari S, Tudorache I, Sturz G, Winterhalter M, Hilfiker A, Haverich A (2006) Flow-dependant re-endothelialization of tissue-engineered heart valves. J Heart Valve Dis 15:287–293
2. van der Zijpp YJ, Poot AA, Feijen J (2003) Endothelialization of small diameter vascular prostheses. Arch Physiol Biochem 111:415–427
3. Ahmann KA, Johnson SL, Hebbel RP, Tranquillo RT (2011) Shear stress responses of adult blood outgrowth endothelial cells seeded on bioartificial tissue. Tissue Eng Part A 17:2511–2521
4. Ferrans VJ, Spray TL, Billingham ME, Roberts WC (1978) Structural changes in glutaraldehyde treated porcine heterografts used as substitute cardiac valves. Transmission and scanning electron microscopic observations in 12 patients. Am J Cardiol 41:1159–1184
5. Bozzola JJ, Russell LD (1999) Electron microscopy, 2nd edn. Jones and Bartlett, Sudbury, MA
6. Dykstra MJ (1992) Biological electron microscopy: theory techniques and troubleshooting. Plenum, New York
7. Reidy MA, Schwartz SM (1983) Endothelial injury and regeneration. IV. Endotoxin: a non-denuding injury to aortic endothelium. Lab Invest 48:25–34
8. Mohtai M, Yamamoto T (1987) Smooth muscle cell proliferation in the rat coronary artery induced by vitamin D. Atherosclerosis 63:193–202

Chapter 28

Genotypic and Phenotypic Analysis of In Vivo Tissue Regeneration in an Animal Model

Christopher W. Genheimer

Abstract

Determining the in vivo response to cellular therapies is important in evaluating the effectiveness of regenerative medicine therapies. Such treatment modalities leverage the treated individual's ability to elicit the body's innate healing response to repair/regenerate damaged tissues or organs. Detailed within this chapter is the process of evaluating the host tissue response to a candidate cell therapy through analysis of key transcript and protein targets.

Key words Tissue regeneration, Regenerative medicine, Tissue engineering, Stem cell, Progenitor cell, qRT-PCR, Western blot, Regenerative response induction, Assay development

1 Introduction

This chapter examines molecular characterization of the in vivo regenerative response in a rodent model of chronic kidney disease (CKD) as a specific example to illustrate methods that may be applied to any animal, organ system, or chronic disease where the intent is to study tissue regeneration. Regeneration of complex solid organs (e.g., kidney) involves the defined reconstitution of multiple specialized cell types organized within highly complex three-dimensional micro-architectures. The regenerative response of the kidney to acute injury is generally understood to be mediated by dedifferentiation of the resident tubular cell population with concomitant acquisition of a stem/progenitor cell phenotype, followed by proliferation and reacquisition of tubular characteristics (1). Numerous studies on the regeneration of renal architecture and function following acute kidney injury point to tubular epithelial cells as central in the restoration of function (2, 3). Tubular cells can be separated from a primary kidney cell isolate prepared from the medulla, cortex, and cortico-medullary junction compartments. Upon intra-renal administration to rodents with CKD-like physiology, such cells increase host survival and enhance

Joydeep Basu and John W. Ludlow (eds.), *Organ Regeneration: Methods and Protocols*, Methods in Molecular Biology, vol. 1001, DOI 10.1007/978-1-62703-363-3_28, © Springer Science+Business Media New York 2013

renal functionality (4–6). Such therapeutically bioactive cell populations may be valuable components of products developed to augment organ function in patients with CKD.

In the current methodology, we applied quantitative gene and protein expression analyses of several known stem cell markers, including SOX2, CD24, CD133, UTF1, NODAL, and LEFTY1 (7–12), to characterize the in vivo regenerative response in explanted kidneys of 5/6-nephrectomized (N_x) rats at 1, 12, and 24 weeks post treatment. We have identified the pluripotency factor SOX2 as a robust transcriptional marker induced within 12 weeks of renal cell treatment with significant up-regulation through 24 weeks post treatment. Furthermore, we have shown differential protein expression profiles between untreated and renal cell-treated rats, with the most significant up-regulation of regeneration-associated proteins at 12 weeks post treatment. Based on these data, we have developed the regenerative response index (RRI) to quantitatively evaluate the acquisition of a pluripotent state associated with stem/progenitor cell phenotype during the induction of a regenerative response within the kidneys of rats with CKD. These data are consistent with a relative increase in the proportion of kidney cells acquiring a stem/progenitor cell phenotype, bringing forward tubular cell dedifferentiation and/or stem cell recruitment as one possible mechanism of action underlying the therapeutic effects of renal cell therapies. By applying our findings more broadly, molecular assays that incorporate the assessment of SOX2 and RRI may be used to monitor the in vivo response to any candidate cellular therapy.

2 Materials

Prepare all solutions using ultrapure water and analytical grade reagents. All reagents should be stored at room temperature unless indicated otherwise.

2.1 Buffers and Reagents

1. Ultrapure sterile water.
2. Liquid nitrogen.
3. Bradford reagent.
4. MES running buffer (pH 7.3): 50 mM MES, 50 mM Tris base, 0.1% SDS, 1 mM EDTA.
5. Blocking buffer: 50 ml TBS-T, 4% w/v low-fat powdered milk.
6. ECL Advance chemiluminescent reagent.
7. TaqMan® Primer/Probes (see Table 1).
8. TaqMan® Gene Expression Master Mix.

Table 1
qRT-PCR primer/probes

Rat TaqMan primers for progenitor cell markers		
Gene	**Abbreviation**	**TaqMan Cat #**
POU class 5 homeobox 1	POU5F1/Oct4A	Rn01532129_gl
Nanog	Nanog	Rn01462825_ml
RNA exonuclease 1	Rex1	Rn01408442_gl
SRY (sex determining region Y)-box 2	Sox2	Rn01286286_gl
v-myc myelocytomatosis viral oncogene	c-Myc	Rn00561507_ml
Musashi	MSI1	Rn00596059_ml
Podocalyxin	PODXL	Rn00593804_ml
Telomerase reverse transcriptase	TERT	Rn01409452_gl
GATA binding protein 4	GATA4	Rn00595169_ml
Undifferentiated embryonic cell transcription factor 1	UTF1	Rn01498190_gl
Nodal homolog from mouse	NODAL	Rn01433623_ml
Snail homolog 2 from Drosophila	SNAI2	Rn00709370_ml
Left–right determination factor 1	LEFTY1	Rn01412531_gl
v-Kit Hardy-Zuckerman 4 Feline sarcoma viral oncogene	KIT/CD117	Rn00573942_ml
Melanoma cell adhesion molecule	MCAM/CD146	Rn00576900_ml
Prominin 1	PROM1/CD133	Rn00572720_ml
Nerve growth factor receptor	NGFR/CD271	Rn00561634_ml
CD24	CD24	Rn00562598_ml
Cadherin-11	CDH11	Rn01536921_gl
Retinoic acid receptor alpha	RARA	Rn00580551_ml
Peptidylprolyl isomerase B	PPIB	Rn00574762_ml

9. Primary and secondary antibodies (see Table 2).
10. NuPAGE® Novex 10% Bis-Tris SDS-PAGE Gels.
11. Phosphate-buffered saline (PBS): 3 mM potassium chloride, 1.5 mM potassium phosphate monobasic, 138 mM sodium chloride, 8 mM sodium phosphate dibasic.
12. EBC lysis buffer: 50 mM Tris base (pH 8.0), 120 mM sodium chloride, 0.5% NP40, and 1 protease inhibitor tablet/10 ml of sterile water.

Table 2
Primary and secondary antibodies

Primary abs				
Vendor	**Cat #**	**Antibody description**	**[Ab] μg/ml**	**MW (kDa)**
R&D Systems	MAB7461	Mouse anti-human lefty-A long and short isoforms	1	40
Abcam	AB19898	Rabbit anti-human, mouse, and rat CD133	1	110
Millipore	MAB4337	Mouse anti-human and rat UTF1	1	36
Abcam	AB55676	Mouse anti-human NODAL	1	40
Cell signaling	2748	Rabbit anti-human, mouse, and rat SOX2	1	35
Becton Dickinson	551133	Mouse anti-rat CD24	1	78

Secondary abs			
Vendor	**Cat #**	**Antibody description**	**Ab dilution**
Vector Labs	PI-2000	Peroxidase Horse Anti-Mouse IgG Antibody	1:60,000
Vector Labs	PI-1000	Peroxidase Goat Anti-Rabbit IgG Antibody	1:60,000

13. Tris-buffered saline with Tween-20 (TBS-T): 137 mM sodium chloride, 3 mM potassium chloride, 25 mM Tris base (pH 7.4), and 0.1% Tween-20.

2.2 Equipment

1. Mortar and pestle.
2. Microfuge tubes.
3. Liquid nitrogen dewar.
4. iBlot system (Invitrogen).
5. Microcentrifuge.
6. Spectrophotometer.
7. XCell SureLock® Mini-Cell (Invitrogen).
8. ABI-Prism 7300 Real-Time PCR System.
9. ChemiDoc™ XRS Molecular Imager and Quantity One® software.
10. Image J v1.4 software.

2.3 Commercial Kits

1. RNeasy Mini Plus Kit (Qiagen).
2. SuperScript® VILO™ cDNA Synthesis Kit (Qiagen).
3. iBlot Gel Transfer Stacks, Nitrocellulose (Invitrogen).

3 Methods

Carry out all procedures at room temperature unless specified otherwise.

3.1 RNA Isolation and cDNA Synthesis

1. Obtain sufficient liquid nitrogen to fully immerse tissue to be analyzed.
2. Place tissue to be analyzed into mortar and immerse with liquid nitrogen (see Note 1).
3. Pulverize tissue using pestle until only a fine powder remains and aliquot into microfuge tubes. Store in multiple tubes at −80°C to allow for multiple molecular assays to be performed as needed.
4. Extract total RNA from 30 mg of tissue using the RNeasy Plus Mini Kit (or equivalent) following the manufacturer's instructions and store excess RNA at −80°C.
5. Determine RNA concentration spectrophotometrically.
6. Determine the volume of RNA needed for 1.4 μg based on RNA concentration established in step 5.
7. Generate cDNA using the volume of RNA calculated in step 4 and the SuperScript® VILO™ cDNA Synthesis Kit (or equivalent) following the manufacturer's instructions.
8. Following cDNA synthesis, dilute each sample 1:6 by adding 200 μl of diH_2O to bring the final volume to 240 μl and store at −20 or −80°C (see Note 2).

3.2 qRT-PCR

1. Thaw cDNA and TaqMan® Primer/Probes (Table 1) at room temperature.
2. Prepare sufficient reaction master mix for the number of samples being analyzed by adding 10 μl TaqMan® Gene Expression Master Mix (2×) and 1 μl TaqMan® Primer/Probe (20×) for each reaction (see Note 3).
3. Aliquot 11 μl of master mix into the appropriate wells of a 96-well PCR plate (a 386-well plate may be used if your thermal cycler is compatible). See Fig. 1 for a diagram of a typical plate layout.
4. Add 9 μl of diluted cDNA from step 8 (Subheading 3.1) to the appropriate wells of the 96-well plate (the total reaction volume should be 20 μl).

	Test Genes										Endogenous Control	
	CD24		CD133		NODAL		LEFTY1		SOX2		PPIB	
	1	2	3	4	5	6	7	8	9	10	11	12
A	Sample #1	Sample #9	Sample #1	Sample #9	Sample #1	Sample #9	Sample #1	Sample #9	Sample #1	Sample #9	Sample #1	Sample #9
B	Sample #2	Sample #10	Sample #2	Sample #10	Sample #2	Sample #10	Sample #2	Sample #10	Sample #2	Sample #10	Sample #2	Sample #10
C	Sample #3	Sample #11	Sample #3	Sample #11	Sample #3	Sample #11	Sample #3	Sample #11	Sample #3	Sample #11	Sample #3	Sample #11
D	Sample #4	Sample #12	Sample #4	Sample #12	Sample #4	Sample #12	Sample #4	Sample #12	Sample #4	Sample #12	Sample #4	Sample #12
E	Sample #5	Sample #13	Sample #5	Sample #13	Sample #5	Sample #13	Sample #5	Sample #13	Sample #5	Sample #13	Sample #5	Sample #13
F	Sample #6	Sample #14	Sample #6	Sample #14	Sample #6	Sample #14	Sample #6	Sample #14	Sample #6	Sample #14	Sample #6	Sample #14
G	Sample #7	NTC	Sample #7	NTC	Sample #7	NTC	Sample #7	NTC	Sample #7	NTC	Sample #7	NTC
H	Sample #8	Control	Sample #8	Control	Sample #8	Control	Sample #8	Control	Sample #8	Control	Sample #8	Control

Fig. 1 qRT-PCR plate layout. An example of a 96-well plate showing analysis of five test genes (CD24, CD133, NODAL, LEFTY, SOX2) compared to an endogenous control (PPIB). NTC = no template control (i.e., negative control)

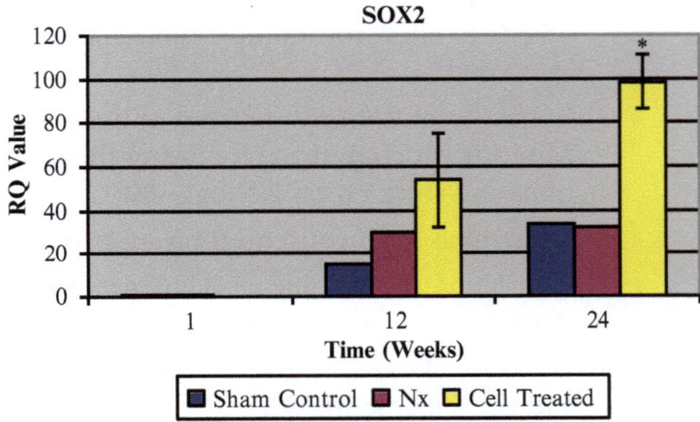

Fig. 2 SOX2 is a transcriptional marker of regenerative response induction. Time course for transcription of SOX2, a marker associated with maintenance of the pluripotent state in sham control, 5/6 nephrectomized kidney (N_x), 5/6 nephrectomized kidney with cell-based treatment. *Bar* indicates standard error; *asterisk* (*) indicates significance at *p*-value <0.05

5. Briefly centrifuge the plate to pellet the reagents (see Note 4).
6. Load the 96-well plate into an ABI-Prism 7300 Real-Time PCR System (or equivalent) and follow manufacturer's instructions for running a plate.
7. After completion of the run, perform data analysis using the comparative Ct method (see Note 5). See Fig. 2 for an example of typical data analysis.

3.3 Protein Extraction

1. Thaw an aliquot of pulverized tissue from step 3 (Subheading 3.1) and add 250 μl of EBC lysis buffer.
2. Allow the sample to lyse for 15 min at room temperature with rocking (see Note 6).

3. Centrifuge the sample for 10 min at 13,000 RPM (microcentrifuge) in order to pellet the cellular debris and collect the supernatant.
4. Determine the protein concentration in each supernatant by Bradford assay.

3.4 SDS-PAGE and Western Blot

1. Using 30 μg of protein, prepare the samples for SDS-PAGE by adding 7.5 μl of NuPAGE LDS Sample Buffer (4×) and 3 μl of 2-mercaptoethanol (10×), and bring up to 30 μl with diH_2O.
2. Vortex the samples gently and place at 95°C for 5 min to fully denature the proteins.
3. Load the entire 30 μl of sample into a well of NuPAGE® Novex 10% Bis-Tris Gels. Be sure to leave at least one lane available for molecular weight markers (see Note 7).
4. Fill the XCell SureLock® Mini-Cell with 800 ml of MES running buffer and electrophorese for 40 min at 200 V. Stop the electrophoresis once the tracking dye has reached the bottom of the gel.
5. Remove the gel from the SureLock® Mini-Cell and pry the plates open using a spatula (see Note 8).
6. Transfer the gel to a nitrocellulose membrane using the iBlot Gel Transfer Stacks and iBlot system following the manufacturer's instructions. Using this system the transfer process only takes 7.5 min. While the transfer is taking place, prepare blocking reagent by adding 4% w/v low-fat milk to TBS-T. Excess blocker can be made and stored at 4°C for later use.
7. Upon completion of the transfer, discard the gel and filter paper and place the nitrocellulose membrane in a clean container. Block the membrane with 15 ml of blocker and incubate at room temperature for 2 h with rocking (see Note 9).
8. Prepare the primary antibody solutions by adding 2% w/v low-fat milk to TBS-T and diluting the primary antibodies as shown in Table 2. Discard the blocker and add the primary antibody. Incubate overnight at room temperature with rocking. Discard the primary antibody and wash the membrane three times/10 min each with TBS-T.
9. Prepare the secondary antibody solutions the same way as in step 8 using the dilutions shown in Table 2. Incubate at room temperature for 1.5 h with rocking.
10. Discard the secondary antibody solution and wash the membrane three times/10 min each in TBS-T followed by two 10 min washes in diH_2O. The final two washes are to ensure the residual detergent is removed prior to developing.

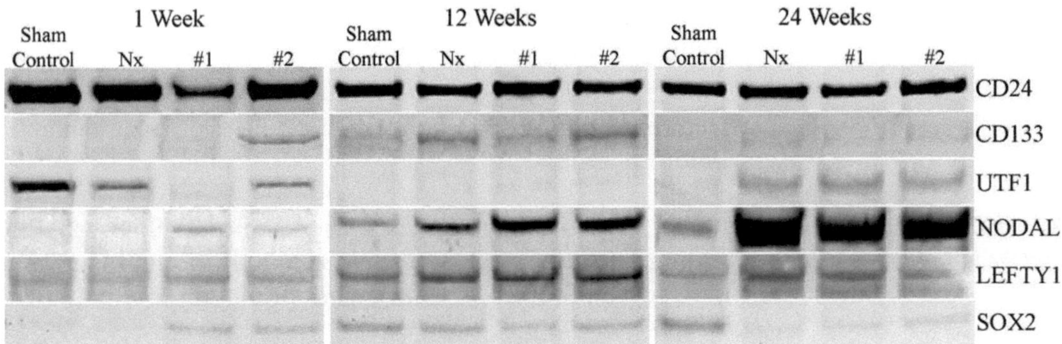

Fig. 3 Western blot analysis of stem markers CD24, CD133, UTF1, NODAL, LEFTY1, and SOX2, in sham control, 5/6 nephrectomized kidney (N_x), and cell-based treatment rats (#1 and #2) at 1, 12, and 24 weeks post treatment. Lanes were normalized by total mass of protein loaded

11. Discard the last water wash and place the membrane on a piece of saran wrap (the saran wrap should be at least 4 times larger than the membrane). Prepare the ECL Advance chemiluminescent reagent by mixing parts A and B 1:1 and apply enough reagent to cover the membrane. Allow the membrane to incubate for 5 min at room temperature. Discard the chemiluminescent reagent by grasping a corner of the membrane with forceps and gently picking upward while soaking up the excess reagent into a kimwipe. Place the membrane back onto the saran wrap and fold over a piece to cover the membrane. Photograph the membrane using a chemiluminescent reader such as the ChemiDoc™ XRS molecular imager with Quantity One® software (see Fig. 3).

3.5 Regenerative Response Induction Calculation

1. Use Image J v1.4 software for quantitation of the western blot data to determine the regenerative response (see Note 10).
2. Load a picture of the western blot into Image J and invert the image to ensure all colors are in gray scale. Next, create a rectangular box that is large enough to fully encompass the band of the target protein and measure the band intensity. Repeat this process for each sample, reusing the same box to ensure the area analyzed remains constant. Once all samples have been analyzed export the data into Excel for analysis.
3. Determine the intensity per unit area by dividing the measured band intensity by the area that was measured. Repeat this process for each protein marker that was analyzed.
4. Now quantitative comparisons can be made between samples across multiple proteins. Statistical analysis can also be done to determine significance (see Table 3 and Fig. 4).

Table 3
Regenerative response index (RRI)

Sample	Detector	1 week	12 week	24 week
Sham control	CD24	173.32	131.10	156.88
	CD133	29.91	53.85	44.65
	UTF1	141.78	26.10	31.34
	NODAL	40.52	51.95	44.12
	LEFTY1	57.57	73.32	44.67
	SOX2	16.34	23.91	31.74
	RRI	90.47	85.46	81.89

Sample	Detector	1 week	12 week	18 week
Nx	CD24	159.67	109.93	179.25
	CD133	34.62	61.96	49.33
	UTF1	62.37	26.34	50.09
	NODAL	37.11	94.47	166.03
	LEFTY1	50.00	89.21	76.41
	SOX2	15.04	21.23	19.98
	RRI	82.26	97.87	140.56

Sample	Detector	1 week	12 week	24 week
Cell treated	CD24	129.82	125.41	170.10
	CD133	43.25	64.17	41.59
	UTF1	39.68	27.07	48.24
	NODAL	35.17	152.00	135.69
	LEFTY1	53.14	110.29	57.27
	SOX2	18.20	19.27	19.95
	RRI	62.89	135.61	112.61

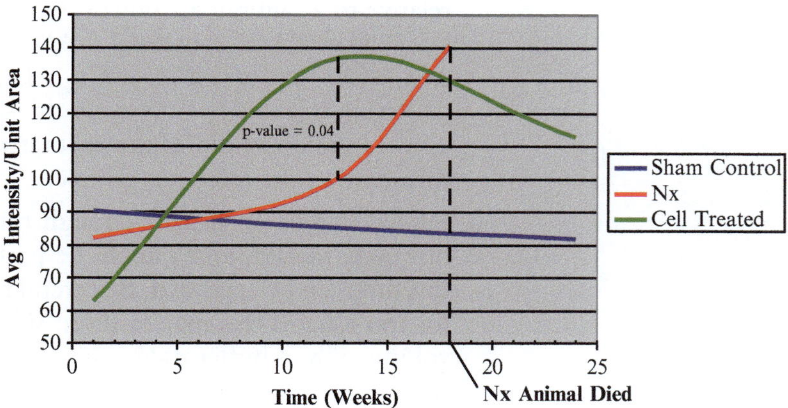

Fig. 4 Time course for acquisition of a stem cell phenotype during regenerative response induction. Time course of regenerative response index (RRI) over 24 weeks in sham control, 5/6 nephrectomized kidney (N_x), and 5/6 nephrectomized kidney with cell-based treatment (average of cell-based treatment #1 and #2)

4 Notes

1. Be sure to handle the liquid nitrogen with proper safety equipment. Use a ladle or metal cup to pour the liquid nitrogen into the mortar adding enough to just cover the piece of tissue. It may be necessary to add liquid nitrogen several times as it will evaporate quickly in the mortar.
2. This dilution can vary depending on the type of analysis being conducted. We have found that the 1:6 dilution gives sufficient sample to complete 26 reactions; however, should more reactions be required, a 1:10 or 1:15 dilution can be used without any signal loss.
3. If you are running ten samples for each primer/probe, then you will need 100 μl master mix (2×) and 10 μl of primer/probe (20×). Mix these reagents together in a separate 0.5 ml microfuge tube and then aliquot 11 μl into the appropriate wells of the reaction plate. Be sure at least one of your primer probes is an endogenous control. It is perfectly acceptable to use more than one endogenous control per plate. All of our experiments were carried out using peptidylprolyl isomerase B (PPIB) as the endogenous control.
4. Centrifuge the plate to ensure all the reagents are together at the bottom of the well. Failing to do this can result in false negative or unusual amplification curves. It is not necessary to mix the reagents in the well as mixing will occur naturally during the amplification cycle.
5. The comparative Ct method is a mathematical calculation to determine relative quantity of a gene target. In this calculation, the gene target is normalized to an endogenous reference and relative to a calibrator (control) sample. The equation is given by $RQ=2^{-\Delta\Delta Ct}$ where $\Delta\Delta Ct=\Delta Ct_T-\Delta Ct_{Cal}$, where $\Delta Ct_T=Ct_T-Ct_R$ and $\Delta Ct_{Cal}=Ct_{Cal}-Ct_R$. RQ = relative quantity, Ct = cycle threshold, Ct_T = cycle threshold of the test sample, Ct_{Cal} = cycle threshold of the calibrator sample, and Ct_R = cycle threshold of the reference sample.
6. Fifteen minutes is usually enough time to lyse most samples; however, should your sample be particularly difficult to lyse, this time can be extended. Be careful not to allow lysis to occur for too long as this may result in significant protein loss. The volume of lysis buffer used may also be adjusted depending on the sample and will have a direct effect on the final protein concentration. Therefore, if you need really concentrated protein, use 50–100 μl of lysis buffer.
7. We routinely use the Novex® Sharp and MagicMark™ molecular weight markers from Invitrogen. This allows us to have a

visible ladder (Novex® Sharp), to monitor electrophoresis and visualize western transfer, and a chemiluminescent ladder (MagicMark™), for sizing our western blots. These molecular weight markers can be mixed together without interfering with one another, in the event only one ladder well is available. Also, when using the MagicMark™ chemiluminescent ladder, do not load more than 2.5 μl as this will result in overexposure of the ladder during the blot developing step.

8. Place the plate and gel into a container of sterile water to help ease the gel off of the plate without tearing. This also makes handling the gel easier when assembling the transfer.
9. The membrane may block overnight at 4°C if needed.
10. Image J v1.4 is a freeware image analysis program that is available from the NIH at the following web address: http://rsbweb.nih.gov/ij/download.html.

References

1. Bonventre JV (2003) Dedifferentiation and proliferation of surviving epithelial cells in acute renal failure. J Am Soc Nephrol 14:S55–S61
2. Guo JK, Cantley LG (2010) Cellular maintenance and repair of the kidney. Ann Rev Physiol 72:357–376
3. Lin F, Moran A, Igarashi P (2005) Intrarenal cells, not bone marrow-derived cells, are the major source for regeneration in postischemic kidney. J Clin Invest 115:1756–1764
4. Kelley R et al (2010a) Bioactive renal cells augment kidney function in a rodent model of chronic kidney disease. International Society for Cell Therapy conference. www.tengion.com/news/documents/Kelley%202010%20ISCT%20podium%20FINAL.pdf
5. Kelley R et al (2010) A tubular cell-enriched subpopulation of primary renal cells improves survival and augments kidney function in a rodent model of chronic kidney disease. Am J Physiol Renal Physiol 299:F1026–F1039
6. Presnell SC et al (2010) Isolation, characterization, and expansion (ICE) methods for defined primary renal cell populations from rodent, canine, and human normal and diseased kidneys. Tissue Eng Part C Methods 17:261–273
7. Sagrinati C et al (2006) Isolation and characterization of multipotent progenitor cells from the Bowman's capsule of adult human kidneys. J Am Soc Nephrol 17:2443–2456
8. Sagrinati C et al (2008) Stem-cell approaches for kidney repair: choosing the right cells. Trends Mol Med 14:277–285
9. van den Boom V et al (2007) UTF1 is a chromatin-associated protein involved in ES cell differentiation. J Cell Biol 178:913–924
10. Chambers I, Tomlinson SR (2009) The transcriptional foundation of pluripotency. Development 136:2311–2322
11. Shen MM (2007) Nodal signaling: developmental roles and regulation. Development 134:1023–1034
12. Tabibzadeh S, Hemmati-Brivanlou A (2006) Lefty at the crossroads of "stemness" and differentiative events. Stem Cells 24:1998–2006

Chapter 29

Histological Evaluation of Tissue Regeneration Using Biodegradable Scaffold Seeded by Autologous Cells for Tubular/Hollow Organ Applications

Elias A. Rivera and Manuel J. Jayo

Abstract

The accurate interpretation of histological outcomes is a critical endpoint in preclinical studies. Thus, the toxicologic pathologist plays a vital role in conducting a comprehensive microscopic evaluation that would ultimately help defining the safety and functionality in Tissue Engineering/Regenerative Medicinal (TERM) products. In spite of many advances in regenerative medicine, there are no specific guidelines for the histological assessment of TERM products (Jayo et al. Toxicol Pathol 36:92–96, 2008). In this chapter, we describe the methodology designed to facilitate the detection of structural and functional changes when conducting a histological assessment including tissue collection (test article extraction), sampling, processing and fixation, special stains, statistical analysis, and morphometry.

Key words Tissue regeneration, Histology, Functionality, Safety assessment, Tubular organs, Special stains, Morphometry

1 Introduction

Tissue Engineering/Regenerative Medicinal (TERM) products constitute a rapidly emerging technology, whose relevance to clinical application depends on preclinical testing and a strong scientific rationale and understanding of basic tissue and cellular responses. The key to success in generating a sound histopathology report relies on the proper interpretation and characterization of tissue/cellular reparative regeneration versus reparative healing response (1, 2). In compiling this chapter, we utilized previous GLP and non-GLP study materials and learnings, thus focusing our attention to the methodology applied in order to capture the macroscopic and microscopic changes and adequately correlate their final outcome to tissue regeneration. The primary objective was to provide the study pathologist with a complete guide designed to facilitate the detection of structural and functional changes when

Joydeep Basu and John W. Ludlow (eds.), *Organ Regeneration: Methods and Protocols*, Methods in Molecular Biology, vol. 1001, DOI 10.1007/978-1-62703-363-3_29,

conducting a histological assessment to determine the safety and functionality of a TERM product in order to meet regulatory requirements. Our discussion will encompass a description of tissue collection, sampling, tissue processing, macroscopic and microscopic assessment, histology parameters, scoring and grading scheme, immunohistochemical staining techniques, morphology, and statistical analysis. All the preclinical GLP study materials and methods utilized in this chapter were conducted in compliance with 21 CFR (Code of Federal Regulation), Part 58 Good Laboratory Practice (GLP).

2 Materials and Methods

2.1 Study Design

A tubular scaffold is comprised of biodegradable polymers; polyglycolic acid (PGA) felts and poly (lactic-co-glycolic acid) polymers (PLGA) seeded with autologous smooth muscle cells (SMC) were utilized to form a neo-urinary conduit (NUC) construct or Neo-Bladder Augment (NBA). Multiple studies were conducted ranging from 3-month to 6-month duration, included total cystectomized Yorkshire swine and partial (trigone-sparing) cystectomized canine (mongrel) models, respectively. The swine were implanted with the NUC, and the canines were implanted with the NBA. Regardless of the surgical procedure, the peritoneum was wrapped around the construct (implant) to provide a vascular source and watertightness. In the NUC, the ureters were reconnected to the proximal end to channel urine to an outflow skin stoma at the distal end. In the NBA, the ureters were left intact and the implant was anastomosed at the margins of remaining bladder tissue (the trigone).

2.2 Tissue Collection

In vivo data collection and analysis such as adequately collecting necessary tissues, physical history, and functional chemistries must be conducted by a trained professional. At necropsy, the pathologist should perform visual inspection of the test article in situ and record a concise description of gross findings, and properly document the findings by gross photography. Specimen identification should contain standard calibrated label showing the study number, animal number, and date. Extraction/explantation of test article requires careful consideration to avoid unintentional lacerations/cuts that may compromise the specimen sampling. Thus, adhesions and fibrous tissue surrounding the test article should be gently removed by blunt dissection. Ideally, the specimen should be gently rinsed with a saline solution to remove excess blood or debris prior to submerging into fixative. At the minimum, the following tissues should be collected for histological evaluation: kidneys, ureters, implant, and lymph nodes (lumbar and mesenteric). However, evaluation of the kidneys, ureters, implant, stoma, and thoracic, abdominal, and pelvic cavities and their organs and tissues should be performed. If any gross lesions, adhesions, and/or organ changes are present, they should be evaluated, photographed, and collected for histological assessment. Furthermore, it is recommended that all major organs/tissues such as heart, lungs, liver, brain (including pituitary), thyroid and parathyroids, and bone (sternum) be weighed and collected/saved for potential future microscopic analysis.

2.3 Tissue Fixation

Prior to selecting the type of fixative, it is important to consider the staining technique to be tested, especially the sensitivity of the antigen for immunohistochemistry protocols. In our study, a 10% neutral-buffered formalin solution was used, as it is considered the best general fixative for specimens because it preserves the widest range of structures and requires a relatively short incubation time (3). As a general rule, 24–48 h is required for optimal preservation of the specimen at a tissue to fixative ratio of 1:10. Fixation can be accomplished by immersion or perfusion. In tubular or hollow organs, perfusion fixation is preferable since it would reduce shrinking, keep anatomical orientation, and prevent collapsing of the lumen that would otherwise potentially compromise adequate representation (sampling) of the specimen. In addition, one may consider securing the specimen to a paraffin board during the fixation process, as it would significantly reduce fixation artifacts pertaining to the anatomical orientation of the explant.

2.4 Tissue Sampling

Regardless of the paradigm of implant application, whether the treatment involves bladder augmentation or total replacement using Neo-Bladder Augment, Neo-Bladder Replacement, or Neo-Urinary Conduit technologies, it would be appropriate to collect samples of native tissue interface proximal and distal of the test article. In addition, collection of draining lymphatics particular to the anatomy, as well as tissue/organs that drain or function above and below the evaluation site may need to be collected and examined. In some instances, gross and sub-gross photography driven anatomical histological sections are collected. Protocol-driven collection of fresh, frozen, and fixed tissues is highly recommended. Necropsy personnel (prosector, technician, or pathologist) should be conscious when handling the extracted specimen (test article) and be gentle when scraping or washing any surface “debris” because it may be integral to the regenerative process and or represent biointegration of scaffold. In the NUC studies, depending on the length of the test article, it was bisected transversely into two pieces (proximal and distal), and then further bisected longitudinally (parallel with the urine outflow), close to the medial plane to yield two pieces (left and right sides), as shown in Fig. 1. Typically, 2–3 sections were chosen and embedded in macro-cassettes, labeled Prox., Mid., and Distal. Additional sections were obtained from the

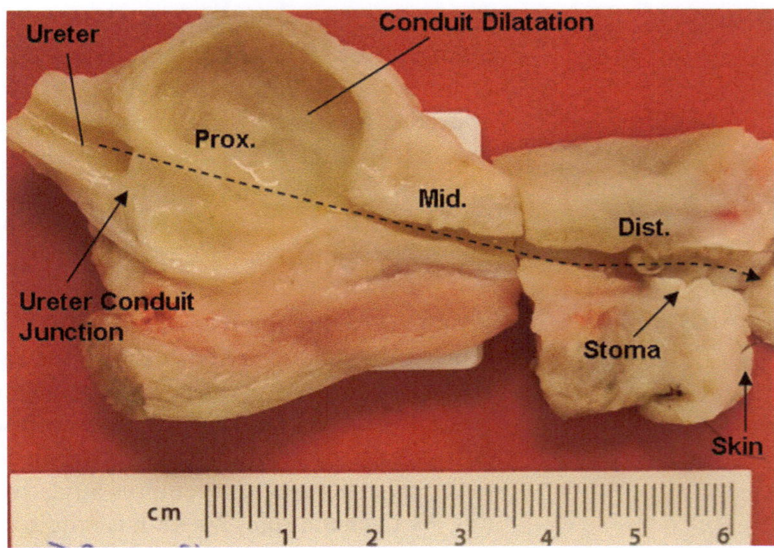

Fig. 1 Macrophotograph of trimmed neo-urinary conduit, longitudinally bisected showing the ureter–conduit junction and the various sampling sites (proximal, middle, and distal). Urine flows from the ureters through conduit's body and out the stoma (indicated by the *blue-dashed arrow*)

left and right ureter conduit junction, left and right ureters (mid. level), left and right kidney, and lymph node (lumbar and mesenteric). In the NBA studies, the complete bladder was removed (trigone, anastomotic site, and neo-bladder), the ureters were ligated, and the urethra was appropriately catheterized to allow it to be fixed under pressure with 10% neutral-buffered formalin (NBF). After fixation, the bladder was cut in half along a cranial/caudal line creating four (4) quadrants of bladder tissue. A total of 12 sections (3 from each quadrant) were collected for histological evaluation. When the surgical interface was apparent, at least two sections were collected from each quadrant to capture the native bladder and neo-bladder interface. In addition, samples of organs described above were collected and saved for future evaluation.

2.5 Histology Processing

Traditional routine hematoxylin and eosin (H&E) and any staining procedure that is related to healing or unwanted repair such as Masson's trichrome for fibrous tissue and Periodic acid-Schiff (PAS) for mucin should be considered at the minimum. However, depending on the organ and the location harvested, cells may produce variable enzymes that can be detected by non-processed fresh or frozen enzymatic protocols. Consider antibodies for inflammatory cells if modulation of inflammatory cells is part of your mechanism of action (MOA) and immunohistochemical staining to characterize tissue layer differentiation. Otherwise, limit evaluation to what is normal during embryo, fetal, and adult for the tissue being assessed.

2.6 Immunohistochemistry

Immunohistochemistry (IHC) staining is a method of detecting the presence of specific proteins in cells or tissues. Tissue samples such as paraffin-embedded tissue sections, frozen sections, and cytospin preparations are prepared in an appropriate manner for immunohistochemical staining. The following immunohistochemical staining was performed with antibodies directed at *cytokeratin AE1/AE3* (epithelium marker), *cytokeratin 7* (urothelium marker), and *calponin* (smooth muscle marker). The positive control tissue elements for calponin consisted of smooth muscle cells in 10% formalin-fixed, paraffin-embedded samples of normal porcine urinary bladder. Negative control tissue elements for calponin consisted of stroma in 10% formalin-fixed, paraffin-embedded samples of normal porcine urinary bladder. Positive control tissue elements for AE1/AE3 and cytokeratin 7 consisted of epithelial cells in 10% formalin-fixed, paraffin-embedded samples of normal porcine urinary bladder tissue.

2.7 Reagents and Solutions

The following reagents and solutions were used in the preparation of immunohistochemistry staining. All solutions were prepared using ultrapure (distilled) water. The name and date of preparation, expiration date, storage information, and preparer's initials were recorded on labeling.

Wash buffer (TBS with Tween 20), 30% hydrogen peroxide, serum-free protein block, proteinase K, antibody diluents, mouse antihuman AE1/AE3, mouse antihuman CK7, mouse antihuman calponin, rabbit anti mouse immunoglobulin, Envision + rabbit HRP, DAB + chromogen and substrate buffer, hematoxylin extra, 100% alcohol, glacial acidic acid, ammonium hydroxide, xylene, mounting medium, and negative control mouse IgG1.

2.7.1 Solutions

1. *95% alcohol*: 950 ml of 100% alcohol + 50 ml of distilled water. Store at room temperature.
2. *Wash buffer 10×*: 1000 ml + 9000 ml distilled water. Store at room temperature.
3. *3% hydrogen peroxide*: 1 ml of 30% hydrogen peroxide + 9 ml distilled water. Store at 4°C.
4. *Mouse antihuman cytokeratin AE1/AE3 antibody working solution*: Dilute the stock solution of the cytokeratin AE1/AE3 antibody in antibody diluent (see Note 1) to a dilution of 0.302 mg/l (1:400). Adjust volumes according to experimental trial. Store at 4°C.
5. *Mouse negative control IgG1 working solution*: Dilute the stock solution of the IgG1 negative control solution in antibody diluent to a dilution of .0.302 mg/l (1:331). Store at 4°C.
6. *Rabbit anti mouse immunoglobulin solution*: Dilute the stock solution of the rabbit anti mouse immunoglobulin in antibody diluent to a dilution of 1:400. Adjust volumes according to experimental trial. Store at 4°C.

7. *DAB+chromogen*: The solution is one drop of DAB + chromogen per 1 ml of DAB + substrate buffer. Amount is specified by the Dako Autostainer and dependent on amount of slides tested per trial. Store at 4°C.
8. *5% glacial acetic acid*: 25 ml of glacial acetic acid + 525 ml of distilled water. Store at room temperature.
9. *Ammonium water*: 2 ml of ammonium hydroxide + 548 ml of distilled water. Store at room temperature.

3 Immunohisto-chemistry Staining Protocols

3.1 Cytokeratin AE1/AE3 Protocol

Detection of cytokeratin AE1/AE3 on paraffin-embedded tissues can be performed using a mouse monoclonal IgG1 kappa antibody specific to cytokeratin AE1/AE3 structural epitopes along with a peroxidase enzyme-conjugated polymer labeling system. Cytokeratins are generalized as the most fundamental markers of epithelial differentiation. Antibody AE1 reacts with subfamily-A cytokeratins. Antibody AE3 reacts with subfamily-B cytokeratins.

3.1.1 Procedure

Room temperature: 18–26°C.

When performing the cytokeratin AE1/AE3 IHC stain, the maximum number of slides tested in a humidity chamber was 30. Within the 30 slides, there will be positive and negative controls. The negative controls are stained with a mouse negative control IgG1 isotype solution.

3.1.2 Manual Staining Procedure for Cytokeratin AE1/AE3

Incubation steps are performed in a humidity chamber; double rinses are performed in a Coplin jar.

1. Xylene (double rinse) 4–10 min per rinse.
2. 100% alcohol (double rinse) 1–5 min per rinse, 95% alcohol (double rinse), 1–5 min per rinse.
3. Tap water rinse (single rinse), 1–5 min.
4. Wash buffer rinse (double rinse), 3–5 min per rinse.
5. 3% hydrogen peroxide (single rinse), 5–10 min.
6. Wash buffer rinse (double rinse), 3–5 min per rinse.
7. Proteinase K (single rinse), 5 min.
8. Wash buffer (double rinse), 3–5 min per rinse.
9. Protein Block (single rinse), 5–10 min.
10. Drain slides.
11. Mouse antihuman AE1/AE3 (1:400), 30–35 min.
12. Wash buffer rinse (double rinse), 3–5 min per rinse.
13. Rabbit anti mouse immunoglobulin (1:400), 30–35 min.

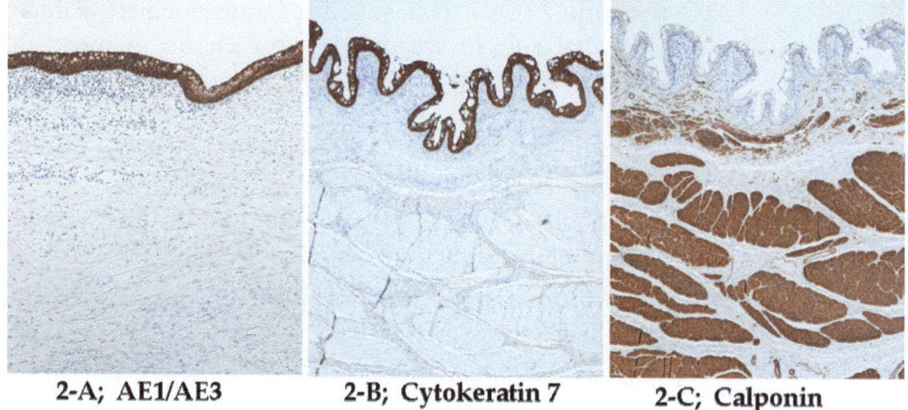

Fig. 2 (**a**–**c**) Acceptance staining criteria for immunohistochemistry staining of positive controls for epithelium (AE1/AE3), urothelium (CK7), and smooth muscle cells (Calponin), where the intermediate filaments stain *dark brown* and nuclei *blue*

14. Wash buffer rinse (double rinse), 3–5 min per rinse.
15. Envision + rabbit HRP, 30–35 min.
16. Wash buffer rinse (double rinse), 3–5 min per rinse.
17. DAB + chromogen, 5–10 min.
18. Distilled water (double rinse), 2 min per rinse.
19. Hematoxylin (dips), 4–5 dips.
20. Distilled water (single rinse), 1 min.
21. 1% glacial acetic acid water (single rinse), 30 s.
22. Distilled water (single rinse), 1 min.
23. 0.05% ammonium water (single rinse), 1–2 dips.
24. Distilled water (single rinse), 1 min.
25. 95% alcohol (single rinse), 1 min.
26. 100% alcohol (double rinse), 1 min per rinse.
27. Xylene (double rinse), 1 min per rinse.
28. Mount and coverslip.

3.1.3 Acceptance Criteria (see Fig. 2a)

AE1/AE3 positive intermediate filaments (staining dark brown). Nuclei (staining light blue).

False-positive results: Nonspecific or endogenous streptavidin reacting with the chromogen substrate.

3.2 Cytokeratin 7 Protocol

Detection of cytokeratin 7 on paraffin-embedded tissues can be performed using a mouse monoclonal IgG1 kappa antibody specific to cytokeratin structural epitopes along with a peroxidase enzyme-conjugated polymer labeling system. Cytokeratins are generalized as the most fundamental markers of epithelial differentiation.

Cytokeratin 7 (CK7) is classified as an intermediate filament, which plays a vital role in creating a cytoskeleton in most eukaryotic organisms (4).

3.2.1 Solutions

1. *95% alcohol*: 950 ml of 100% alcohol + 50 ml of distilled water. Store at room temperature.
2. *Wash buffer 10×*: 1,000 ml + 9,000 ml distilled water. Store at room temperature.
3. *3% hydrogen peroxide*: 1 ml of 30% hydrogen peroxide + 9 ml distilled water. Store at 4°C.
4. *Mouse antihuman cytokeratin 7 antibody working solution*: Dilute the stock solution of the cytokeratin 7 antibody in antibody diluent (Dako) to a dilution of 0.5825 mg/l (1:800). Adjust volumes according to experimental trial. Store at 4°C.
5. *Mouse negative control IgG1 working solution*: Dilute the stock solution of the IgG1 negative control solution in antibody diluent to a dilution of 0.5825 mg/l (1:171). Store at 4°C.
6. *Rabbit anti mouse immunoglobulin solution*: Dilute the stock solution of the rabbit anti mouse immunoglobulin in antibody diluent to a dilution of 1:200. Adjust volumes according to experimental trial. Store at 4°C.
7. *DAB + chromogen*: The solution is one drop of DAB + chromogen per 1 ml of DAB + substrate buffer. Amount is specified by the Dako Autostainer and dependent on amount of slides tested per trial. Store at 4°C.
8. *5% glacial acetic acid*: 25 ml of glacial acetic acid + 525 ml of distilled water. Store at room temperature.
9. *Ammonium water*: 2 ml of ammonium hydroxide + 548 ml of distilled water. Store at room temperature.

3.2.2 Procedure

Room temperature: 18–26°C.

When performing the Cytokeratin 7 IHC stain, the maximum number of slides that will be tested at a time is 30 in a humidity chamber. Within the 30 slides, there will be positive and negative controls. The negative controls are stained with a mouse negative control IgG1 isotype solution.

3.2.3 Manual Staining Procedure for Cytokeratin 7

Incubation steps are performed in a humidity chamber; rinses are performed in a Coplin jar.

1. Xylene (double rinse), 4–10 min per rinse.
2. 100% alcohol (double rinse), 1–5 min per rinse.
3. 0.95% alcohol (double rinse), 1–5 min per rinse.
4. Tap water rinse (single rinse), 1–5 min.
5. Wash buffer rinse (double rinse), 3–5 min per rinse.

6. 3% hydrogen peroxide (single rinse), 5–10 min.
7. Wash buffer rinse (double rinse), 3–5 min per rinse.
8. Proteinase K (single rinse), 5 min.
9. Wash buffer (double rinse), 3–5 min per rinse.
10. Protein Block (single rinse), 5–10 min.
11. Drain slides.
12. Mouse antihuman CK7 (1:800), 30–35 min.
13. Wash buffer rinse (double rinse), 3–5 min per rinse.
14. Rabbit anti mouse immunoglobulin (1:200), 30–35 min.
15. Wash buffer rinse (double rinse), 3–5 min per rinse.
16. Envision + rabbit HRP, 30–35 min.
17. Wash buffer rinse (double rinse), 3–5 min per rinse.
18. DAB + chromogen, 5–10 min.
19. Distilled water (double rinse), 2 min per rinse.
20. Hematoxylin (dips), 4–5 dips.
21. Distilled water (single rinse), 1 min.
22. 1% glacial acetic acid water (single rinse), 30 s.
23. Distilled water (single rinse), 1 min.
24. 0.05% ammonium water (single rinse), 1–2 dips.
25. Distilled water (single rinse), 1 min.
26. 95% alcohol (single rinse), 1 min.
27. 100% alcohol (double rinse), 1 min per rinse.
28. Xylene (double rinse), 1 min per rinse.
29. Coverslip and mount.

3.2.4 Acceptance Criteria (see Fig. 2b)

CK7 positive intermediate filaments (staining dark brown).
Nuclei (staining light blue).

False-positive results: Nonspecific or endogenous streptavidin reacting with the chromogen substrate.

3.3 Calponin Protocol

Calponin is a small protein filament associated with smooth muscle contraction. Detection of calponin on paraffin-embedded tissue can be performed using a mouse monoclonal IgG1 kappa antibody specific to calponin structural epitopes along with a peroxidase enzyme-conjugated polymer labeling system.

3.3.1 Solutions

1. *95% alcohol*: 950 ml of 100% alcohol + 50 ml of distilled water. Store at room temperature.
2. *Wash buffer 10×*: 1,000 ml + 9,000 ml distilled water. Store at room temperature.
3. *3% hydrogen peroxide*: 1 ml of 30% hydrogen peroxide + 9 ml distilled water. Store at 4°C.

4. *Mouse antihuman calponin antibody working solution*: Dilute the stock solution of the calponin antibody in antibody diluent (Dako) to a dilution of 0.86 mg/l (1:100). Adjust volumes according to experimental trial. Store at 4°C.
5. *Mouse negative control IgG1 working solution*: Dilute the stock solution of the IgG1 negative control solution in antibody diluent to a dilution of 0.86 mg/l (1:116). Store at 4°C.
6. *Rabbit anti mouse immunoglobulin solution*: Dilute the stock solution of the rabbit anti mouse immunoglobulin in antibody diluent to a dilution of 1:400. Adjust volumes according to experimental trial. Store at 4°C.
7. *DAB+chromogen*: The solution is one drop of DAB+chromogen per 1 ml of DAB+substrate buffer. Amount is specified by the Dako Autostainer and dependent on amount of slides tested per trial. Store at 4°C.

3.3.2 Procedure

Room temperature: 18–26°C.

When performing the calponin IHC stain, the maximum number of slides that will be tested at a time is 30 in a humidity chamber. Within the 30 slides, there will be positive and negative controls. The negative controls are stained with a mouse negative control IgG1 isotype solution.

3.3.3 Manual Staining Procedure for Calponin

Incubation steps are performed in a humidity chamber; rinses are performed in a Coplin jar.

1. Xylene (double rinse), 4–10 min per rinse.
2. 100% alcohol (double rinse), 1–5 min per rinse.
3. 95% alcohol(double rinse), 1–5 min per rinse.
4. Tap water rinse (single rinse), 1–5 min.
5. Wash buffer rinse (double rinse), 3–5 min per rinse.
6. 3% hydrogen peroxide (single rinse), 5–10 min.
7. Wash buffer rinse (double rinse), 3–5 min per rinse.
8. Proteinase K (single rinse), 5 min.
9. Wash buffer (double rinse), 3–5 min per rinse.
10. Protein Block (single rinse), 5–10 min.
11. Drain slides.
12. Mouse antihuman calponin (1:100), 30–35 min.
13. Wash buffer rinse (double rinse), 3–5 min per rinse.
14. Rabbit anti mouse immunoglobulin (1:400), 30–35 min.
15. Wash buffer rinse (double rinse), 3–5 min per rinse.
16. Envision+rabbit HRP, 30–35 min.

17. Wash buffer rinse (double rinse), 3–5 min per rinse.
18. DAB + chromogen, 5–10 min.
19. Distilled water (single rinse), 1–5 min.
20. Hematoxylin, 4–5 min.
21. Distilled water (single rinse), 1–5 min.
22. Buffer rinse (single rinse), 1–5 min.
23. Distilled water (single rinse), 1–5 min.
24. 95% alcohol (single rinse), 1–5 min.
25. 100% alcohol (double rinse), 1–5 min.
26. Xylene (double rinse), 1–5 min.
27. Mount and coverslip.

3.3.4 Acceptance Criteria (see Fig. 2c)

Calponin-positive protein filaments (staining dark brown).

Nuclei (staining light brown).

False-positive results: Nonspecific or endogenous streptavidin reacting with the chromogen substrate.

4 Statistical Analysis

Depending on sample numbers, various statistical analyses can be used on raw and derived data. Raw data may be the scores while the derived data may be the percent change in score from baseline. Paired and non-paired *t*-tests, chi-square, nonparametric tests, and ANOVA *t*-tests can be used to evaluate within and among groups.

5 Tissue Evaluation

While the primary focus of our discussion pertains to the microscopic assessment, it is important to correlate the macroscopic observations to the microscopic findings. In that context, evaluation should be tailored to the tissue, application, implantation procedures, and MOA. Using a list of expected and unexpected histological changes a histological grading system can be used to apply severity, intensity, and magnitude scores that support efficacy or safety. Those changes that are determined to correlate to clinical and chemical findings may help you to classify the changes listed into an either replacing or regenerating pathway. An example of macroscopic observations and microscopic correlate is shown in Table 1.

Table 1
Macroscopic findings with microscopic correlates

Animal no.	Tissue	Macroscopic observation	Microscopic correlate
	Kidney, left	Hydronephrosis, moderate	Inflammation, chronic, diffuse, pyelonephritis
	Kidney, right	Hydronephrosis, mild	No correlation
	Ureter, left	Hydroureter	Dilatation, mild
	Ureter, right	Normal	Normal
	Bowel	One, focal adhesion to NUC	Adhesions conduit to intestine present
	Uterus	Adhesions to ureters and NUC	Adhesions uterus to NUC and ureters present

6 Histological Assessment

6.1 Grading Scheme

Most of the literature uses a 0–4 scale, but directionality and magnitude can be a problem for evaluator when applying to a change that is supposed to change with neo-tissue formation and increase with time versus a change that high level may be unwanted with time such as inflammation and an extreme change away from baseline. In this study we utilized the 0–4 grading scheme and focused our evaluation to functional and safety parameters and correlated in vivo clinical, gross, and subgross findings to histological data to assess evidence of efficacy and safety. The microscopic assessment and grading was focused on the process of taking place as a whole, i.e., the regenerative event associated with the restoration of function and structure of native organ by regenerative healing rather than reparative healing, which is associated with fibrosis and scarring of connective tissue. Thus, grading criteria to the parameters evaluated were developed using a combination of similar grading schemata in referenced literature (2, 3, 5). The grading schemes of all histological parameters were designated as follows:

Grade 0 (Normal): This score corresponds to an absence of histological change; no apparent histological change.

Grade 1 (Minimal): This score corresponds to a small histological change. The tissue involvement was considered minor, small, or infrequent. The score reflected a focal, multifocal, or diffuse distribution, in which approximately <10% of the tissue was involved.

Grade 2 (Mild): This score corresponds to a noticeable, but not prominent histological change. The tissue involved was considered small, but consistently present. The score reflects a focal, multifocal, or diffuse distribution in which approximately 10–25% of the tissue is involved.

Grade 3 (Moderate): This score corresponds to a histological change that was a prominent feature of the tissue and consistently present. The score reflects a focal, multifocal, or diffuse distribution in which approximately 26–50% of the tissue is involved.

Grade 4 (Marked/Severe): This score corresponds to a histological change that was overwhelming and persistent. The change may or may not a have an adverse effect on organ function, depending on the nature of the finding. The score reflects a focal, multifocal, or diffuse distribution in which approximately >50% of the tissue is involved.

6.2 Immunohistochemistry Grading

All slides (including negative and positive controls) were initially judged for adequacy of tissue elements and staining. Intensity of staining was graded as follows: ± (equivocal), 1+ (weak), 2+ (moderate), 3+ (strong), 4+ (intense), or neg (negative).

7 Functional Parameters

Evaluation and interpretation of functional parameters was developed to examine the ongoing process by assessing two primary tissue responses: reparative regeneration and reparative healing. These two processes are very similar at 1-month post implantation but remarkably different by 3 months as their structural and functional characteristics display distinctive proportions of the stromal (fibroblasts, vessels, and nerves) and parenchymal (epithelial and smooth muscle) components. At 3 months, regenerative healing would show greater stromal organization, consisting of marked neovascularization with smoothly interwoven connective tissue elements and significantly improved organization of parenchymal components comprised of smooth muscle-like bundles and less dense collagenous tissue. In contrast, reparative healing would show less vascularization, within a dense collagenous tissue and less organized spindloid cell infiltrate (3).

The histological assessment and corresponding tissue changes pertaining to the regenerative and healing processes were scored using the following parameters: epithelialization, luminal surface and muscular layer fibrovascular response, myofibroblastic reaction, peritoneal surface integration, lumina surface and muscularis scaffold fibers scores. An example of individual animal scoring format is shown in Tables 2 and 3.

7.1 Epithelialization

This is a major component of the regenerative process, and therefore an increase in epithelial or urothelial cell coverage is expected. Typically, higher epithelialization scores are most often correlated to mucosal hyperplasia and thus scores should be based on the presence, thickness, and extent of epithelial coverage, which may

Table 2
Histological characterization of functional and safety parameters

Conduit site	Dorsal			Ventral			Ureteral junction		Conduit junction
	Proximal	Middle	Distal	Proximal	Middle	Distal	Left	Right	Stoma skin
Functional parameters									
Epithelialization									
Luminal surface fibrovascular response									
Muscular layer fibrovascular response									
Myofibroblastic response									
Peritoneal surface integration									
Safety parameters									
Tissue necrosis									
Luminal surface fibrous response									
Muscular layer fibrous response									
Overall inflammatory response									
Scaffold degradation									
Calcification and/or heterotopic bone formation									

be variable depending on the organ application. Furthermore, characterization of epithelial coverage should be performed using IHC staining techniques, i.e., AE/AE3 for epithelium and cytokeratin 7 for urothelium (Figs. 3, 4, and 5).

7.2 Luminal Surface Fibrovascular Response

This score is based on the presence and extent of neovascularization and corresponding inflammatory reaction. As part of the regenerative process, numerous angiogenic capillaries are expected to be prevalent and extend toward the luminal surface to form the lamina propria. This process is initially facilitated by the peritoneal wrap. The formation of new capillaries that begin to infiltrate the loose to variably dense connective tissue stroma in conjunction with inflammatory cells, consist primarily of macrophages and giant cells,

Table 3
Special stains characterization

	Dorsal			Ventral			Ureteral junction		Conduit junction
Conduit site	**Prox.**	**Mid.**	**Dist.**	**Prox.**	**Mid.**	**Distal**	**Left**	**Right**	**Stoma skin**
Section location									
Epithelium (P = present; A = absent)									
AE1/AE3 labeling									
CK-7 labeling									
Mucin (PAS stain)									
Squamous epithelium (P = present; A = absent)									
AE1/AE3 labeling									
CK-7 labeling									
Mucin (PAS stain)									
Smooth muscle layers/bundles									
Calponin									

although neutrophils and lymphocytes may be variably present. Depending on the duration of the study, higher scores would reflect early time of explantation (30–90 days) and lower scores are expected beyond 90 days, as neoangiogenesis should have subsided, to form normal lamina propria.

7.3 Muscular Layer Fibrovascular Response

This score is based on the presence of fibrovascular reaction, intensity of inflammation, vascular response, and extent of neo-organ wall composition of fibrovascular stroma. At time close to implantation, higher scores are indicative of an increased regenerative response. At later time point, i.e., greater than 90 days, inflammation, neoangiogenesis, and overall fibrovascular reaction should have subsided, and the neo-organ's wall architecture should display urinary-like tissue components, resembling that of native urinary tissue (Fig. 6).

7.4 Myofibroblastic Reaction

This score is based on the presence and intensity of the reaction and extent of smooth muscle-like repopulation of the neo-organ wall. In the early regenerative process, smooth muscle-like spindloid cells with or without fibroblastic characteristics should begin to populate the implanted tissue. These cells form the stroma necessary for tissue structure (support) and architecture (shape),

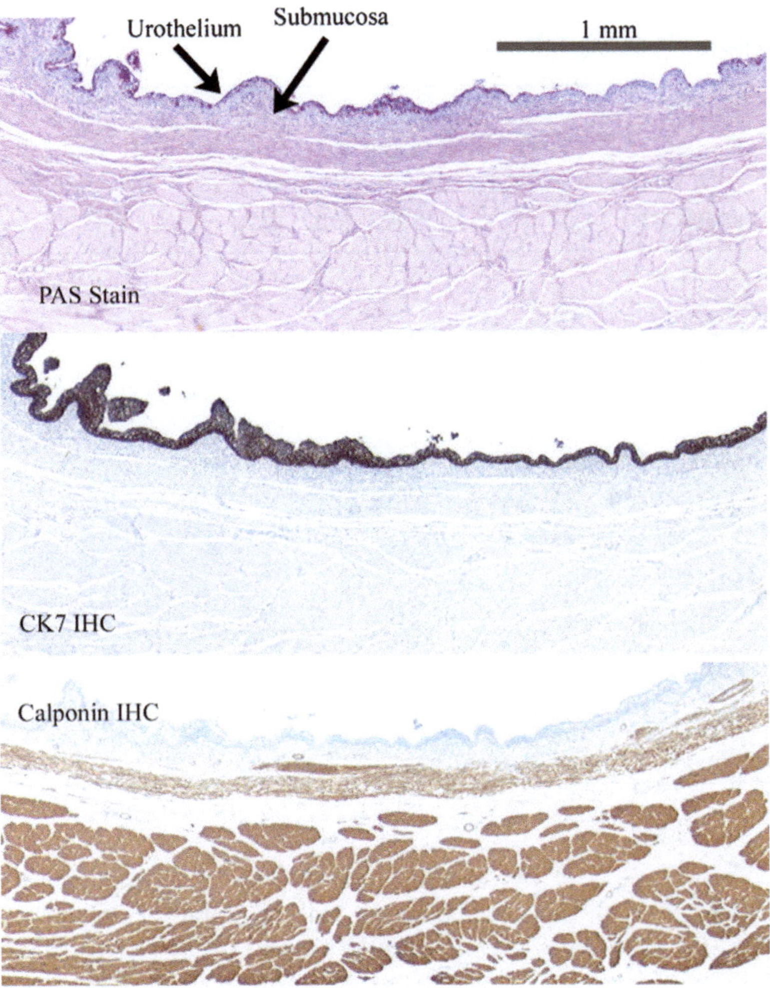

Fig. 3 Representative sections from the proximal end of the conduit. *Top panel*, mucin granules are observed in the urothelium (*pink/purple* with Periodic acid-Schiff stain). *Center panel*, the epithelium is staining positive with CK7 immunohistochemistry (IHC), indicative of urothelial-type epithelium. *Bottom panel*, smooth muscle layer are staining positive by calponin (IHC staining)

which will occupy the space of degraded scaffold biomaterial. At a later time point, it is expected that these myofibers continue to form, extend, and ultimately organize into smooth muscle bundle and layers (Fig. 7).

7.5 Peritoneal Surface Integration

This scored is based on the presence of peritoneum (wrap), and its integration to the external surface of the construct (neo-organ). As part of the regenerative process, angiogenic vessels from the peritoneal wrap form fibrovascular pedicles that infiltrate into the construct. The vessels provide oxygen, nourishment, and other

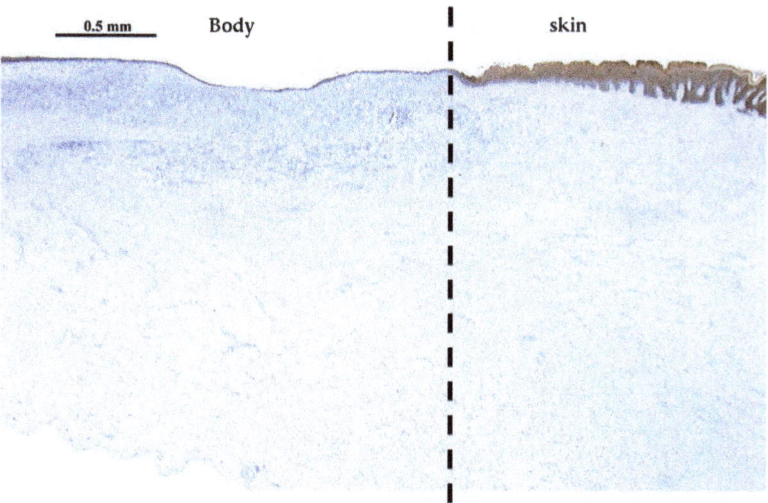

Fig. 4 Histological representation of the conduit's body (mid. and distal portions), showing the junction between the skin (keratinized epithelium) and non-keratinized squamous epithelium, which extends over the conduit's wall. Note the epithelium is staining brown by AE1/AE3, an IHC marker for epithelium

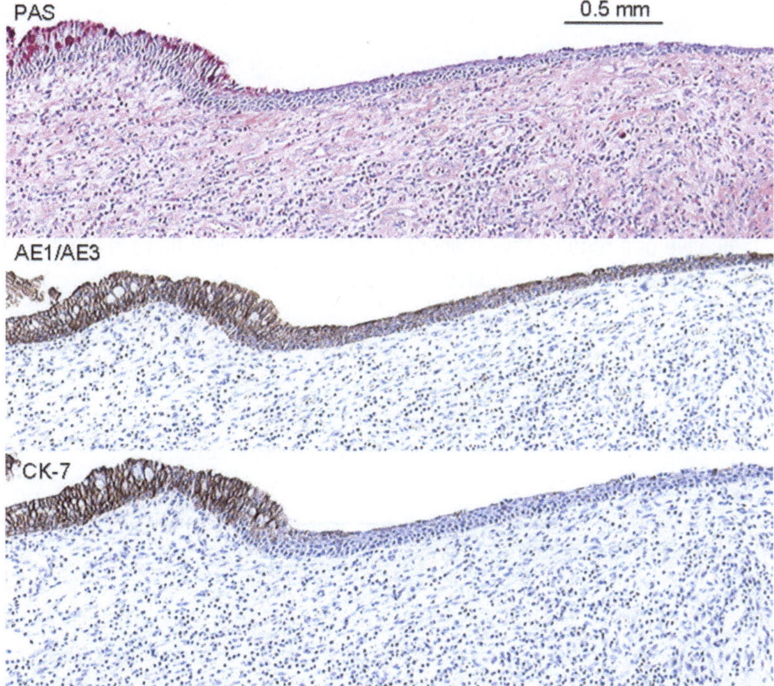

Fig. 5 Histological representation from the midportion of the conduit showing the junction between stoma-derived non-keratinized squamous epithelium (at *right*), and the ureter-derived urothelium-type epithelium (at *left*). *Top panel*, mucin (PAS positive) granules are observed in the urothelium. *Center panel*, AEI/AE3 positive marker for epithelium (staining *brown*). *Bottom panel*, CK7 positive for urothelium at *left* but negative for epithelium at *right*

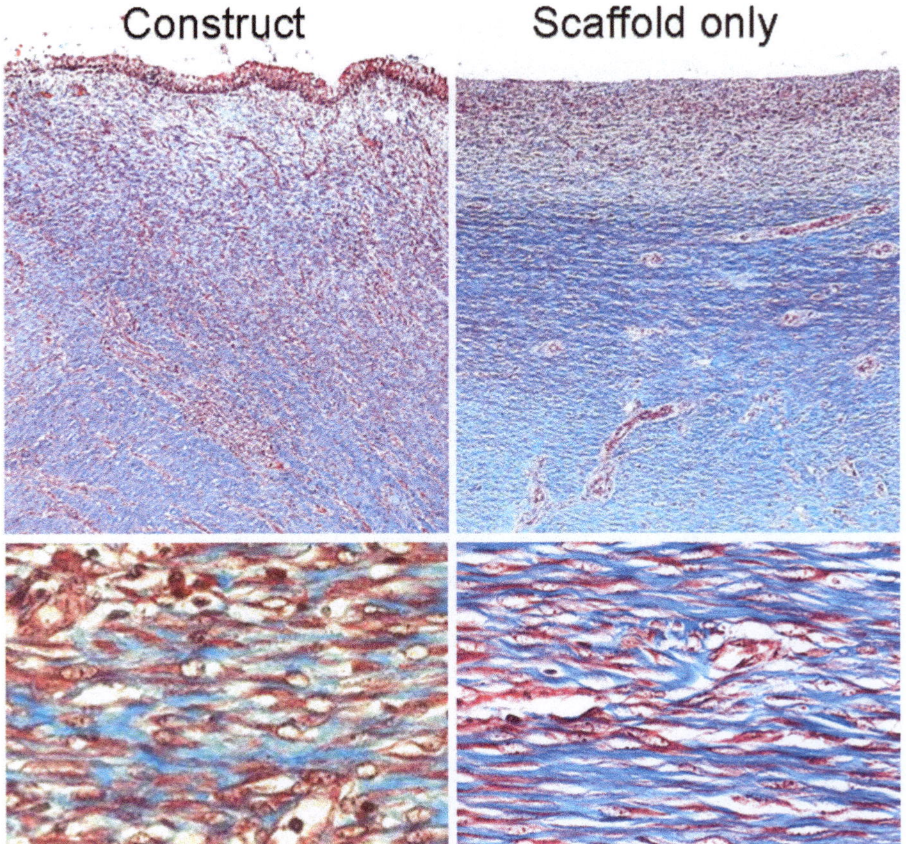

Fig. 6 Histological representation of implant's wall construct versus scaffold only (4 weeks post implantation). The intensity of the fibrovascular response appears greater in the construct versus scaffold only implant which is indicative of a regenerative response. In the scaffold only there is more deposition of fibrous tissue matrix within the wall, which is indicative of a reparative process

inflammatory cells that play an important role in removing the scaffold material and secreting other cell precursors necessary in the reparative process (Fig. 8).

8 Safety Parameters

An overt tissue response is considered to have a negative impact on the safety of the neo-organ being replaced. Whether the inflammatory reaction becomes excessive, scaffold degradation is delayed, or tissue necrosis becomes prevalent, these overt tissue changes are likely to impair the normal healing and regenerative process. Thus, the following parameters were interpreted as having an impact on the regenerative outcome of the construct: tissue necrosis, luminal surface fibrous response, muscular layer fibrous response, inflammatory response, scaffold degradation and calcification, and/or ectopic bone formation.

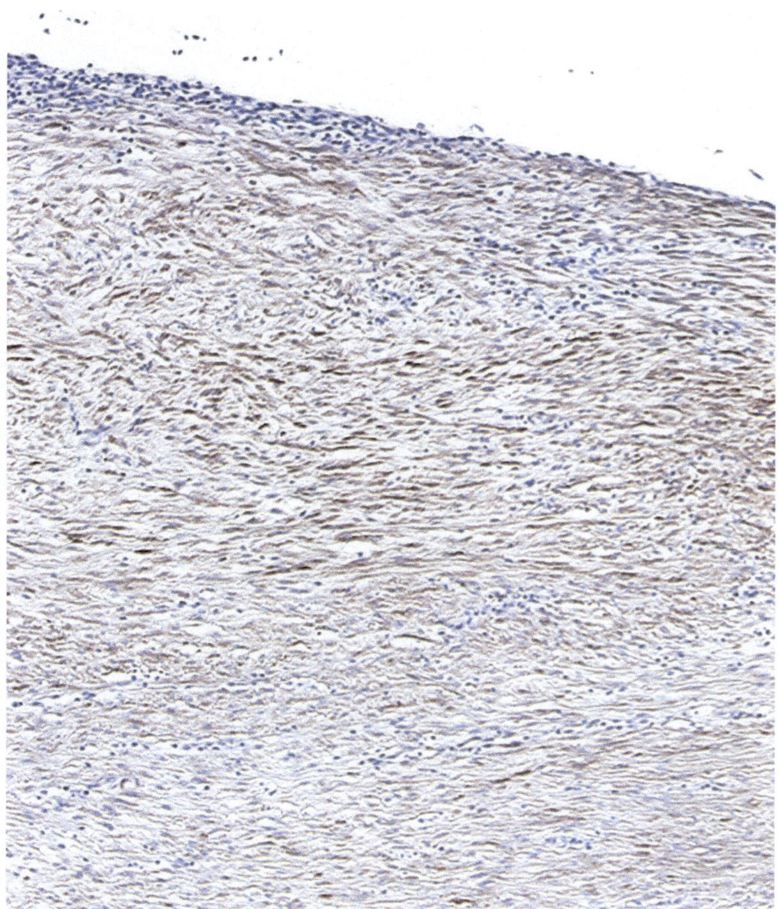

Fig. 7 Histological representation of neo-bladder wall. There are no bundled myofibers; however, there is a marked myofibroblastic response characterized by presence of myofibroblast which are staining *brown* by calponin immunohistochemistry

8.1 Tissue Necrosis

This type of tissue response is not expected as part of either the normal healing or regenerative response, and is certainly nondesirable in the process, as it indicates an adverse environment and is likely to have been caused by local tissue ischemia. The region affected may show other findings such as hemorrhage, edema, inflammation (chronic active or acute, neutrophilic), parenchymal tissue changes such as smooth muscle dystrophic mineralization, atrophy, and fiber disarray.

8.2 Luminal Surface Fibrous Tissue

This score is based on the presence and extent of fibroplasia. As part of the healing process, dense connective tissue is deposited to provide structural support or, in the case of poorly degradable sutures and/or implants, a fibrous capsule may be formed surrounding the material, separating it from the host tissue. This change represents an attempt by the body to isolate or wall off the

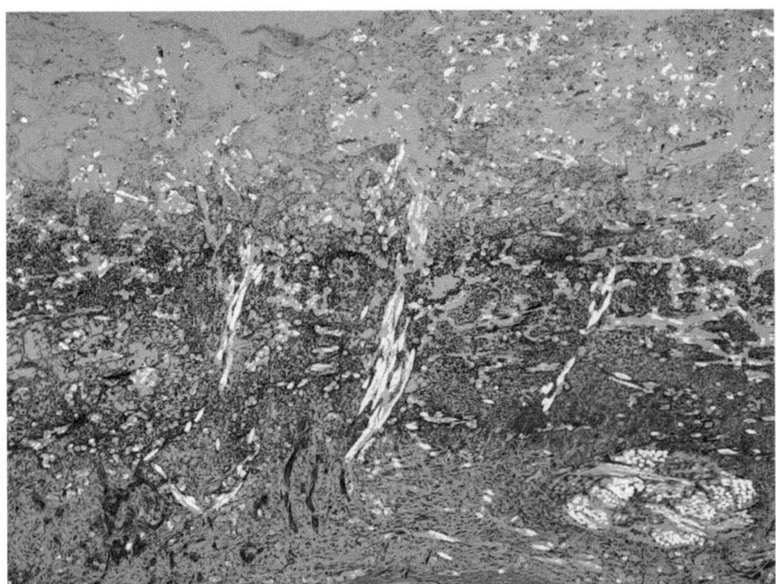

Fig. 8 Histological representation of the construct wall at 6 days post implantation. Note there is marked accumulation of remnant scaffold material (fibers) seen under polarizing field. The luminal surface (L) of the wall is located at the *top* of the panel. Note there is cellular infiltration with intense fibrovascular response in the outer 1/3 of the wall, as the peritoneal wrap provides immediate vascularization

device/implant. It is characterized by excessive deposition of mature, collagen-rich connective tissue with or without associated inflammatory cells (fibroplasia) or an integrated fibrovascular source (Fig. 6).

8.3 Muscular Layer Fibrous Tissue

This score is based on the presence and extent of fibroplasia associated with smooth muscle fiber replacement and is characterized by an excessive fibroblastic deposition of mature, collagen-rich connective tissue with or without inflammatory cells (Fig. 6).

8.4 Inflammatory Response

While at early stages of implantation the inflammatory process is part of the healing response and thus, an anticipated physiologic resorption, a process by which foreign material, i.e., suture and/or scaffold material, is removed from the host tissue; however, any excessive accumulation of acute or chronic inflammatory infiltrate not associated with this process should be interpreted as an adverse event and probably the result of contamination or infection.

8.5 Scaffold Degradation

This score is based on the presence of scaffold fibers observed in the implanted construct and the intensity of inflammatory reaction and extent of tissue composed of degraded scaffold material. Typically, the scaffold polymer fibers should be undergoing

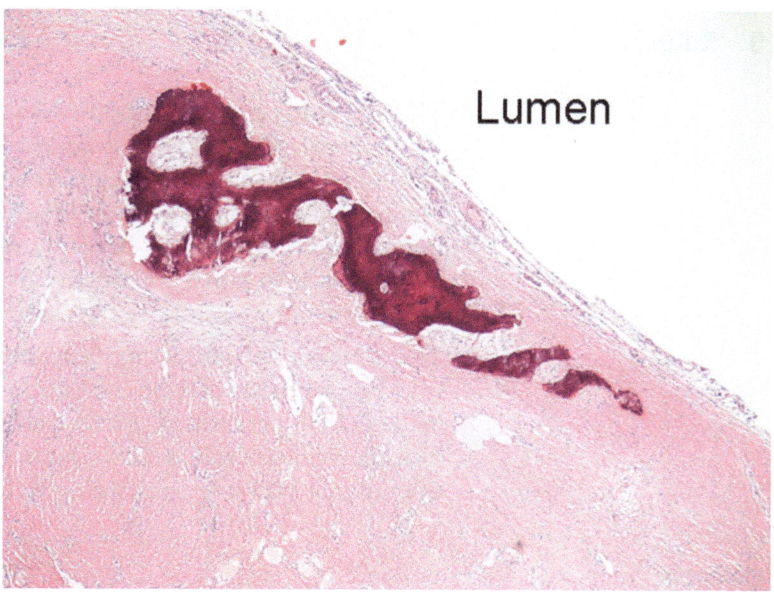

Fig. 9 Histological representation of neo-bladder wall showing subluminal heterotopic bone formation. HE, ×40 original magnification

degradation with or without inflammatory cells (through hydrolysis), but the presence of inflammatory cells would accelerate the fiber degradation and removal of foreign material. A decrease in scaffold fibers is expected with time (Fig. 8).

8.6 Calcification and/or Heterotopic Bone Formation

This score is based on presence of calcification and/or heterotopic bone formation. As part of the healing process, and dependent on oxygen tension and proper conditions, urothelium is capable of producing bone spicules with or without hematopoietic marrow. Furthermore, previous studies have shown focal ossification of peritoneal scars in a swine model is associated with the surgical procedure (trauma) to peritoneum (6) (see Fig. 9).

9 Histo-morphometry

This approach should be considered if there is evidence of tissue proportionalities in the test article when compared to untreated controls. For example if wall thickness of the neo-organ (construct) evaluated after 12 weeks of implantation appears less or thicker than the scaffold only control, it would be prudent to confirm and document this finding by morphometric analysis as it may be used to support claims of safety and efficacy (see Note 2).

10 Notes

1. This is a ready-to-use antibody diluent by DAKO. It provides an appropriately buffered medium for the dilution of both polyclonal and monoclonal antibodies and for the preparation of negative control reagents in IHC. Interaction between antibodies and epitopes can be easily disturbed by extreme pH. Thus, it is imperative to use the proper antibody diluent in order to achieve optimal antibody performance.
2. The accuracy of conducting histomorphometric assessment of bladder-like tissue may be compromised when tissue fixation is inadequately performed. Study pathologist must be aware of fixation artifacts such as shrinking or contracting and dilatation of muscular wall when recording wall thickness.

References

1. Jayo MJ, Watson DD, Wagner BJ, Bertram TA (2008) Tissue engineering and regenerative medicine: role of toxicologic pathologists for an emerging medical technology. Toxicol Pathol 36:92–96
2. Jayo MJ, Jain D, Wagner BJ, Bertram TA (2008) Early cellular and stromal responses in regeneration versus repair of a mammalian bladder using autologous cell and biodegradable scaffold technologies. J Urol 180: 393–397
3. Prophet EB, Mills B, Arrington JB, Sobin LH (1992) Laboratory methods in histotechnology. Armed Forces Institute of Pathology, Washington, DC
4. Moll R (1998) Cytokeratins as markers of differentiation in the diagnosis of epithelial tumors. Plenum, Hermann, NY
5. Morton D, Kemp R, Francke-Carrol S, Jensen K, McCartney J, Monticello T, Perry R, Pulido O, Roome N, Schaffer K, Sellers R, Snyder P (2006) Best practices for reporting pathology interpretations within GLP toxicology studies. Toxicol Pathol 34:806–809
6. Baker IK (1993) The peritoneum and retroperitoneum. Pathol Domestic Anim 2:425–445

Index

Joydeep Basu and John W. Ludlow (eds.), *Organ Regeneration: Methods and Protocols*, Methods in Molecular Biology, vol. 1001, DOI 10.1007/978-1-62703-363-3, © Springer Science+Business Media New York 2013

F

G

H

I

K

L

M

N

O

P

Q

R

S

T

U

W

X

MIX
Papier aus verantwortungsvollen Quellen
Paper from responsible sources
FSC® C105338

If you have any concerns about our products, you can contact us on
ProductSafety@springernature.com

In case Publisher is established outside the EU, the EU authorized representative is:
Springer Nature Customer Service Center GmbH
Europaplatz 3, 69115 Heidelberg, Germany

Printed by Libri Plureos GmbH
in Hamburg, Germany